AF553657

ANGIOSPERMS

ANGIOSPERMS

By

Dr. Pooja

Department of Botany

R.C.C. College

Ghaziabad (U.P.)

DISCOVERY PUBLISHING HOUSE

NEW DELHI-110002

First Published-2004

ISBN 81-7141-788-4

Published by

DISCOVERY PUBLISHING HOUSE

4831/24, Ansari Road, Prahlad Street,
Darya Ganj, New Delhi-110002 (India)
Phone: 23279245 • Fax: 91-11-23253475
E-mail:dphtemp@indiatimes.com

Printed at:

Tarun Offset Printers, Delhi-53

Preface

Angiosperms or taxonomy is a very popular, most attractive and useful branch of biological sciences. It has a long history. It gave birth to other phases of botany but remains exciting, interesting, and important because it deals directly with the fascinating differences among the species of plants that inhabit the earth. The aim of present title is to provide a general overview of the field in most effective and positive manner. It is designed to meet the needs of undergraduate students of Indian Universities that is why only Indian species belonging to different families have been described. The chapters are largely independent so that the teacher may choose the desired sequence of topics. The level of presentation is based on the assumption that the students will have had an introductory biology or botany course. Also, it is hoped that this book will prove stimulating to serious amateurs, teachers, and professionales who specialize in other fields but utilize classification and other taxonomic information about plants.

Several plants found around us have been described in semitechnical language. The description of the various families and genera have been supported with several illustrations. The subject matter and many illustrations have been taken from many journals. So author does not claim any originality of the work.

Author is thankful to all her colleagues, friends, botanists and educationists who have accorded their full co-operation and well wishes during the preparation of manuscript of the present title.

The author tried hard to be accurate and upto date in statement and realises the impossibility of completely avoiding errors therefore, the author will greatly appreciate having his attention called to any questionable statement.

The author expresses her gratitute to Mr. Wasan and staff of M/s Discovery Publishing House for their whole hearted co-operation in the publication of this book.

Author

Contents

1

INTRODUCTION

Systematic botany is the science of identifying, naming, and classifying all plants. It is a challenging field of study. The potential economic uses of plants may not be immediately evident, but we must know which plants are related to one another in order to predict their properties. Wild relatives of our cultivated plants often have genes that can provide the desirable qualities, such as disease resistance, needed by plant breeders for crop improvement. Nonscientists can readily understand evolution and changes by observing the relationships of form and function that are so prominent among flowers.

Definitions

Systematic botany deals with the study of the diversity of plants and their identification, naming, classification and evolution. Since there is no agreement or etymological basis for the distinctions between *systematics* and *taxonomy*, these terms are used interchangeably in this text. However, it should be realized that some authors do differentiate between systematics and taxonomy, the former having the broad definition as just given, with taxonomy restricted to the study of classification.

Taxonomy

It is the arrangement of plants into groups having common characteristics. These groups are then arranged into a system. Similar species of flowering plants are placed in a *genus* (plural, genera); similar genera are grouped into *families*; families with common features are arranged into *orders*, orders into *classes*, into *divisions*. Classification results in the placing into a hierarchy of ranks or categories such as species, genera, families, and so on. In addition to expressing relationships based on common features, classification serves as a filing and information retrieval system and allows easier reference to the organisms comprising the filing system.

Identification or Determination

The recognition of certain characters of flower, fruit, leaf, or stem and the application of a name of a plant with those particular characters is the *identification*. Recognition occurs when the specimen under consideration is similar to a previously known plant. If comparison of the specimen with all similar species reveals that it differs from them, it may be named as a new species. The term *classification* is often wrongly used in this sense and such use should be discouraged.

Taxon

A *taxon* (plural, taxa) is a covenient and useful term applied to any taxonomic group at any rank: for example, species, genus, or family, or to a taxonomic group where rank is uncertain. As a general term, taxon can be used to indicate the rank of the group as well as its components. Taxa are separated from one another and are recognized by the features unique to each taxon in the same way we distinguish our friends from one another or differentiate among breeds of dogs or any other grouping of objects.

Nomenclature

The orderly application of names to taxa in accordance with the International Code of Botanical Nomenclature is called the *nomenclature*. This code provides procedure for selection of the correct name or formulation of a new name.

Description

The *description* of a taxon is a listing of its features or morphological characteristics referred to as *taxonomic characters*. A shortened description consisting of only those taxonomic characters useful in separating a taxon from other related taxa is known as a *diagnosis*, and the characters used in the diagnosis are called *diagnostic characters*.

Flora

The term *flora* refers either to the plants growing in an area surrounded by a geographic or political boundary or to an inventory of the plants or particular area or region. *Floristics* is the descriptive term used to refer to an investigation of the flora. A manual is a book that provides an inventory of the flora and the means of identifying the plants using descriptive keys.

Other Approaches

Various approaches in systematics include *classical taxonomy*, which consists largely of museum research but often includes field work, and *biosystematics*, which involves ecological, cytological, and genetic investigations, and experimental studies of living populations in the field, experimental garden, laboratory, and greenhouse. *Numerical taxonomy* is the treatment of various types of taxonomic data by computerized methods. *Chemical taxonomy*, *chemotaxonomy*, or *chemosystematics* is the use of chemical features plants in developing classifications.

Systematic Botany

Primeval human used food and medicinal plants that grew in their environment, recognizing hundreds of different plants. This early recognition of useful and harmful plants marked the beginning of systematic botany. The development of language made it possible for observations of plants to be accumulated, and this knowledge could be passed from one generation to the next. Today the basic recognition and grouping of plants has developed into a highly complex science concerned with classifying plants into groups based on postulated evolutionary relationships.

Systematic botany includes all activities that are part of the effort to organize and record the diversity of plants and acquaints us with the fascinating differences among the species of plants. The activities of a plant taxonomist are basic to all other biological sciences, since systematics provides an inventory of plants, schemes for identification, plant names, and a system of classification of plants. Not only is systematics basic to other scientific fields, but also it depends on other disciplines for information and data useful in constructing classifications. A sound classification, bringing related plants together, may suggest problems worthy of study by ecologists, plant breeders, pharmacologists, horticulturists, but biochemists.

Objectives

Plant taxonomy has five objectives:

1. To inventory the world's flora.
2. To provide a method for identification and communication.
3. To produce a coherent and universal system of classification.
4. To demonstrate the evolutionary implications of plant diversity.
5. To provide a single Latin "scientific" name for every group of plants in the world, both extant and fossil.

Although the inventory of the earth's flora is largely complete in the northern temperate zone, much remains to be done in the tropics. About 1 million of an estimated 1,500,000 species of all kinds of organisms have been described for the temperate regions, but in the tropics only about 50,000 of an estimated 3 million species of organisms have been described thus for. It has been estimated that there are about 15,000 species of flowering plants in South America that have not been named or made known to science.

The primary tasks of systematics botany-that is, to explore, describe, and classify the flora of the world is thus far form complete. Once names and classification have been provided, three must be methods for persons to identify a taxon as being similar to another known entity. This is accomplished by producing descriptions, keys, catalogs, illustrations, manuals, or other publications that aid in the identification of specimens. Technology is presently available for the use of computers to aid in identification. In the future, computer programs may be routinely used in plant identification. Nevertheless, training, practice and experience remain the best teachers in plant identification. In the century since Darwin, biologists have been able to demonstrate that evolutionary lineages occur among taxa.

The evolutionary development or lineage of a taxon is its *phylogeny*. Phylogenetic relationships exist that show that diverse species and the lineages they represent did not arise spontaneously, but may have had a common ancestral form. Modern classification attempts to use all available informations about plants to develop a phylogeny. Since the fossil record is often fragmentary, especially for flowering plants, information must be gathered form a wide variety of sources to formulate a hypothesis about phylogeny.

Requirement for Names

A method of taxonomy is necessary to allow us to identify plants and animals and to communicate scientifically with others. Indeed, classification is both an information storage and retrieval system without which scientific communication would be impossible. A plant's name is the key that unlocks the door to its total biology.

Ecologists, biochemists, horticulturists, and others must have a reference system for the plants they use in their research. However, systematic botany is not meant only for scientist; it may be used by other people with varied interests and training who are interested in the natural history of plants. The scientific name of a plant communicates the species and genus, and from that the family may be easily determined. In addition, one can assume that there are other individual plants that share certain of the same features. For instance, when a forester identifies a white oak (*Quercus alba*), it can reasonably be assumed that there are other individual white oaks in nature that have similar morphological features, structure, and physiology. The knowledge that a plant is *Quercus alba* automatically predicts that much information will be applicable to the plant.

Phases of Plant Systematics

One of the primary concerns of plant taxonomy is to provide an inventory of plants of the world. This is sometimes called *alpha taxonomy* and is an activity of even preliterate peoples. However, the most active recorded period began with the voyages of discovery by Western Europeans in the 1400s. The peak of botanical exploration, in terms of plants described, was the late 1800s, but exploration and discovery continue today, especially in the tropics. Plants material collected on these early expeditions to the far corners of the earth was sent to botanists in Europe for naming and description. In the late 1700s and early 1800s, botanists were overwhelmed with the amount of plant specimens collected for study. Collections of pressed and dried specimens called *herbaria* (*Singular*, *herbarium*) were developed.

Botanists frequently became interested in particular taxa and exchanged plant specimens with other botanists for extensive study. Some of these collections later formed the nuclei of world-renowned herbaria. By the late 1800s, many botanical centers were established in Europe and North America. Their herbaria grew repidly as access to remote regions of the earth increased. In this phase, many species were described, named, and classified into genera and families for the first time. The floras of given areas became known through herbarium specimens.

Exploration and discovery continue today, especially in the tropics, although new collection from the temperate region are continually made. After adequate herbarium herbarium material is accumulated for given geographic or political region, the second phase caled *synthesis* begin. Classification based on *morphology*, the form and structure of plants, is developed. This presents the means of identifying the plants of that area. Botanists who have the goal of developing improved classification investigate the distribution of characters in taxa. This syntheses phase reached its plateau in the late 1800s and continues with considerable study today.

The foundation of taxonomy was largely developed before the general acceptance of Darwin's evolutionary ideas of the late 1800s and was based on *typology*. The influence of this doctrine is evident in systematics today. The early adherents of typology reasoned that each natural taxon had an idealized pattern or "type". In its extreme form, such reasoning led to the denial of variability within taxa. Today, however, a loosely defined typology is a useful and necessary king of thought in taxonomy. The endless array of plant groups encountered by taxonomists must be organized mentally. Thus taxonomists seek repetitive patterns as aids in recognizing mosses, ferns, gymnosperms, dicots, grasses, sunflowers, oaks, and so on Typology was most influential during the early years of the exploration and discovery phase. It also continues into the synthesis phase.

Another *experimental* phase is the combining of data for interpretation in evolutionary or phylogenetic terms. The emphasis is on understanding the course of evolution and its cause at the specific or generic level. Studies are currently ongoing where workers are concerned with morphological and chemical variation and cytological features such as chromosome number, behaviour, and morphology. Controlled experiments are conducted in genetics. Hybridization, common in plants, is studied. Such research has led to revised concepts of evolutionary theory. Most information obtained form experimental studies is used in improving our knowledge of species and genera and has therefore had an impact on the classification of higher plant families and orders. With experimental methods, variation within species has becomes welldocumented, and the cause of variation among plants may often be deduced.

Critical Problems and Opportunities

The urge among many scientists to follow the newest approaches in the biological sciences in an attempt to exploit the greenest pastures of research has caused, in many universities, an imbalance in the biological sciences to the extent that systematics has been neglected. Fortunately, there seems to be a growing awareness among many biologists that systematics is a fundamental prerequisite for many biological investigations and that the lack of competent taxonomists is a limiting factor in certain areas of research, especially in ecology, plant exploration, and comparative biochemistry. There are many opportunities for students to make a contribution to society thought plant systematics. Many floras and classifications were done over 100 years ago and are out of date.

Classifications change as additional knowledge accumulates, so that some names as well as the arrangement of taxa used in older floras are now incorrect. Since the 1800s many new species have been named, but most have never been critically examined with reference to overall classifications. Much field work remains to be done, especially in the tropics. In view of the rapid destruction of natural vegetation in tropical areas and because of agricultural practices and encroaching civilization, there is an urgent need to explore the flora of these regions. A striking example of the urgency in this respect can be examined in Ethiopia, where in 1954, forest covered about 16 percent of the countryside. By 1975, forest cover was only 4 percent.

In most tropical rain forests, the plants have never been inventoried, yet the few that remain are rapidly being cleared. Many species that inhabit these areas may be lost forever. For example, a wet forest strip of coastal Ecuador was almost completely denuded of forest in the 1960s. A small area of less than 1 km^2 of forest preserve, recently investigated by taxonomists from the Missouri Botanical Garden, remains. This forest preserve is now completely surrounded by banana and oil palm plantations. Some 50 new species of flowering plants have been described from this 1 km^2 preserve. The major timber tree of this coastal region is *Persea theobromifolia*, which was not described until the 1970s. *persea theobromifolia* is a relative of the cultivated avocado (*Persea americana*) and may possibly prove to be a desirable rootstock for it.

Commercial avocado growers have been searching for an understock resistant to root-rot. Further, several species of pileas and gesnerias from the same region have recently been introduced into cultivation as ornamentals and are grown commmercially by the tropical foliage plant industry. It is of critical importance for plant taxonomists to examine the potential economic uses of plants. By way of example, the yeheb nut (*Cordeauxia edulis*), once abundent in Somalia where its pods were used as a staple food, was threatened by extinction in the early 1950s.

Botanists became aware of its precarious situation and so secured the seeds, and it is now cultivated on a small scale in Somalia and Kenya. Because *C. edulis* thrives under extremely low precipitation, the shrub is an important potential food plant for arid regions where the human populations are frequently subjected to famine. The continental United States is richly endowed with a marvelously diverse flora of approximately 20,000 species of vascular plants. Unfortunately, it is estimated that about 1,200 of these species are threatened, approximately 750 are in danger of extinction, and perhaps as many as 100 are already extinct. In the first half of the nineteenth century only about species of plants are known to have become extinct in the United States, but in the first 50 years of the twentieth century, about 45 species of plants succumbed.

To date, a few of the native plant species of North America have been studied in detail in terms of environmental requirements, ecological life histories, reproductive strategies, and seedling establishment. Indeed, such investigations may provide information about why a plant is rare and how populations of endangered species might be increased or preserved. Much remains to be learned about the details of species distribution in North America, including the precise locations of living stands of endangered species, and about the relative and sometimes changing frequencies of individual species. For these reasons there are active floristic research programs today in many individual American states and Canadian provinces.

Historical Background of Classification

The history of classification is an exciting facet of plant systematics. Preliterate people arranged plants by their usefulness, whether edible, poisonous, or medicinal. As the purposes of classification expended, the criteria for classification changed. The groupings of plants of the early hunter-gatherers eventually gave way to a classification reflection plant affinities. Today we group plants by their presumed natural and phylogenetic relationships at the specific, generic, familial, and higher levels.

Beginning of Taxonomic Studies

The discoveries of the use of plants for food and later as medicine began at a very early stage in human evolution. Prehistoric people knew and used nearly all the important crop plants that we cultivate today. Food gatherers modified wild species by selecting plants with such features as tastiness in vegetables or higher yield in grains. Contemporary studies indicate that primitive peoples in remote areas today recognize and have precise names for large numbers for plants in their local environment. Some of these people regularly use plants for fish or arrow poisons, others for drugs to treat wounds or sickness, and still others for narcotic or hallucinatory purposes. Classification by preliterate people is at least partly based on the useful and harmful properties of plants. These groupings often parallel current classification concepts and have been referred to as *folk taxonomies*; that is, classifications that developed within the society through the need of the society and without scientific efforts.

Early Civilizations

Early civilization developed in western areas such as Babylonia and Egypt, where cultivation of crops was feasible. Since agriculture supported these civilizations, botanical lore was of great importance. But it was not until the development of writing and convenient writing materials such as papyrus, made from a Nile River sedge (*Cyperus papyrus*), that experience and knowledge of plants could be easily recorded.

Theophrastus (about 370-285 B.C)

The Greek philosopher Theophrastus was an outstanding naturalist. He was disciple of Aristotle and is often called the "father of botany". After Aristotle's death in 323 B.C., Theophrastus inherited Aristotle's library and garden. Theophrastus is credited with several hundred manuscripts, but only two botanical discourses survive, *Enquiry into Plants and The Causes of Plants*. These are available in translation. His writings summarize what was known of plants at the time and are similar to notes for lactures. Theophrastus classified plants into four major groups; herbs, undershrubs, shrubs, and trees. He described approximately 500 different species of plants. Theophrastus noted many differences in plants, such as corolla types, ovary positions, and inflorescences. He distinguished between flowering and nonflowering plants. He was aware of structural features, such as the pericarp of fruits, and the distinctions among plant tissues. The botanical information contained in the writing of Theophrastus was sound and sophisticated. The original contributions of his work were not improved upon until after the Middle Ages. Many names for plants today are derived from those used by Theophrastus.

Pliny the Eide (A.D. 23-79)

Pliny, a Roman naturalist and writer attempted to compile everything known in the world into an extensive 37-volume encylopedia entitled *Historia naturalis* ("Natural History"). Nine of the volumes were devoted to medicinal plants. Although it contained many fables and was inadequate by today's standards, Pliny's accounts of plants remained for centuries as the only reflection of the achievements in Greek botany by Theophrastus. He must be considered as one of the most significant contributors. It is unfortunate that his studies were cut short by his death during an eruption of Mount Vesuvius

Dioscorides (First century A.D.)

The Roman military surgeon Dioscorides was the most important botanist after Theophrastus. Traveling extensively with the Roman armies, he had firsthand knowledge of plants used as remedies. Apparently interested in improving medical service in Roman empire, dioscorides prepared a truly outstanding book, *Materia medica*, which included descriptions of some 600 species of medicinal plants, 100 of which were not described by Theophrastus. Excellent illustrations were later added. For 1500 years this book was attentively studied. No drug was recognized as genuine unless named in *Materia Medica*. The most beautiful and famous copy was prepared about A.D. 500 for Emperor Flavius Olybrius Anicius as a gift to his daughter, Princess Juliana. It is available in facsimile, and the illustrations can be easily associated with familiar plants. The original manuscript is now in Vienna. Many of the names used by Dioscorides appear today as familiar generic names. *Materia medica* was not a deliberate attempt at classification; however, some plants such as legumes, which today are regarded as closely related, were grouped together. *Materia medica* contained less botany than the works of Theophrastus, but it's usefulness in medicine caused it to be considered the definitive work of plant knowlege until the end of the Middle Ages.

Middle Ages

During the European middle Ages, little progress was made in original scientific study of plants. Wars and the decay of the Roman Empire caused the destruction of much literature. Manuscripts were lost at a faster rate than they could be laboriously copied in the newly founded monasteries. Botanical knowledge was largely confined to the preciously known works of Theophrastus, pliny, and Dioscorides.

Islamic Botany

Originated around A.D. 610 to about A.D. 1100, some classical botanical works were preserved by the Moslem society because the Islamic scholars had a great admiration for Aristotle and other Greek scholars. Since their scientific interests were of a practical nature, pharmacy and medicine of the Islamic people were highly developed. Islamic botanists provided practical lists of drug plants however did not develop original schemes of classification.

Albertus Magnus, "Doctor Universalis" (c. 1193-1280)

The botanical work of Albertus Magnus *De vagetabilis* not only dealt with medicinal plants, as had the previous works by the Greeks and Romans, but it also provided descriptions of plants. These excellent descriptions were based on firsthand observations of the plants. Albertus Magnus made efforts to classify of plants and is believed to have been the first to recognize, on the basis of stem structure, the differences between monocots and dicots. After Albertus Magnus more people looked at living plants in nature, and more significantly, they began to print books about them.

HERBALISTS

In the period of Renaissance, there was revival of scientific spirit, and interest in botany increased. The invention of printing with movable type in about 1440 allowed botanical books to be produced that were available to a wider audience than the former hand-copied manuscripts. Botanical books were produced with descriptions and illustrations made from woodblocks or metal plate engravings. They were intended to be used for indentifying medicinal plants. These books, or *herbals*, were in turn sought and used by the gatherers and diggers of medicinal plants, who were called *physicians* or *herbalists*. The herbals were written for utilitarian purposes; they were often cheaply done and of poor intellectual quality, sometimes containing independent observations but generally little that might be considered real scientific advancements.

German Herbalists

In the sixteenth century Germany was a center of botanical activity. The most outstanding contributions of this period were in the form of herbals by Otto Brunfels (1464-1534), Jerome Bock (1489-1554), Valerius Cordus (1515-1544), and Leonhard Fuchs (1501-1566). Brunfels, Bock, and Fuchs are sometimes called the "German Fathers of Botany". Brunfels was the earliest German Renaissance writer of note on botany. Bock's *Neu Kreuterbuck* contained excellent descriptions and some beginnings toward a system of classification.

Historia plantarum was completed by Cordus in 1540 but was not published until 1561, some 17 years after his death. Cordus in herbal contains descriptions of 446 species in flower and fruit. The botanical descriptions were prepared in an orderly format based on studies of living plants. Fuch's *De historia stirpium*, was better-documented and was the most noteworthy of the herbals of that period. German herbalists summarized much knowledge of plants, but no coherent classification scheme was produced perhaps because relatively few plants were discussed and there was no real need for an elaborate taxonomic scheme at this point in time. The activities of the German herbalists signified a return to personal observations, descriptions based upon actual living

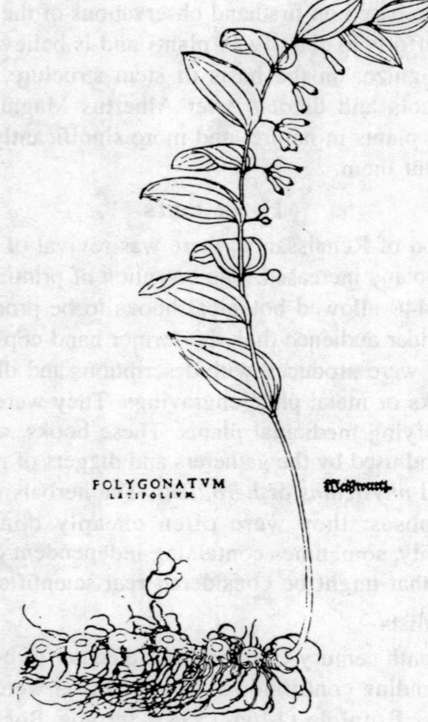

Fig. 1.1. Solomon's seal from Fuchs' De historia stirpium.

plants, and attempts to enumerate the local flora. Indeed, hints of classifications or groupings began to appear; for example, Brunfels placed several species of plantain (*plantago*) together; there were vague recognitions of kinds (species) and groups (genera).

Herbals of Other Countries or Civilizations

English botanical activity of the Sixteenth century was represented by William Turner (1510-1568) and John Gerard (1542-1609), The Dutch were represented by Charles de L' (Latinized as Carolus clusius) (1526-1609), Rembert Dodoens (Dodon-aeus) (1517-1585), and Mathias del l' Obel (Lobelius) (1538-1616). Pierandrea Mattiolia (Matthiolus (1500-1577), an Italian published in numerous editions a popular annotated and illustrated work derived from the writing of Dioscorides. Botanical activity flourishing throughout Europe. Although we have focused on Western Europe, the botanical activity in other civilizations should not be overlooked.

The Aztecs of Mexico developed botanical gardens, cultivated plants for both food and as ornametals, and used medicinal herbs. The *Badianus Manuscript*, an herbal, was compiled in 1552 by two Aztecs. Chinese civilizations was older and much more advanced during the Middle Ages than civilization in Western Europe. The Chinese were printing on paper with movable block type before A.D. 1000. They were very interested in plants and introduced many species into cultivation. Botanical works were apparently produced in China around 3600 B.C.; however the oldest manuscript still in existence dates back to about 200 B.C.

In India agriculture was developed nearly 2000 B.C, and many different crops were cultivated. One interesting Indian botanical work, thought to have been written around the first century, indicates that methods of cultivations were well known. For an additional account of the early history of botany in various cultures the reader is referred to A.G. Morton's *History of Botanical Science* (1981).

Italian Rennaissance

Luca Ghini (1490-1556)

Credit for the development of botany in the first half of the sixteenth century must be given to the Italian, Luca Ghini. He is sometimes credited with the invention of the herbarium, which provided the technological advance needed for the rapid accumulation of knowledge on classifications. Dried herbarium specimens could be studied at anytime, not just during the growing season, and could be used over a period of years, thus providing a "library" of pressed and dried plants. Herbaria were indispensable for the comparison and classification of plants. For example, most Linnaeus' descriptions of non-Swedish plants were from herbarium specimens. Ghini was also instrumental in the establishment of botanical gardens at Pisa and Florence. Gardens provided living plants for study, and the plants grown in them were used to prepare realistic illustrations.

Transition Period

The transition period for the renaissance to the modern period produced many notable workers and much literature, which can be only hinted at in this brief discussion. The aggressive exploration of the New World in the 1600s was responsible for the discovery of many new plants. In Europe the herbalists studied these plants and added them to their herbals. Because of the large number of new species, botanists needed to develop a more precise system for naming and arranging plants. For example in 1542, Fuch was concerned with 500 species, Bauhin in 1623 with 6,000 and Ray Iisted 18,000 species in 1682. Coping with such an avalanche of plants required discrimination of species and clearly demonstrated a need to recognize and describe genera.

Andrea Cesalpino (1519-1603)

Cesalpino (Lat.Caesalpinus), an Italian physician, followed Aristotelian reasoning and logic. Among the 1, 500 plants of *De plantis libri*, which appeared in 1583, he sought a classification using an philosophical rather than a purely utilitarian approach by basing his classification of the features of plants. Cesalpino followed the methodology of downward classification by the logical division, that is, starting with a number of easily recognized classes- or example, trees, shrubs, or herbs-then subdivide into subordinate sets of subclasses defined by single characters. However, a study of his classification shows 32 groups of plants, recognized by him, that appear to be related. It is likely that Cesalpino started with these groups and then searched for characters that would permit him to arrange them in accordance with logical, downward division. Cesalpino profoundly influenced botanists of a later date including Tournefort, Linnaeus, and other who continued with the same methodology.

Caspar Bauhin (1560-1624)

Bauhin, a Swiss botanist, *Published Pinax theatri botanici* in 1623 a list of 6000 plants. *Pinax* provided a much- needed synonymy of plant names by listing for each plant all the names given to it by different botanists. This proved to be a valuable botanical catalogue of plant names. Bauhin used some binomial nomenclature (names consisting of only two words) and appeared to have an understanding of a concept of grouping species into genera. The genera were not described but were defined by the characters of the included species.

John Ray (1627-1705)

John Ray, the son of a British black smith, was a graduate of Trinity college. He published numerous works. However, his two most significant botanical publications were *Methodus plantarum nova* (Ray, 1682) and *Historia plantarum* (Ray, 1686-1704), in three volumes. The last edition of *Methodus*, published in 1703, treated 18,00 species, many of which came from areas other than Europe. However, some inflation in number of species occurred in these works because of Ray's narrow concept of species. Ray developed a system, of classification based on form relationships by grouping together plants that resembled one another.

Ray believed that an acceptable classification was one that joins together those plants that are similar in features and separates those that differ in their characteristics. He attempted to define species in the terms of morphology and reproduction. His classification was the greatest advance in theoretical botany of the seventeenth century.

Ray's system later influenced the thinking of the de Jussieu and de Candolle families. In Ray and Bauhin can be seen the beginnings of *natural calssification*, that is the grouping together of those plants that resemble one another. Today it is said that plants that resemble one another are usually closely related. However, it was not until the time of Darwin that the idea of lineages was firmly established.

Joseph Pitton de Tournefort (1656-1708)

Joseph Pitton was a pupil of Magnol. He was a French botanist and is best known for his *Istitutiones rei herbariae*, published in 1700. *Institutiones* was very popular because of the ease in identifying its 9000 species arranged into 700 genera. Compared with Ray's *Historia*, his system of classification was interior because it was artificial. Its purpose was not to group closely related species but to aid in identification. Tournefort placed great emphasis on the genus. Although the roots of the concept are much older, dating to Aristotle, he is sometimes known as the "father of his genus concept." Tournefort used the concept of genus consistently and provided descriptions of the genera. Form that point on, the concept of genus was well established in classification. There is evidence in his writing that he began to develop a system of group at a higher level than that of the genus-that is aggregates of genera.

Carl Linnaeus (1707-1778) and Linnaean Period

Linnaeus was a great Swedish naturalist and classifier. He was born in Rashult, Sweden and became the renowned botanist of the eigthteenth century. He sometimes called the "father of taxonomy." Today Linnaeus is best remembered for his consistent use of a referable system of nomenclature, that is, the binomial system of nomenclature. Linnaeus entered the University Of Lund in 1727 to study medicine. Dissatisfied with Lund, in 1729 he moved to the University of Uppsala, where he came under the influence of the Dean, Olaf Celsius, who introduced him of Professor Rudbeck, a professor to botany.

On an expedition to Lapland in 1732, linnaeus greatly increased his knowledge of natural history,. In 1735 Linnaeus went to the Netherlands and quickly finished his medical degree at the University of Harderwijk. While in the Netherlands he became the personal physician to the wealthy banker George Clifford, who had extensive botanical and horticultural interests. Clifford became the patron of linnaeus. The three years spent in the Netherland and traveling in Europe were important and creative years in Linnaeus' life. Before returning to Sweden in 1738, he published several books on natural history. He was also able to meet many of the prominent naturalists of the time, including John Frederik Gronovius and Hermann Boerhaave from the Netherland, Professor J.J. Dillen and Sir Hans Sloane in England, and the de Jussieu brother in France.

After establishing a medical practice on his return to Sweden, he became professor of medicine and botany at the University of Uppsala in 1741, a position he held until his death in 1778. His three best-known works are the following: *Systema naturae*, 1735, et seq.,presenting his system of classification in outline form; *Genera plantarum*, 1737, providing descriptions of many genera; and *Species Plantarum*, 1753 a two-volume catalog used for plant identification. These have been reprinted many times and are available in most scientific libraries. Linnaeus was hailed by many of his contemporaries as starting a new epoch in botany. We now view his work as the culmination of an attempt to create a workable system of identifying and classifying plants. Even though the system was useful for identifying plants, natural relationships were not stressed and unlike plants were often grouped together.

Linnaeus divided plants into 24 classes based in large part on the number, union, and length of stamens. Plants with one stamen were placed in class Monandria, with two stamens call Diandria, and then Triandria, Tetrandria, Pentandria, and so on. Classes were divided into orders based on the number of styles in each flower. The strength of Linnaeus's artificial "sexual system" is its simplicity, since botanists could easily use it to apple names to plants. Today we consider the greatest contribution of linnaeus to be his consistent use of a precise and referable system of binomial nomenclature.

"Referable" means that the name of the species immediately indicates the genus into which it is classified. In *Species plantarum*, each plant had a generic name, a polynomial descriptive phrase intended to serve as a definition of the species analogous to our dichotomous keys, and trivial name (i.e., specific epither) printed in the margin. This trivial name was used originally for purposes of indexing. Essentially, it was shorthand for the "real name," the polynomial. Only later was its usefulness recognized. The generic name *Serratula*; the trivial is glauca, and the polynomial descriptive phrase is *foliis ovato-oblongis accuminatis serratis, floribus corynbosis, calycibus subrotundis. Serratula glauca* forms the binomial. Information was also included about previous publications and illustrations, and about specimens in his herbarium and where the specimens were collected.

The conveniece of using the trivial name in combination with the generic name was immediately obvious, and the binomial system of nomenclature was widely accepted. Linnaeus was not the first to use to names for each species; Bauhin and other had done so, but not consistently. Linnaeus's nomenclatural scheme survives intact to this day. Linnaeus recognized the limitations to be expected when only one set of characteristics is used (e.g., stamen number), but the convenienced for identifying plants was great. This convenience may have even impeeded botanical progress toward the development of a natural classification. His taxonomic scheme was purely artificial and was basically devoid of informational content.

It was the influence of the nomenclature that caused the taxonomic system to impede the progress of botanical calssification. In any analysis of the work of Linnaeus, the conditions of the eighteenth century and the confused state of botanical nomenclature must be considered. Linnaeus made use of the works of Bauhin, Caesalpino, Ray, Tournefort, and others and put together a syntheses of their ideas. Linnaeus and his followers are best associated with a botanical era, that of the artificial sexual system of classification. Linnaeus brought order to the overwhelming array of literature, and consensus and simplicity back into taxonomy and nomenclature. He is remembered for his energetic enthusiasm . Linnaeus' *Genera Plantarum* listed and briefly described the plant genera, continuing the tradition of Bauhin and Tournefort. *Species plantarum provides* much information on each species, but no generic descriptions.

Linnaeus' genera were often well-characterized groups of species; he obviously recognized such groups by visual inspection and elaborated logical division later on. Linnaeus' Teaching abilities resulted in his added importance as a naturalist. He attracted hundreds of enthusiastic students. Among his better-known students were Peter Kalm, Fredrick Hasselquist, Peter Thunberg, and Daniel Solander. His collections were sold by his widow in 1783 to J.E. Smith, an English botanist and one of the founders of the Linnean Society of London. The herbarium of Linnaeus is now housed in London at the Linnean Society in Burlington House, Piccadilly. (The name of the society is the "Linnean." The term used to describe Linnaeus' methods is

Linnaen.") The herbarium sheet have been photographed and are available in microfiche in many research libraries. Much has been written on the life and works of Linnaeus, and more information may be found in references cited at the end of this chapter (Blunt 1971; Daniels et al., 1976; Frangsmyr, 1983; Fries 1923; Stafleu, 1971; Stearn, 1957).

Natural Systems

By the end of the 1700s, most botanist realized that there were "natural affinities" among plants. Opposition developed to the artificial sexual classification scheme of Linnaeus because his system often placed unlike plants together (e.g., cacti and cherries). This opposition was especially strong in France, where the sexual system had never been fully accepted. There was gradual abandonment of (1) the use of single characters to classify plants, and (2) the selection of characters based on theory rather than experience or experimentation. The development of a classification system reflection natural relationships—which could also be used to aid in identification-became a major focus of botanical activity. A "natural system" of classification implies that plants presumed to be related are cataloged together. In its original context, the natural system was designed to reflect God's plan of creation and not one of lineages.

Michel Adanson (1727-1806)

The French naturalist Adanson traveled in tropical Africa, where the flora and fauna made his aware of the failings of one-character taxonomy. He questioned the validity of logical division and the sexual system and stated that the only natural method was to consider all parts of the plant. He advocated the use of as many characters as possible (Mayr, 1982). Adanson was the first botanist to attempt to determine how many characters should be used and whether there should be preference for certain characters. He realized that characters may vary in their taxonomic significance and proposed the weighting (Placing emphasis) of characters a *Posterori.*

The science of taxonomy has shown that some characters are in fact more valuable than others ; yet modern taxonomists try to use as many characters as possible. Adansonian methods have found advocates in recent years in numerical taxonimists who use computers to develop classifications from all the measurable features of plants. Adanson described groupings similar to our modern orders and 58 families (a category suggested earlier by Ray) as set forth in *Familles des plantes* (1763).

J.B.P. de Lamarck (1744-1829)

de Lamarck was the french biologists. He is generally best known for his unsuccessful attempts to provide an explanation for the idea of evolution, but he is also well known to taxonimists for his treatment of the plants of France. In *Flore francaise*, Published in 1778, he set forth his procedures for determining which plant precedes another in a natural series, and he stated rules for the grouping of species and the treatment of orders and families. Equally as important, de Lamarck developed as analytical method for identification similar to our modern keys found in various manuals.

Botanists of de Jussieu Family

There were four botanists in the de Jussieu family : the brother Antoine (1688-1758), director of the Jardin des Plantes in Paris: Joseph (1704-1779), explorer and collector in South America; and Bernard (1699-1777), founder of the royal Botanical Garden at Versailles; and their nephew Antoine-Laurent (1748-1836) founder in 1793 of the Musee d' Histoire Naturelle de Paris. Sometimes before the French Revolution, a decision was made to arrange the plants in the garden at Varsailles according to the Linnaean system. The garden was never arranged in this manner because Bernard de Jussieu felt that plant that "looked alike" (i.e., were presumed to be related) should be grouped together. This was not always true with Linnaeus' artificial sexual system, where unlike things were grouped together because they had the same number of stamens. It should, however, be remembered that the idea of evolution and lineges was not established until Darwin's time.

In 1763 Antoine-Laurent de Jussieu joined his uncle and began to work on natural groupings of plants in the garden. In 1789, during the French Revolution, Antoine-Laurent de Jussieu published *Genera plantarum secundum ordines naturales disposita*, a work resulting from his experiences in arranging the garden. The publication of *Genera plantarum* caused the idea of natural systems to be widely accepted. Antoine-Laurent de Jussieu recognized 100 carefully characterized groups of plants that are now called *families* (he called them *orders*). His system divided plants into three main groups—that is, the acotyledones (cryptogams and a few misplaced monocotyledons), monocotyledones (monocotyledons), and dicotyledones (dicotyledons and gymnosperms)- and it forms the immediate progenitor of our modern systems of classification. That this treatment was remarkable is indicated by the acceptance of most of the groupings today, along with the features by which they are recognized. The de Jussieu system was superior to the artificial sexual system of Linnaeus and was fundamental to further progress toward a natural system of classification. The philosophy of the natural system was firmly established in the scientific community by Antoine-laurent de jussieu. It would continue until the idea or organic evolution became firmly entrenched in the late 1800s.

de Candolle Family

The reputation of the Swiss-French de Candolle family was established by Augustin Pyramus de Candolle (1778-1841). Although born in Geneva, de Candolle received his botanical training in Paris and was greatly influenced by the French botanists of this day. In 1813 he published *Theorie elementaire de la botanique*, which set forth the principles of plant taxonomy. His classification scheme differed somewhat from the de Jussieu system, but it was a natural calssification scheme dividing the plants into two major groups: the cellulares (nonvascular) and the vasculares (vascular plants). A major effort of A. P. de Candolle from 1816 to his death in 1841 was the *Prodromus Systematis naturalis regni vagetabilis*, an attempt to classify and describe every known species of vascular plant. Although never finished, it remains the only worldwide treatment of certain groups. The first seven volumes were written by A. P. de Candolle himself, while the next ten were written by specialists and were edited by his son, Alphonse (1806-1983). Included in the *Prodromus* were all the dicotyledons (monocotyledons were not treated) accounting for 58,000 species in 161 families; de Çandolle placed the gymnosperms among the catkin-bearing dicotyledons and began the dicotyledons with the family Ranunculaceae (buttercups and their relatives).

S. L. Endlicher (1805-1849)

Over 6800 genera of plants were classified in Endlicher's *Genera plantarum*. This notable German botanist separated the Thallophyta (algae and fungi) from the higher plants (cormophyta). He began the dicotyledon families with those having apetalous flowers.

A. T. Brongniart (1801-1876)

In 1843, Brongniart advocated the division of plants into Cryptogamae (nonseed-bearing) and Phanerogamae (the seed-bearing plants). Brongniart also suggested the need for the careful comparison of living and fossil form of plants, significant aspect of modern plant systematics.

George Bentham (1800-1884) and Sir Joseph Delton Hooker (1817-1911)

Attention of many botanists during the first half of the nineteenth century was focused on the development and improvement of the natural systems of classification. This era was climaxed by the publication of Bentham and Hooker's *Genera Plantarum* in 1862-1883. Bentham was a self-trained English botanist, and Hooker was director of the Royal Botanic Gardens at Kew, near London. *Genera plantarum* is a three - Volume work in Latin, giving names and descriptions of all genera of seed plants. The classification system by which it is arranged is clearly derived from the system of de Jussieu and de Candolle. Bentham and Hooker divided the dicotyledons into : (1) Polypetalae (free petals), (2) Gamopetalae (fused petals), and (3) Monochlamydeae (apetalus). The dicotyledons were followed by the gymnosperms and monocotyledons. In all, 200 families and 7569 genera were treated.

Genera plantarum was one of the last great works in which the species concepts were based on the idea that species are fixed entities, unchanged through time and placed on earth by the Creator. Although it appeared after Darwin published his *The Origin of Species*, it is pre-Darwinian in concept, despite the fact that Hooker and Bentham were both supporters of Darwin. Nevertheless, their delimitation of families and genera is based on natural relationships, reflecting the heritage of de Jussieu. The reason that Genera Plantarum is so useful, even today, is that Bentham and Hooker prepared the generic descriptions from observations made of the plants themselves rather than by copying descriptions from the literature. Their generic descriptions are accurate, complete, and precise; the large genera are divided into sections and subsections; and the geographical range is given. Their system of classifications was widely accepted and used in the British colonial floras and elsewhere. British herbaria, such as Kew and the British museum, are still arranged according to it.

Botany in the Nineteenth Century

Many exciting and fascinating discoveries were made by botanists during the nineteenth century. The amazing Scotsman, Robert Brown, explored Australia and other exotic places, investigated floral morphology, recognized the naked ovule as the fundamental distinction between gymnosperms and angiosperms, and reported observations of the nuclei of cells. Wilhelm Hofmeister, the brilliant German botanist of the mid- 1800s, determined the life cycles of cryptogams (plants that are not seed-bearing) and formulated the concept of alternation of generations. The epoch-making researches of Eduard Strasburger in the late 1800s demonstrated nuclear fusion in the ovules of gymnosperms and angiosperms and laid the foundations of plant cytology. He failed to notice, however, the extraordinary "triple fusion" of the polar nucleus with the second sperm nucleus of the pollen, a phenomenon peculiar to flowering plants, which was eventually discovered in 1898 by the Russian, S. G. Navashin. These important developments, and others brought about by technological improvement in microscopes, gave systematist and greater understanding of diversity and provided the information needed to adjust their classifications and to reflect biological entities.

Influence of Darwin's Theory of Evolution of Systematic

By the middle of the nineteenth century, the stage was set for the idea of evolution. The observations and collections of world travelers and the endless flow of specimens back into Europe made it impossible to ignore knowledge regarding the fossil record and variation in species of plants and animals. The history of the theory of evolution is as fascinating as that of the history of plant systematics; however, we shall consider only its impact on plant systematics. Charles Darwin (1809-1882) and A. R. Wallace (1823-1913) wrote a paper entitled "On The Tendency of Species to Form Varieties and Species by Natural Means of Selection," which was read before the Linnean Society of London of July 1, 1858.

On November 24, 1859, Darwin's *The Origin of Species* was published, and every copy of the first edition was sold that very day. Biology was irrevocably changed. By the twentieth century, the concept that species are not unchanging creations was widely accepted. Rather than endless generations of species created all alike, species are dynamic, variable population systems that change with time and form lineages of closely related organisms. This idea provided a conceptual framework that had great impact on systematics. It should be pointed out that Darwin won his spurs as a taxonomist by eight years of study on the classifications of barnacles. Darwin wrote that his work on the taxonomy of barnacles was of great value and provided insight to him when he had to discuss the principles of classification in *The Origin of Species.* In this chapter he recognized certain rules of classification, namely, that (1) organisms have lineages, (2) characters constant over large groups have value in developing classifications, (3) taxa should be based upon correlated complexes of characters, and (4) characters should be weighted based on a *posteriori* information.

Transitional Phylogenetic Systems

As the impact of evolutionary theory became apparent, taxonomists began to integrate evolutionary concepts into their classifications. There were no abrupt departures from pervious systems, since the post-Darwinian system were firmly based on plant morphology. Taxonomists consciously tried to arrange the natural groups of plants in an evolutionary sequence proceeding from the simplest to the most complex. Botanist struggled with the problem of structures, which seemed simple but in reality resulted from the reduction or fusion of more complex ancestral features. Attempts were made to develop classification system that were evolutionary but that may seem naive to present-day taxonomist. Such system are said to be *phylogenetic*. It was soon recognized that the development of phylogenetic classification had several practical problems.

In the flowering plants it is usually impossible to reconstruct past evolutionary pathways because of the absence of a clear-cut fossil record. Difficulty arises in assuming that each taxon is monophyletic; that is, the group has a common origin. Finally, there is the difficulty of arranging a branching system such as a flora within the confines of a linear sequence in a book, where the first taxa treated are the least specialized and the last are the most advanced and presumably derived.

August Wilhelm Eichier (1839-1887)

Germany was the leading center for the study of plant morphology during the second half of the nineteenths century. Classification at that time was exclusively morphological, and German plant morphologists began to consider their data in light of evolutionary theory. A number of worker made contributions. Among the most important was Eichler, who in 1883 developed what was to become a widely accepted classifications of the plant kingdom. He divided the plant kingdom into the nonseed plants (the Cryptogamae) and the seed plants (the Phanerogamae), as previous botanists had already done. The former group contained algae, fungi, bryophytes, and seedless vascular plants, whereas the latter was divided into gymnosperms and angiosperms; angiosperms were further separated into monocotyledons and dicotyledons. The dicotyledons were subdivided into two groups, the Choripetalae (plants with flowers having free petals) and Sympetalae (those with fused petals); the old apetalae (flowers without petals) being distributed among the two groups. Eichler did not accept, or at least did not understand, the idea of secondary reduction (i.e., form complex back to simple flowers), which affected the placement of certain groups in his classification scheme.

Adolf Engler (1844-1933) and Kari Pranti (1849-1893)

Engler was a professor of botany at the University of Berlin and director of the Berlin Botanical Garden from 1889 to 1921. He proposed a system of classification based on that of Eichler, differing only in detail. Among with his associate, Prantl, he published *Die naturlichen Pflanzenfamilien* in many volumes from 1887 to 1915. This monumental work includes keys and descriptions for all the plant families. *Die naturlichen Pfilanzenfamilien* is abundatly illustrated, and the morphological, anatomical, and geographical literature are summarized. Engler and Prantl prepared many of the treatments, but a number were contributed by specialists. The size and scope of *Die naturlichen Pflazenfamilien* give it practical far greater than its value as a calssification device.

Engler's taxonomic scheme was widely accepted because of the detail and elaborate form of its presentation. Most non-British herbaria and many floras still follow his sequence in arranging plants. Engler did not intend for the precise linear sequence of his system to be understood as evolutionary. In the Englerian system, Plants were grouped into a number of divisions, many of which were algal group. In this scheme, the monocotyledons preceded the dicotyledons; the latter comprising two group: the Archichlamydeae (Chloripetalae or Polypetalae), which preceded the Metachlamydeae (Sympetalae or Gamopetalae). The first group of dicotyledons began with the catkin-bearing plants and their relatives, called the Amentiferae. Engler believed that simple, unisexual flowers were primitive, and in his earlier work he rejected the idea of secondary reduction. His system of classification has undergone continual revision by his followers and has been published many times in successive editions of *Syllabus der Pflanzenfamilien.* Volume 2 of the twelfth and latest edition, edited by Melchior and treatind the angiosperms, appeared in 1964. Through the years the general outline has changed little, although some families have been shifted within the classification. The present Englerian system is generally regarded as obsolete since most botanists do not agree with its assumptions about the primitive flower.

Intentional Phylogenetic Systems

Taxonomic schemes that try to reflect evolution are said to be *phylogenetic*. By the turn of the twentieth century, the impact of Darwinism and genetics was fully felt in plant systematics, and systems of classification were proposed that were intentionally phylogenetic. One of the advantanges of phylogenetic systems is that they are rich in informational content, since the identity of a plant implies kmowledge of its affinities and evolutionary relationships. Actually, the natural system and the phylogenetic systems are similar in many respects, and the families and genera differ little in content. The difference is in the placement of families and order, as well as what is considered to be the primitive flower. The Englerian school considered simple, unisexual flowers to be primitive. Examples of an Englerian primitive flower would be something like the unisexual flowers found in willow (*Salix*) or cottonwood (*Populus*). "Primitive = simple" in the Englerian context is a concept derived from reason, while in phylogenetic context, "primitive" means those things that were here first. However, without facts from the fossil record some of the latter reasoning may well be circular.

Charles E. Bessey (1845-1915)

A professor of the University of Nebraska, Bessey received part of his American to make a major contribution to the theory of plant classification. In 1894 Bessey, greatly influenced by the evolutionary ideas of Darwin, Published an intentionally phylogenetic taxonomic system based upon the principles of organic evolution. He considered the flowering plants to be monophyletic, that is, derived from one evolutionary line and representing a continuum of lineages. His system was based on a set of *dicta* or concept of primitive features found in ancient plants, versus the advanced characters of more recently evolved plants. His system is basically a modified Bentham and Hooker calssification with emphasis on primitive ranalian dicotyledous stock giving rise to the other dicotyledons and to the monocotyledons. His final classification was published in outline form shortly after his death in 1915. Among the lower plants, Bessey's system of classification was similar to Engler's scheme; however, it was different in that the pteridophytes and gymnosperms were subdivided into a number of divisions. Further, the dicotyledons were arranged in a quite different classification and began with the Magnoliaceae-Ranunculaceae complex.

Hans Hallier (1862-1932)

In 1905 the German botanist Hans Hallier published a phylogenetic classification based on many of the same principles used by Bessey. Hallier began the flowering plants with Ranunculaceae-Magnoliaceae complex and derived the monocotyledons from the dicotyledons. Since the time of Bessey and Hallier, most classifications of the flowering plants have used either a Bessey-Hallier system or a modified Englerian scheme. Changes were largely modifications at the level of family and order and in proposed lineages.

John Hutchinson (1884-1972)

The British botanist Hutchinson was associated with the Royal Botanic Gardens at Kew. In his *Families of Flowering Plants* (1973) and *Genera of Flowering Plants* (1964-1967), he proposed a system of classification resembling Bessey's but differing in several fundamental point. Essentially, hutchinson derived the flowering plants from hypothetical proangiosperms (i.e., the transitional plants from gymnosperms to angiosperms) and divided them into three lines: the Monocotyledones, Herbaceae Dicotyledondes, and Lignosae Dicotyledones. The primitive monocotyledons were derived from the primitive herbaceous dicotyledons, their point of origin being in the Ranales. Hutchinson considered the woody versus herbaceous habit of fundamental importance among the dicotyledons. The woody line was derived from the woody Magnoliales and the herbaceous line from the herbaceous Ranales. Hutchinson's treatment of individual families and genera, and especially that of the monocotyledons, is excellent. The division of the dicotyledons into woody and herbaceous lines is considered unfortunate and unnatural because it often places closely related families some distance apart. This has detracted from the overall value of his work.

Contemporary System of Classififcation

At present, the classification of the plant kingdom continues to be modified as new information become available. In recent years, the classification of flowering plants has benefited from data from paleobotany, biochemical systematics, and ultrastructure using scanning and transmission electron miscroscopes. Used in conjunction with information derived from traditional sources such as comparative anatomy and gross morphology, these kinds of data further refine the classification. The American plant taxonomist Robert Thorne proposed a synopsis of his classification in 1968 and somewhat more detailed exposition in 1976. Armen Takhtajan (1980) of the Soviet Union and Arthur Cronquist (1981) of the New York Botanical Garden presented detailed outlines of their classifications.

The systems of Thorne, Takhtajan, and cronquist are Besseyan in origin and tradition. G. L. Stebbins (1974) has developed a classification of the flowering plants in the Thorne-Takhtajan Cronquist tradition but with some modifications. These recent classifications continue to improve, and no doubt the future will see more revisions as botanists attempt to develop phylogenetic classifications based on evolutionary principles using newly available information. Among such attempts is the work of the Danish botanist, Rolf Dahlgren (1977, 1980, 1981), who bases his classification on the distributions of a large number of characters drawn from many biological disciplines including embryology, chemistry, and anatomy. Dahlgren follows the widely held modern view that none of the extant groups of flowering plants is ancestral to any other present-day group. Thus, the Magnoliaceae-Ranunculaceae assemblage is not ancestral to the other flowering plants but has simply retained a large number of primitive features. In England, K. R. Sporne (1976) has focused his efforts on developing a classification using character correlation among angiosperms. Sporne presents his system in the form of a circular diagram with the derived groups radiating outward and the groups located near the center of the circle, regarded as primitive.

2

PLANT NOMENCLATURE

The assignment of names to plants is called *nomenclature*. Just as all known objects are given names or should be given names, plants also are given names for two main purposes: (i) as an aid to communication and (ii) to indicate relationship. It involves principles governed by rules developed and adopted by the International Botanical Congresses. The rules are formally listed in the International Code of Botanical Nomenclature (Voss, 1983) and are often referred to simply as the "Code." The ultimate goal of this precise system, as embodied in the Code, is to provide one correct name for each taxon. The rules of nomenclature are subdivided into *articles*, which must be adhered to, and *recommendations*, which are optional. Although classification schemes may change with time, the scientific names of plants are relatively stable. The plant retains its name although the family or higher taxonomic categories are changed.

Much effort has been devoted to establishing procedures for naming taxa and for changing names that were incorrectly assigned. Nomenclature and classification are different but inseparable. The placement of a plant or group of plants in the classification scheme may be determined by knowing its name. When the generic name of plant is known, it is possible with the proper bibliographic aids to determine the family to which that genus is usually assigned. Such a bibliographic tool is *Dictionary of the Flowering plants and Ferns* (Willis, 1973).

BASIS OF SCIENTIFIC NAMES

The Present system of nomenclature is the result of historic series of changes that gradually became formalized. The oldest plant names we now use are the common names used in ancient Greece and Rome.

Today all plant names have a Latinized spelling or are treated as Latin regardless of their origin. The custom originates from medieval scholarship and the use of Latin in most botanical publications until the middle of the nineteenth century. The assignment of names was relaitvely unstructured until the seventeenth century when the number of plants known to botanists began to increase greatly. This resulted in a need for a more precise naming system for plants.

During several centuries before 1753, names were often composed of three or more words. These names are called *polynomial*. For example, in the herbal of Clusius (1583), the name *Salix pumila angustifolia altera* is used for a species of willow. This complex name-description system was not workable because it was cumbersome and not readily expandable. In 1753 with Linnaeus's *Species plantarum*, the *binomial* format was substitued for the polynomial. This two-word format made naming more convenient and provided a readily expandable system. Our present formal nomenclature began with the publication of Linnaeus's *Species plantarum* (1753). Since 1753 nomenclature procedures have become standardized through periodic legalistic revision so that plants are not named haphazardly.

Common Names

Many plants, because they are useful in some way or other, have common names or *vernacular* names or *local* names. Widely distributed plants have a large number of common names. Pansy (*Viola tricolor* L,) for example, is grown in most of the European and American gardens and has about 50 common English names. In a multilingual country like India, almost all useful plants have local names which differ from language to language and even from dialect to dialect. Not only this, but in *Ayurveda*, the ancient Indian literature on medicinal plants, the mango (*Mangifera indica* L.) is known by over 50 different names, all in the Sanskrit language. The common names obviously have a limited usage and for wider use throughout the country or the whole world, these have proved to be unsatisfactory. Since they are neither universal nor methodical, they may be misleading and inadequate.

Scientific Names

To overcome the difficulties raised by common names, botanists have given scientific names to all the known plants. These are methodical and thus provide means for international communication. It was agreed by botanists of the world that scientific names should be in Latin. There are various reasons for this. The most important fact is that this language is not being used by any country or nation at present. Secondly, at one time it was a widely used language throughout the European countries and a lot of botanical literature has been written in Latin. It is from Latin that most of the European languages have evolved. Although it is now a "dead language". There was obviously no proposal for Sanskrit to be used for naming plants although much more botanical literature is written in this language. However, since Europe, and particularly Greece, has dominated the whole world in the field of science during 1600 to 1850 A.D. Latin is now a language of botany and other allied sciences. It has been found to be suitable for descriptive phases of natural science. The script, however, is Roman.

Composition of Scientific Names

The genus Name and specific epithet together form a binomial called the *species name*. The term *species name* is often erroneously used to refer to the specific epithet alone, but the species name consists name of both the generic name and the specific epithet. A complete scientific name must be followed by the third element, the name of the person or persons who formally described the plant. For example, the complete scientific name for white oak is *Quercus alba* Linnaeus; the genus is *Quercus*, the specific epithet *alba*, and author citation Linnaeus. The author element of a name is often abbreviated, and "L." is normally used for the authority is place of "Linnaeus." To be correct, the species name of white oak is not "*alba*," but is *Quercus alba* L. Therefore a complete scientific name of a species consists of three elements: (1) the genus (plural, genera), (2) the specific epithet, and (3) the author citation.

Generic names

The generic name is a singular Latinized noun or a word treated as a noun. It is always written with an initial capital latter. After a generic name has been spelled out at least once, it may be abbreviated by using the initial capital latter; for example, "*Q.*" for *Quercus*. Generic names may not consist of two words unless they are joined with a hyphen. Latin inflectional endings are used for both generic names and specific epithets. Section 3 of the International Code of Botanical Nomenclature deals with what makes a generic names. Stearn's *Botanical Latin* (1983) is an excellent reference for the mechanics and grammar of botanical Latin. The name may be taken from any source, and it may commemorate some person of distinction. Genera such as *Linnaea* for Linnaeus of *Jeffersonia* for Thomas Jefferson are commemorative.

Many ancient common names, such as *Asparagus* and *Narcissus*, were converted into generic names directly from Greek. Features of plants, such as the liverlike leaves of *Hepatica*, gave generic names to still others, the word *Hepatica* being derived from the Latin word for liver. Information about a plant is sometimes expressed in a generic name because it indicates in a general way the kind of plant under consideration. With familiar genera we can recognize the plant by their generic names, for example, *Rosa* as rose and *Pinus* as a pine, both of which are ancient colloquial names.

Specific epithets

Specific epithets may be derived from any source and may honor a person, or they may be derived from an old common name, a geographic location, or some characteristic of the plant, or they may even be composed arbitrarily. The specific epithet is often an adjective illustrating a distinguishing feature of the species. Specific epithets consisting of two words must be hyphenated, as in the case of *Capsella bursa-pastoris* (L.) Medic. The specific epithet usually agrees with the gender of he generic name if the specific epithet is an adjective. If the specific name is an adjective placed in a genus that has the masculine ending *-us*, a species might be spelled *albus*, but if it is a genus with a feminine spelling, it would be spelled *alba*. In spite of its-us ending, *Quercus* is feminine for the purposes of botanical Latin: thus, *Quercus alba*. It is customary to treat all trees in botanical Latin as feminine, as was usually the situation in classical Latin.

A specific epithet may also be a noun in apposition carrying its own gender. When the noun is in apposition, it is normally in the nominative case-for example, *Pyrus malus* for the common apple. When a specific epithet is named after a person and ends in a vowel or *-er*, the latter *-i* is added (e.g., *glazioui*), but if it ends in consonant, the latters *-ii* are added (e.g., *ramondii*) (Recommendation 73C of the Code). When named for female, it ends in *-iae* or *-ae*; e.g., *luciliae*. Specific epithets derived from geographical names usually are terminated by *-ensis*, *-(a)nus*, *-inus*, *-ianus*, or *-icus*; examples are *quebecensis*, *philadelphicus*, and *caroliniaus* (Recommendation 73D). The Code recommends that all specific epithets be written with a small initial latter, but capital latter may be used when epithets are derived from a person's name, from former generic names, or from common names. Both the generic name and the specific epithet are customarily underlined when written or typed; when printed, they are in italics or boldface. The author citation is never underlined.

Author

The name of the person or persons following the genus and specific epithet indicates the author. It is a source of historical information regarding the name of the plant (Clausen, 1938). By giving the author's name, one may discriminate among names. The author citation may be abbreviated; for example, "L" for Linnaeus or "Michx." for Andre Michaux. Frequently a name will have two authors, with the first in parentheses. For example, with *Vernomia acaulis* (Walter) Gleason, the positioning of these two authors show that this species was first described by Walter, who supplied the epithet *acaulis*.

Walter put it in a genus other than *Vernonia*, and at some later point Gleason transferred this species to *Vernonia*. When the rank of a taxon is changed or when a species is transferred from one genus to another, the name of the describing author is placed in parentheses and is followed by the name of the person who made the change. Transfer are sometimes necessary in taxonomic studies when new information suggests that taxonomic boundaries be realigned. Name changes should be made only after careful consideration of taxonomic relationships and must follow the requirements of the International Code of Botanical Nomenclature (Weatherby, 1946).

Rules of Nomenclature

The increased number of plants known to European botanists in the eighteenth century required the development of order and stability in plant nomenclature. The first elemental rules of naming plants were proposed by Linnaeus in 1737 again in 1751. In the latter part of the eighteenth century, *Priority*, or the use of the oldest name, was recognized as the cornerstone of nomenclature, ensuring that each plant had a unique name. Botanists who did not adhere to this principle created confusion in the naming of plants. A. P. de Candolle in his *Theorie elementaire de la botanique* (1813), set forth a detailed set of rules on the process of assigning names.

Later, the rules of A. P. and A. de Candolle evolved into our present International Code of Botanical Nomenclature. Numerous plants were inescapably named tow or more times by accident, so in the years following Linnaeus a complex synonymy developed. Steudel in 1812 and 1840 and 1840-1841 published an index of plant of names, *Nomenclator botanicus*, which listed all names known to have been assigned to plants. This was useful for checking names and synonyms. It was the forerunner of *Index kewensis*. In 1867 the First International Botanical Congress was convened in Paris by Alphonse de Candolle, the son of A.P. de candolle. Botanists from many countries met and adopted a set of rules for nomenclature. Most of the rules had been proposed by Alphonse de Candolle.

The rules were an excellent beginning, but practical applications revealed some inherent deficiencies. The need to modify the rules became evident in the late 1800s when botanists at Kew Gardens in England, at the Botanical gardens at Berlin, at the New York Botanical Garden, and elsewhere began to stray from the rules adopted in Paris. Botanical Congresses were held in 1892, 1905, 1907, and 1910 in an attempt to resolve nomenclatural problems and establish internationally acceptable rules. Nomenclatural problems were standardized on a worldwide basis with general agreement reached by the International botanical Congress of 1930. Subsequent Congresses have been held on a regular basis and have offered only minor modification in the rules. A detailed history of the development of the Code is discussed in Lawrence (1951) and Smith (1957). Many of the terms used in the Code are elaborated upon and easy to find in Mc Vangh, Ross, and Stafleu (1968). Developments and changes in the Code are often discussed in the journal *Taxon*, published by the International Association for Plant Taxonomy.

Principles

Today botanists throughout the world use the International Code of Botanical Nomenclature, which is written in English, French, and German (Voss, 1983). A set of nomenclatural principles forms the philosophical basis of the Code :

1. "The nomenclatures of a taxonomic group is base upon the priority of publication." This very important principle provides that the correct name is the earliest properly published name that conforms to the rules. Earliest published names take precedence over names that conforms to the rules. Earliest published names take precedence over names of the same rank published later. Priority for plant nomenclature being May 1, 1753 for vascular plants and some other groups, but not for all plants. This was the date of Publication of the first edition of Linnaeus' *Species plantarum*.
2. "Scientific names of taxonomic groups are treated as Latin regardless of their derivation." This rule requires that generic and specific epithets, as well as other names, be Latin or treated as if they were Latin. The Code must be consulted for details on selecting the proper grammatical endings for names of all taxa.
3. "Botanical nomenclature is independent of Zoological nomenclature. The Code applies equally to names of taxonomic groups treated as plants whether or not these groups were originally so treated." The first part of this is of practical concern; that is, the

Code provides solely for the Nomenclature of plants. The same name that has been assigned to a plant may also be used by zoologists for naming an animal. For example, *Cecropia* refers both to a moth and to a large genus of tropical trees in the family Moraceae.

4. "Each taxonomic group with a particular circumscription, position, and rank can bear only one correct name, the earliest that is in accordance with the rules, except in specified cases."
5. "The application of names of taxonomic group is determined by means of nomenclatural types." The "type" principle provides that each species name must be associated with a particular specimen, the nomenclatural type. The type for a genus is a species, for a family it is a genus, and so on.
6. "The rules of nomenclature are retroactive unless expressly limited."

Rules adopted by the Congresses operate to affect nomenclatural matters carried out before the passing of the rules. Botanist should consult the latest International Code of Botanical Nomencalture when confronted with solving current nomenclature problems.

Procedures

Detailed procedures based upon these principles are divided into Rules and Recommendation. The Code states, "The objectice of the Rules is to put teh nomenclature of the past into order and to provide for that of the future: names contary to a rule cannot be maintained." The Recommendations dealing with minor points provide guidence and uniformity in naming plats. However, names that are contratry to the Recommendations cannot be rejected for that reason.

Ranks of Taxa

The formal taxonomic hierarchy is a system of categorical ranks with associated names (Scott, 1973). Generally, the species is the basic unit of classification. Each species belongs to a series of taxa of consecutively higher rank. The International code of Botanical Nomenclature provides the series of ranks with names the are the hierarchical categories. The ranks, in descending sequence, provided by the Code are shown in Table 2.1, along with an example of each. The Code, in effect, defines the categories only by listing their sequence. It may not be necessary to use all the categories provided by the Code for a small order, family, or genus, but the sequence of categories must not change. However, certain categories (i.e., species, genus, family) are essential if nomenclature is to function. The categories commonly used in the flowering plants are the class, subclass, order, family, genus, species, and sometimes either subspecies or variety or even sometimes both. Categories such as subfamily, tribe, subgenus, section, and so on may be used and are frequently necessary in large and complex groups.

Table 2.1. Series of Ranks Provided by the international code of botanical nominclature

Ranks of taxa	*Example*	*Endings of ranks above genus*
Division	Magnoliophyta	-phyta
Class	Megnoliopsida	-opsida
Subclass	Asteridae	-idae
order	Asterales	-ales
Suborder		-inales
Family	Asteraceae (or Compositae)	-aceae
Subfamily		-oideae
Tribe	Vernonieae	-eae
Subtribe	Vernonineae	-ineae
Genus	*Vernonia*	
Subgenus		
Section	Lepidoploa	
Subseciton	Paniculatae	
Series	Verae	
Subseries		
Species	*Verninia angustifolia* Michx.	
Subspecies	*V. angustifolia* ssp. *angustifolia*	
Variety		
Subvariety		
Form		
Subform		

In actual practice, species are grouped into genera and genera into families and so on through the sequence of categories. Each rank in turn is more inclusive than the lower categorise. This categorization gives order and accessibility to the classification of plants and provides a meaningful system of information input or retrievel. The Code requires standardized grammatical endings for the categories from division down to subtribe. However, an exception is the use of certain family names which have been sanctioned by the Code because of old, traditional usage. These names do not end in the usual family ending of *-aceae*. The names of these families, along with their alternative names, are Palmae (Arecaceae), Gramineae (Poaceae), Cruciferae (Brassicaceae), Leguminosae (Fabaceae), Guttiferae (Clusiaceae), Umbelliferae (Apiaceae), Labiatae (Lamiaceae), and Compositae (Asteraceae). Botanists are authorized by the Code to use either of these alternatives. Some manuals use the older names and others use the *-aceae* names. Family names ending with -aceae are based on generic names; for example, *Brassica* is the base of Brassicaceae. To follow correct usage, family names are treated grammatically as plural nouns.

Type Method

Names are established by reference to a nomenclatural type. Taxonomists use the type method as a legal device to provide the correct name for a taxon. The nomenclatural type of a species, a *type specimen*, is a single specimen or the plants on a single herbarium sheet. The type specimen for the species *Vernonia alamanii* DC. is located in the de Candolle Herbarium in Geneva, Switzerland. The type of genus is a species; for example, the type of the genus *Vernonia* is *V. noveboracensis* (L.) Michx. The type of family is a genus; for instance, the genus *Aster* is the type genus for the family Compositae (Asteraceae). The nomenclatural type is not necessarily the most representative of a taxon; it is the speciemen or specimens with which the name of that taxon is permanently associated, whether it is a correct name or a synonym.

The type specimen in no way reflects the typological concept of an idealized specimen. The type specimen has nothing at all to do with variation nut indicates the attachment of a name to a particular specimen. When a species new to science is collected, several things must be done : (1) it is given a name; (2) a Latin diagnosis or description is prepared; (3) a type is designated; and (4) the name and description are published. All these must be done in accordance with the Code. An example of a publication of the name and description of a new species. In this description, a type was designated and deposited in the New York Botanical Garden Herbarium and a Latin description was provided. The Code designates several kinds of types.

The *holotype* for the name of a species is the one specimen used or designated by the author in the original publication as the nomenclature type. If a holotype was designated by the author, it may not be rejected; and any type. If a holotype was designated by the author, it may be rejected; and any type chosen after the original publication cannot be regarded as the holotype. Today it is essential that a holotype be designated for a newly described species and deposited in an established public herbarium. In the earlier days when the rules of nomenclature were not standardised or when the type method was not followed, the authors of new taxa did not designate types. In such cases, or when the holotype has been lost or destroyed, a *lectotype* or *neotype* may be designated as a substitute for it.

A *isotype* is any duplicate of the holotype; it is always a specimen. Plant specimens are generally collected in sufficient numbers (minimum 4 specimens) out of which when one is selected as a holotype others are considered as isotypes. But if the author has not designated a single holotype and has used all or more than one specimen, then these are called as *syntypes*. A *lectotype*, as stated above, is a specimen or other element selected from the original material to serve as nomenclatural type when no holotype was designated at the time of publication or as long as it is missing. A *neotype* is a specimen or other element selected to serve as nomenclatural type as long as all of the material on which the name of the taxon was based is missing.

Obviously, a lectotype or a neotype is designated by others and not by the author himself. It is therefore necessary to select them carefully. The Code specifies that lectotype must be selected either from isotypes or syntypes if such exist. If neither is available, a neotype may be selected from any other suitable source. Further, it is strictly recommended by the Code that the original material, especially the holotype, of a taxon be deposited in a permanent responsible institution and that it is scrupulously conserved. When a living material is designated as a type (for Bacteria and Fungi only) appropriate of it should be immediately preserved. The principle of typification does not apply to names of taxa above the rank of family. In the literature one finds occasionally mention of *cotypes* (the term identical with isotype) and *topotypes* (specimens collected subsequently from the same locality as that of the holotype). These, however, are not accepted by the present code.

An additional type, a *paratype*, is sometimes used, which is a specimen other than an isotype or a syntype cited by the author while describing a taxon. In most cases, where no holotype was designated, there will also be no paratypes, since all the cited specimens will be syntypes. However, in cases where an author has cited two or more specimens as types, the remaining cited specimens are paratypes and not syntypes. The Code has proveded a guide to the determination or selection of the nomenclatural types of previously published taxa. The early botanist did not desgnate tyes as is done today. To these early botanist, species were based upon all specimens, illustrations, and descriptions within the limits of the species. These elements or everything associated with the name at first publication are known collectively as the *Protologue*. Recommendation 7B of the Code suggests that when elements of the protologue are heterogeneous, the lactotype should be selected to preserve current usage.

Priority of Names

Priority is concerned with the precedence of the date of valid publication and determines the acceptance of one of two or more names that are otherwise acceptable. A name is said to *legitimate* if it is in accordance with the rules and *illegitimate* if it is contrary to the rules. The rule of priority states, "For any taxon form family to genus inclusive, the correct name is the earliest legitimate one with the same rank, except in cases of limitation of priority by conservation". The Code contains several limitations on the principle of priority, "The principle of priority does nor apply to names of taxa above the rank of family". To avoid disadvantageous changes caused by strict application of priority, some specific, generic, and family names are conserved by action of the International Botanical Congresses.

Conserved names are referred to as *nomina conservanda*. This means that some names, even though they are not the oldest legitimate names, are used in preference to the older names. Occasionally, family or generic names, perhaps published in obscure publications or otherwise not used, will have priority over well-known names despite not having been in regular use. Adoption of such generic names usually requires formally transferring specific epithets to the resurrected genus. To avoid the confusion this would cause, names are conserved by decisions of an International Botanical Congress. Conservation of specific names is restricted to names of species of major economic importance. The conservation of specific names was authorized only recently at the international Botanical Congress held at Sydney, Australia in 1981.

Valid Publication of Names

To become a part of the legal botanical names of taxa must meet certain requirements which are explicitly stated by the Code. "Publication in effected, under this Code, only by distribution of printed matter (through sale, exchange, or gift) to the general public or at least to botanical institutions with libraries accessible to botanists generally.

It is not effected by communication of new names at a public meeting, by the placing of names in collections or gardens open to the public, or by the issue of microfilm made from manuscripts, typescripts or other unpublished material. Offer for sale of printed matter that does not exist does not constitute publication" (Article 29). Currently, publication of handwritten descriptions or descriptions printed in nursery catalogs or seed exchange lists in not considered to be effective publication. A plant name is not effectively published if printed on a label attached to herbarium specimens even if the specimens are widely distributed.

Effective publication refers to the place and form of publication of the names of plants. The botanical community must communicate plant names in widely distribution scientific literature. For valid publication, a name must be *effectively published* in the form specified by the Code. It must be accompanied by a description or a reference to a previously published description for that taxon. Since 1935 all diagnoses of new taxa (algae and fossil plants are excepted) must be written in Latin to be validly published. The *diagnosis* is a statement by the author giving the distinguishing features of the taxon. The description itself need not be in Latin, although it is recommended.

Author's Name

The name of a taxon should include a citation of author or author who originally described that taxon : for example, *Vernonia arkansana* DC., for A. P. de Candolle; *Vernonia* Schreb., For J. D. C. von Schreber; and the tribe Vernonieae Cass., for Henri Cassini. There are many sources of explanations of abbreviated names of authors, including the *Manual of the Vascular Plants of Texas*, Gray's *Manual of Botany* and the *Draft Index for Author Abbreviations* (1980) prepared at the Royal Botanic Gardens, Kew, England. The author citation expedites locating the original plant description, which helps determine the type and date of publication for the taxon. Sources providing references to original descriptions are *Index kewensis* on an international basis and the *Gray Herbarium Index* for New World plants. Either source provides references to the original description. Another function of the author citation is to identify the name. Through unfortunate error an author may publish a name that is preoccupied; that is, the specific epithet may have been used for another taxon in the same genus. The author citation allows the distinction between the two names. Of course, only the earlier name is legitimate.

Author citations can aid botanists in tracing the transference of species form one genus to another. For example, the author citation for

Vernonia noveboracensis (L.) Michx. Reveals that Andre Michaux transferred to the genus *Vernonia* a species originally described in another genus by Linnaeus. The original species name used by Linnaeus was *Serratula Noveboracensis* L. Since *Serratula* L. was published in 1753 and *Vernonia* Schreb. in 1791, it appears as though *Vernonia* violates the rule of priority. When names are published by two authors, the author citations are linked by either & or et (Latin, "and")-for instance, *Opuntia pollardii* Britt. et Rose, for N. L. Britton and J. N. Rose. The author citation *Carex Stipata* Muhl. ex Willd, indicates that the name was ascribed to G. H. E. Muhlenberg but was published by K. L. Willdenow, who attributed the name to Muhlenberg. When a name with a description supplied by one author is published in a work by another author, the word it should be used to connect the names of the two authors, for example *Viburnum ternatum* Rehder in Sargent.

Retention and Rejection of Names

The Code has rules outlining the proper procedures for selecting the correct name when taxa are divided, transferred, or rejected. That is, a genus might be divided into two genera or a species transferred from one genus to another genus; or it a name is illegitimate, it is rejected. A brief synopsis of the major points of the most important rules concerning retention, choice, and rejection of names is presented here. A change in the diagnostic limit separating the taxon from its nearest relatives is not justifiable cause for a change in the name of the taxon. For example, change in the concept of diagnostic characters of a genus or species is not a reason to change a name. If a genus is divided into two or more genera, the original name is retained for the genus that includes the nomenclatural type species for the genus.

Likewise, when a species is divided into two or more species, the original specific epithet must be retained for the species that includes the type specimen. This same rule applies to infraspecific taxa, that is, subspecies and varieties. When a species is described in one genus and later transferred to another genus, the specific epithet, if legitimate, must be retained. *Chrysocoma acaulis* Walt. , 1788, is now treated as *Vernonia acaulis* (Walt.) Gleason, 1906. Chrysocoma acaulis Walt. is the *basionym* of *Vernonia acaulis* or the name-bringing synonym associated with Walter's type specimen, which is located in the Museum of National History in Paris. For many years this species was called *Vernonia oligophylla* Michx., 1803.

In 1906 Gleason properly recognized that there was an older specific epithet and basionym for this taxon, *C. acaulis* Walt., 1788. If a taxonomist considers two previously distinct species to be the same species, the earlier epithet must be selected for the newly combined species. The latter name is then considered to be a taxonomist synonym. In such cases the identity of the basionym is important in determining the correct name for a species. When a subspecies or variety is transferred to another genus or species, it is no different than transferring specific epithets, for epithets hold their priority within rank. If it is transferred without change of rank, the original epithet must be retained, unless there is some nomenclatural barrier.

When two or more taxa of the sme rank are united, the oldest legitimate name or epithet is selected. For example, if the genera *Sloanea* L., 1753, *Echinocarpus* Blume, 1825, and *Phoenicosperma* Miq., 1865, are united, Sloanea L. is the oldest name and would be correct. The other two names become taxonomic synonyms. When identifying plants, one may notice that many manuals will cite one or more synonyms for certain species treated in that flora. This practice is helpful because familiar names may become synonyms. A fine example of this is Radford et al. , *Manual of the Vascular Flora of the Carolinas* (1968). Following the description of *Solidago graminifolia* (L.) Salisbury, they cite as synonyms : *Euthamia graminifolia* (L.) Nutt. -S; *S. graminifolia* var. *nuttallii* (Greene) Fernald -F, G. From this information you can compare treatments of the taxa in different manual.

Vernonia leiocarpa, the correct name for the species, was published by A. P. de Candolle in *Prodromus*, in 1836. The type was collected in Mexico by Karwinski s. n. (Latin for *sine numero*, meaning "without [collection] number") G-DC indicates that the type is located in the de Candolle Prodromus herbarium at the Conservation Botanique de Geneve, Switzerland. Each herbarium is assigned an abbreviation by the International Association for Plant Taxonomy. These abbreviations are found in Holmgren, Keuken, and Schofield (1981). The type specimen was viewed on IDC microfiche of the de Candolle Prodromus hebarium. The exclamation point(!) is an abbreviation for *vidi* (Latin, "I have seen it"), and it indicates that the author of the revision has seen specimen cited. In 1891 Kuntze transferred the specific epthet to the genus *Cacalia*, making the combination *Cacalia leiocarpa* (DC.) Kuntze.

In 1906 Gleason Made the combination *Eremosis Leiocarpa* (DC.) Gleason, by transferring *V. leiocarpa* DC. to *Eremosis*. Gleason also described *Eremosis melanocarpa* Gleason, based on the type collected in Guatemala by Heyde and Lux. 3416 is the collection number of the particular specimen of Heyde and Lux. This holotype is located in NY (New York Botanical Garden) and was examined by the author of the revision, as indicated "!" Isotypes examined are located in F (Field Museum), Gh (Gray Herbarium), MO (Missouri Botanical Garden), and US (United States National Museum). Blake did not recognize Eremosis and transferred the specific epthet *melanocarpa* to *Vernonia*. The author of this taxonomic revision believed *V. melanocarpa* to be a taxonomic synonym of *V. leiocarpa*. A review of the formal taxonomic treatment provides a taxonomic history of the entity. There is a growing an useful practice to give all the names based on the same type specimen in one paragraph, thus using a paragraph for each basionym, its nomenclatural type, and its taxonomic synonyms.

The *basionym* is the epithet with which the type is associated. Another important rule of the Code is the following : "A legitimate name or epithet must not be rejected merely because it is inappropriate of disagreeable, or because another is preferable or better known, or because it has lost its original meaning". Therefore, the name *Scilla peruvina* cannot be rejected because it grows in the Mediterranean area rather than in Peru. *Vernonia Crinita* Raf. is better known than the older *V. arkansan* DC., but *V. crinita* Raf. Cannot be retained just because it is better known. A name is a *later homonym* if it is spelled like a name previously and validly published for a taxon of the same rank based on a different type. Different genera or different species within a genus cannot have the same name. If they do, the earlier name is legitimate and the later name is a later homonym. *Tepeinanthus* Boiss, ex Benth, 1848, is a later homonym of *Tapeinanthus* Herb., 1837. *Astragalus rhizanthus* Boiss. , 1843, is illegitimate because it is a later homonym of *Astragalus rhizanthus* Royle, 1835.

The Code deals with other matters, including spelling of names and epithets. The names of plants must be spelled as they were originally published unless there was a spelling or typographic or typographic error. A frequent source of confusion is the naming of infraspecific taxa. If a subspecies or veriety is described in a species not previously divided into infraspecific taxa, there is automatically a "type" subspecies or variety. This bears the same epithet as the species but is not followed by an author citation. This means that a species with infraspecific taxa must have at least tow subspecies or varieties. For example, the two subspecies taxa must have at least two subspecies or varieties. For example, the two subspecies or *Vernonia obtusa* (Blake) Gleason are *Vernnonia obtusa* subsp. *obtusa* and *Vernonia obtusa* (Blake) Gleason subsp *parkeri* S. B. Jones (Subsp. *obtusa* is not followed by an author citation.) The Code requires that the epithet be repeated and the oritinal type specimen be the type of subsp. *obtusa*. The logic for this is simple. Creation of a subspecies (or variety) automatically creates two subspecies: (1) the entity that the author has in mind when erecting the new subspecies, and (2) the remaining materials within the species.

The former receives the new name (e.g., *Vernonia obtusa* subsp. *Parkeri* S. B. Jones), while the latter is signified by repetition of the specific epithet (e.g., *Vernonia obtusa* subsp. *obtusa*). Names such as *Vernonia obtusa subsp. Obtusa subsp. obtusa* are termed autonyms or automatically established names. Hybrids between different species in the same genus or between closely related genera are sometimes described and named. In order to be validly published, the names of hybrids follow the same rules as those that relate to names of nonhybrids. Additional rules and recommendations needed for the naming of hybrids are found in the Code in Appendix 1, Names of Hybrids. Some of the more important points include the following : Hybrids between two species or the same genus are shown either by a formula- for instance, *Salix aurita* L. x *S. caprea* L., or if desired, by a formal name-for example, *Quercus* X *beadlei* Trel., a hybrid of *Q. alba* L. X *Qmichauxii* Nutt. It is permissible under the Code to refer to them by formula or to give them a name (binomial). The opinions or plant taxonomists differ as to whether hybrids should or should not be named.

Cultivated Plants

Article 28 of the Code deals with nomenclature for cultivated plants. Plants brought in form the wild and cultivated retain the names applied to the same taxa in their native habitat. Horticultural plants that are produced in cultivation through hybridization, selection, or other processes and that are worthy of being named receive cultivar names. The term *cultivar* denotes an assemblage of cultivated plants that is clearly distinguished by any characters (morphological, physiological crytological, chemical, or other), and that following reproduction (sexual or asexual), retains its distinguishing characters (Brickwell, 1980). *Cultivar* is derived from the terms *cultivated variety*.

It should be noted that the concept of the cultivar is not analogous to the botanical veriety (varietas), the latter being a category below that of species and governed as such by the international Code of Botanical nomenclature. Cultivar names are written with a capital initial latter. They are either preceded by the abbreviation cv. (meaning cultivar") or placed in single inverted commas, for example, *Hosta* 'Decorata.'

Cultivar names may be used after generic, specific, or common names. Examples of cultivars are *Camelia japonica* cv. Purple Dawn and *Citrullus* cv. Crimson Sweet (or watermolon cv. Crimson Sweet, or *Citrullus lanatus* cv. Crimson Sweet,). The use of the term variety to refer to cultivers is improper, but the usage was traditional for a long time and is only now going out of style. Detailed information on specialized nomenclatural situations dealing with cultivated plants may be found in the International Code of Nomenclature for Cultivated Plants. The nomenclature for cultivated plants must follow this Code. Some of he more important general rules governing cultivar names are the following:

1. New cultivar names must now be in modern languages and not be Latin names. For example, the Latin '*albo-marginata*' cannot be given to a new cultivar. The only exception to this are names of botanical taxa reduced to cultivar ranks.
2. It is recommended that cultivar names be registered with a registration authority to prevent duplication or misuse of cultivar names. Fro example, the American Hosta Society acts as the registration authority for cultivars of the genus *Hosta*.
3. Since January 1, 1959, new cultivar names must not be the same as a botanical or common name or a genus or a species. Thus, names such as *Hosta* cv. 'Rose' would not be permitted.
4. If the botanical name of a species is changed, the cultivar name remains unchanged. For instance, it the scientific name of the tomato should be changed to conform with the Botanical Code, the cultivar names, such as 'Batter Boy' or 'Ultra Girl,' remains unchanged.
5. New cultivar names published after January 1, 1959, require a published or duplicated description that may be given in any language and dated at least to the year.
6. Two or more cultivars in the same cutivar class are not permitted to bear the same name. A cultivar class is usually a genus but may also be a species, a crop type, or a group of cultivers. For example, since there is already a *Hosta* 'Decorata,' a second *H.* 'Decorata' could not be named.

Conclusion

The International Code of Botanical Nomenclature deals with the terms used to denote the ranks of taxa as well as with the scientific names applied to plants. There are valid reasons for the occasional but necessary changes of familiar plant names. Examples and problem for practice with application of the Code may be found in St. John (1958) and Benson (1962). The use of the case method with these problems is an excellent way for a potential taxonomists to develop a working knowledge of the Code. Davis and Heywood (1963) observe, "To most systematist, however, nomenclature is a time-consuming necessity that comes between them and the plant. Nevertheless, it is one of the tools of the taxonomists' trade and for that reason its principles must be mastered." It should be emphasized that for detailed knowledge of the rules of nomenclature, the Code itself must be consulted.

3

PRINCIPLE OF PLANT TAXONOMY

The basis of taxonomy is the similarities and dissimilarities and dissmilatities among organisms. Historically, taxonomy has been a descriptive science based on the variation and form of morphological characters. The classification schemes of the taxonomists of the 1700s and 1800s placed similar-appearing organisms together in species, comparable species into genera, and genera with resemblances into families. Taxonomists sought the "ground plan" of the Creator. With the advent of Darwinism in the late species relatedness and evolution were incorporated into classification. Darwin's evolutionary theory provided the concepts that species represent lineages and that species within a genus have evolutionary affinities with one another. Genera and families represent progressively older divergences in the lineages. Before Darwin, it was believed that each species was created individually be God and remained unchanged through time. This is the idea of *special creation*, a hypothesis now rejected by the scientific community. Because similar appearance often reflects a close relationship, the Darwinian concept of lineages did not radically alter previous classification schemes. It did lead to the development of deliberate phylogenetic classification schemes.

Phylogenetic classifications presume to express to express genetic relationships. In both theory and practice, genera, families, and so on should reflect the notion that the members of each group share a common origin. This concept had little effect on the species, genus, or family levels of classification, but the arrangement of families changed drastically from a linear sequence to a branching format designed to reflect evolutionary relationships. Today most taxonomic treatments are implicitly phylogenetic. They attempt to recognize and bring together related groups of plants. New research methods have been developed, and taxonomists now use not only gross morphology but also chemistry, anatomy, ultrastructure, and a variety of sophisticated techniques to determine relationships that result in the classification of taxa. Changes in classification occur because of the discovery of new information and in the reinterpretation of existing information.

CATEGORIES

A *hierarchy* is an orderly array composed of a series of inclusive levels called *categories*. The categories are the ranks to which taxa are assigned; for example, species, genus, family and so on. The higher categories, such as family or order, are more inclusive than the lower categories of genus of species. The other categories are regularly used to reflect evolutionary relationships in very large and complex group. The categories are defined by the Code according to their position relative to other categories. The Code does not define or explain what is meant, in a biological sense, by a species, a genus, or a family. The classifications used in botany are based on the fundamental category of the species. Species are grouped into the higher category of genus, and then genera are grouped into families. Species can be subdivided into further subspecies or varieties.

Monophyletic

A *monophyletic* taxon is one derived from one ancestral population system, whereas a *polyphyletic* taxon has been derived from two or more ancestral population systems or taxa. It has become a common principle of systematics that taxa should be monophyletic. If assemblages of plants seem to be polyphyletic, they should be divided into monophyletic groups. Consequently, monophyletic categories at all ranks in the hierarchy express evolutionary lineages and relationships. Because of the sparse fossil record, it is often difficult to determine the origin of the higher ranks in the angiosperms, accounting for differences in contemporary classification systems. In actual practice, determination of the monophyletic line represents a hypothesis inferred from the study of present-day members of the group.

Species

A group of individual plants that is fundamentally alike is generally treated as a *species*. Ideally, a species should be separated by distinct morphological differences from other closely related species. This is necessary in order to have a practical classification that can be used by other. However, it is sometimes difficult to delimits a species precisely. If each minor variation in plant populations were to be made the basis of species distinction, there would be no end to the number of species. For systematic purposes, difference in features should not be allowed to obscure resemblances. A species is a concept that cannot be defined in exact terms and is not absolute and inelastic. Rigorous definitions of species are not possible because the criteria may change with the characteristics of each group.

Much of the disagreement about what is actually a species is vague or minor and tends to obscure the large amount of agreement on what constitutes a species. In developing concepts of species, specimens should be regarded as samples of living, reproducing populations of genetically related individuals. Many different kinds of species have various strategies of reproductive isolation that reduce or prevent interbreeding. For example, many species are sexual, a few are asexual; some have arisen by polyploidy, changes in chromosome number, or other mechanisms. Since species represent lineages, system of living populations may be found at various stages or levels of morphological divergence from one another in reproductive isolation. The importance of a concept of divergence was stressed by fosberg (1942), who observed that many kinds of evolutionary processes produced numerous sorts of species at various stages of divergence from one another.

The *biological species concept* envisions the species as distinct population system of central importance in nature and in evolutionary biology. According to this concept, biological species are kept separate from closely related species inhabiting the same general region by reproductive isolating mechanisms that prevent or greatly reduce gene exchange between them. Two such reproductively isolated species can coexist in the same general area and not lose their identity. In actual taxonimic practice, the biological species concept is difficult to apply to plant species. A diploid and a tetraploid cannot effectively exchange genes; but for practical purposes, different ploidy levels within a morphological species are sometime regarded as representing the same species.

When many plant species come into contact, they are in some cases able to exchange genes and may produce fertile hybrids. First-generation hybrids between plant species range in fertility from completely fertile to partially or completely sterile. For this and other reasons, the ability or lack of ability to produce hybrids and exchange genes cannot be used as a general criterion for the erection of species boundaries in most plants. Mishler and Donoghue (1982) suggest that a variety of species concepts in necessary to capture adequately the complexity of the variation patterns in nature. They urge systematists to develop a pluralistic approach to the criteria establishing species and urge specialists to develop species concepts for their particular taxonomic group reflecting the biology of the group. This is not to say that species are not meaningful groups, but simply the recognition that no one criterion can be used to delimit species in all groups of organisms.

Infraspecific taxa

A species embraces the variation within its populations. To manage some of the recurring variation, three infraspecific categories are used by plant systematists to provide formal taxonomic recognition of variation within species. These are *subspecies*, *variety* (Latin, *Varietas*), and *form* (Latin, *forma*). The categories *Subspecies* and variety are applied to populations of species in various stages of differentiation.

A common process of evolution and speciation in many plants is the gradual divergence of a former homogeneous species or population system into two or more diverse population system. Their divergence is usually related to adaptation to differing geographical areas or climates, or to local but sharply distinct ecological habitats. During the process

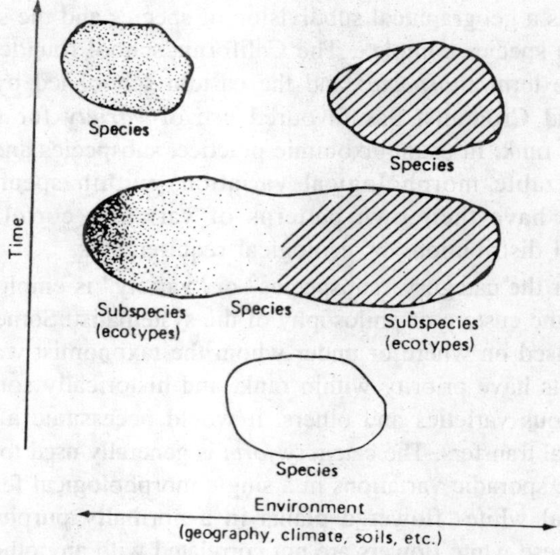

Fig. 3.1. Differentiation of a single species in a relatively homogenous environment into subspecies in differing environments, and eventually into two species following further differentiation of the populations.

or becoming adapted to different habitats, the local populations become genetically distinct. This genetic differentiation is reflected in both morphology and physiology. Often these differentiated populations, or *ecotypes*, occupy adjacent ranges where they interbreed and intergrade at the points of contact.

Consequently, there can be gradual or sharp discontinuities in the variation patterns between the divergent populations, depending on the environmental gradient. These ecoptypes most often form the basis or subspecies or variety. Fosberg (1942) suggests that the stages of divergence can be indicated by the infraspecific category in which the population are placed. This gives the taxonomist a great deal of flexibility by providing several infraspecific ranks that may be used to fit various evolutionary situations. Variety was the first infraspecific category to be used for plants. To Linnaeus, it was primarily an environmentally induced variation. The term *subspecies* was introduced early in the 1800s by Persoon (1805-1807) and was considered a major variation in the species. Gray (1887) regarded the variety as a considerable change in the ordinary appearance of a species.

In 1926 Hall argued that the term *Variety* had so many uses and had been applied to so many different things, from races to cultivars, that its use should be discontinued. Other, such as Fernald, regarded the variety as a geographical subdivision of species and the subspecies as parts of a species complex. The California school founded by Hall has used the term *subspecies* and the eastern school led by Fernald, Gleason, and Cronquist has favoured use of *Variety* for the major infraspecific rank. In usual taxonimic practice, subspecies and varieties are recognizable morphological variations within species. Their populations have their own patterns of variation correlated with geographical distributions or ecological requirements.

Whether the category "subspecies" or "variety" is employed often depends on the custom or philosophy of the systematist. Sometimes this is largely based on where or under whom the taxonomist was trained. Also, epithets have priority within rank, and historically some groups have numerous varieties and others, it would necessitate a swarm of nomenclatural transfers. The category *form* is generally used to recognize and describe sporadic variations in a single morphological feature such as occasional white- flowered plants in a normally purple-flowered species. If these white flowers are not correlated with any other features of the plants, including geographical and ecological distributions, they might be described as a form. Little taxonimic significance is attached to the minor and random variation upon which forms are normally based. For this reason, few modern systematists use the rank of form. It is encountered most frequently in older taxonimic treatments or in treatments of horticultural interest.

Genera

Like species, the genus represent a concept. From a parctical standpoint, the *genus* is an inclusive category whose species have more characteristics in common with each other than with species of other genera within the same family. Genera, therefore, are aggregates of closely related species. Lawrence (1951) writes that from an evolutionary viewpoint, the genus is an important concept because the sum of all characteristics is used to group closely related species, allowing the genus to be viewed as a unit. The function of the genus concept is to together species in a phylogenetic manner by placing the closest related species together within the generic classification. The genera in most families, for example, Magnoliaceae, show striking differences in features, whereas genera in other families, for instance, compositae, are separated on less obvious characters. A possible reason for this difference is that extinction of species and genera has been greater in the older families (Magnoliaceae), resulting in well-marked distinction in feature among genera. There is no size requirement for a genus. A genus consisting of but one species is said to be *monotypic. Leitneria* (Leitneriaceae), consisting of a single species, *L. floridana*, is monotypic, (Cronquist, 1968). Some genera, such as *Senecio* of the Compositae family, contain 2000 to 3000 species.

Genera must be delimited following a worldwide study of their species. If the deciduous and the evergreen rhododendrons of North America are compared, they appear to be two distinct genera. Some taxonomists in the past have placed the deciduous species in the genus *Azalea*. With wider acquaintance of *Rhododendron* on a worldwide basis, the apparent discontinuities break down. As broadly and ordinarily interpreted, the genus *Rhododendron* includes both the deciduous azaleas and the evergreen rhododendrons of North America. The collective group of species must be distinct from the other species groups to support two genera.

If the species of two genera are not readily separable, most taxonomists would agree that it is more convenient to have larger units of classification. If necessary, the ranks of subgenus or section may be preferable to handle the classification and at the same time express relationships. When a genus is split into two or more genera, nomenclatural changes are required for certain of the species. Such changes may be avoided by the use of the infrageneric ranks because they do not cause changes in binomials. The present tendency in plant systematics is to have a broad, conservative view of generic concepts and limits. This is expressed in a relatively stable generic nomenclature. The category of genus is an old and useful concept that serves both the professional and the amateur botanist. Most colloquial or common names parallel generic concepts-for example, oak, pine, rose, buttercup, sunflower.

Families

Much of what has been said about the concept of the genus applies equally to the concept of the family. Ideally, families should be monophyletic and present both a biologically meaningful treatment and a practical taxonomy. Both reproductive and vagetative features are used to characterize families (Cronquist, 1968). The parts of the flower and fruits of flowering plants are numerous and varied, and they are subject to little environmental modification. Consequently, they provide more characters for the definition of families than do vegetative features. The category of family is more inclusive than either the genus or the species and there may be exception to certain characters.

Families are often delimited by ovary position, kinds of pistils and stamens, carpel number, fruit type, symmetry of the flower, leaf arrangement, leaf morphology, and habit. The use of certain characters to delimit families does not preclude the use of that same character to delimit taxa at higher of lower ranks in the hierarchy. For example, the type of fruit is often used to separate families, but in the Rosaceae it is delimits only subfamilies. Some families are recognized easily even by amateurs because of certain distinguishing features including the head (inflorescence) of Compositae, the schizocarp (fruit) and umbel (inflorescence) of Umbelliferae, and the winged schizocarps (fruit) of Aceraceae.

Some families are well defined and are relatively homogeneous, such as the Composite, Cruciferae, Gramineae, Umbelliferae, and Sarraceniaceae. Leguminosae are a large family characterized by a unique type of fruit and seed, yet family members are not homogeneous. The diversity of the family Leguminosae is reflected by the differing classifications that have appeared form time. Some divide the legumes into three families, while other maintain one family with three subfamilies. Taxonomists agree on where the lines should be drawn, but they disagree on the rank to be assigned. A great diversity of both reproductive and vegetative features may be found in families such as Ranunculaceae, Berberidaceae, and Rosaceae. Families, like genera, do not have a fixed size.

The Compositae, one of the largest families of flowering plants, has around 22,000 species; the Leitneriaceae has but a single genus and one species. The size and diversity of plant families are the consequences of their evolutionary history, age, and radiation, coupled with our perception of the taxonomy of the group. Taxonomic schemes are by definition subjective; the whole point of taxonomy is to convey information through a derived scheme. When distinction are sharp among closely related families, there is no problem in defining the family. But as Cronquist (1968) wrote regarding whether *Phryma* should be placed in the Verbenaceae or in the Phrymaceae, "It is purely a matter of taste whether the evident relationship (with the verbenaceae) or the obviously different gynoecium should be emphasized in determining the status of *Phryma*, but assignment of genera to poorly defined families is usually a matter requiring educated judgement."

Thirteen years later Cronquist (1981) reversed his stand and included *Phryma* in the Verbenaceae noting, "The unusual gynoecium of *Phryma* is in pattern with other unusual types in the Verbenaceae. I see no value in keeping *Phryma* as a separate family if other peripheral group are to be retained in the Verbenaceae." Subjective classification is often criticized by biologists whose training does not include a knowledge of the principles of systematics or taxonomic methods.

It may be defended, however, on the grounds of the *predictive value* of classification developed using these principles. Predictive value suggests that because of their common ancestry the members of a monophyletic group will share some features not apparent at the time the classification was originally made. For example, if one wishes to look for new antitumor alkaloids similar to those found in certain species in the Rutaceae, the best place to look would be in other species of Rutaceae. Examples like this are common and have demonstrated the predictive value of classifications developed by systematists. Most taxonomists favour broadly conceived family concepts that lend stability to classification.

Although there is no clear discontinuity of features between the Labiatae and the Verbenaceae, the two are different, and custom maintains the two groups as separate families. In some instances, tradition is a taxonomic character. The Rosaceae may be broken into several families based on good logical evidence. However, the overall similarities within the family, rather than its intrafamily differences, has caused most taxonomists to treat the Rosaceae as a single family. Again, there is no set definition for a family; yet, there is relatively little disagreement on family classification. This suggests among systematists with the delimitation of most families.

Order

Orders include one or more families. The monophyletic requirement for establishing order must be broadly interpreted and should not be too restrictive when developing a workable classification. Order are much more difficult to define and delimit than families. The divergence in the lineages, as reflected in orders, occurred early in the evolutionary history of the flowering plants and left few fossil records. Orders and other higher categories are characterized by an aggregate of characters. Generally, there are greater number of exceptions at the level of order than at the lower ranks. Most taxonomists that orders should be monophyletic, but all do not agree as to their size or to the families included in each.

Bessey (1915) used many orders that often contain only a few families. However, Hutchinson (1973) uses many orders that often contain only a few families. The classification used in this text takes a somewhat broad approach by arranging the flowering plants into 83 orders. Kubitzki (1977) provides an excellent series of papers on the classification of higher categories.

Typological Concepts of Taxa

Systematists often remember the essential characteristic of a species, genus, of family and represent that taxon in their mind by an image embodying its salient features. This symbolism exemplifies what is meant by typological concepts of taxa. Taxonomists familiar with the flora of an area or traveling in new areas unconsciously use typological methods in recognizing the hundreds of species, genera, and families of the flora. For example, acorns, five-angled pith, and buds clustered at tips of twigs provide a recognition pattern for oaks (*Quercus*). Taxonomists also identify plant material by comparison with previously identified specimens or by using keys, figures and descriptions. Since these tools represent only a sample of the populations, a taxonomist build up typological pictures of species, genera, and families from experience with the plants. As Davis and Heywood (1963) have observed, typology is a valid and necessary process in the identification and classification of plants when intelligently used. Only when its use obscures the natural variational populations is it a liability to systematics.

CHARACTERS

The Characters of an organism are all the features or attributes possessed by the organism that may be compared, measured, counted, described, or otherwise assessed. Form this definition, it is clear that the differences, similarities, and discontinuities between plants or taxa are reflected in their characters. The characters or taxon are determined by observing or analyzing samples of individual and recording the observation, or be conducting controlled experiments. Herbaria contain millions of plant specimens providing a ready source of samples for descriptive plant systematics. Plant taxonomy uses morphological and anatomical characters for the purposes of classification. Structures are observed with the eye, hand lens, or light microscope or by using scanning electron microscopy (SEM). Modern instrumentation allows comparative studies of physiological rates, such as photosynthesis, and analysis of chemical compound produced by the plant.

For classification purposes, there is increasing use of evidence from fields as cytology, biogeography, paleobotany, palynology, phytochemistry, population biology, molecular biology and ultrastructure. But as Cronquist (1975) noted, "For the present at least, Comparative study of modern forms, from the eyeball to the SEM level, remains the mainstay of major taxonomic interpretation." He also consluded the because of the time and effort required obtain molecular data "....morphology, in the broad sense, will continue to reign taxonomically supreme for some years to come." Turner (1977), however, disagrees and argues that Cronquist neglects the recent application of chemical data for purposes of classification. Characters have many uses, including the following : providing informaion for construction of taxonimic systems, supplying characters for construction of keys for identification, furnishing features useful in the description and delimitation of taxa, and enabling scientists to use the predictive value of classificaion.

Characters are such things as leaf width, stamen number, corolla length, locule number, and placentation. Some descriptive or numerical term must occur with the character to imply the *character state*. For example, a leaf width of 5 centimetres and basal placentation are character states. When used in description, delimitation, or identification, characters are said to be *diagnostic* or *key characters*. Characters of constant nature, which are used to help define a group, are referred to as *synthetic characters. Quantitative characters* are those features that can be readily counted or measure, whereas *qualitative characters* are such things as flower colour, odor, leaf shape, or pubescence. Quantitative characters are normally easier to obtain and communicate than qualitative characters.

Successful observations of qualitative features provide a challenge to plant taxonomists. Another term often used by systematists is *good character*. This implies that the character is genetically determined, largely unaffected by the environment, and relatively constant throughout the populations of the taxa. The more constant a character, the greater is its reliability and resulting importance. In order to provide character states, there must be differences in the features of flowering plants. Vegetative parts of angiosperms, such as leaves, stems, and roots, are relatively large and easy to observe, but the generally provide fewer characters for classification than reproductive structures, such as sepals, petals, stamens, pistils and so on.

Clones of genetically identical flowering plants, when grown in diverse environments tend to be relatively uniform in characters of the flower and fruit but highly variable in overall size of the plants and leaves. Vegetative features of flowering plants, especially size and shape, tend to be more influenced by environment factors than are reproductive features. As a result, certain Vegetative characters are less reliable and less useful than reproductive characters in taxonomy. A character that is diagnostic for separating two genera in one family may be a key character for families in another order. Some species are delimited by leaf arrangement, while leaf arrangement is a synthetic character of certain families. Characters become important after they are proved valuable by experience.

The relative value and utility of characters are empirically determined, and there is not such thing (in a practical sense) as a "fundamental" character. A rule of long taxonomic precedence require that taxa at all ranks be established on the basis of a correlation of several characters, thus avoiding the incongruities of classfication constructed on the basis of a single character. Ideally, classifications should be based on an understanding of all characters form every population of each species under consideration. But on a practical basis, this is not possible. Some geographic areas may be better sampled than others, and some characters have simply never been examined. There is not enough time to examine every possible character.

Classification

Although the theory of evolution had little immediate impact on the practice of taxonomic research at the level of species, genus, and family, it did impact on taxonomic theory. Systematists began to attempt to develop systems of classfication designed to reflect evolutionary lineages. It was reasoned that such classifications should designed to reflect evolutionary lineages. It was reasoned that such classifications should reflect that course of evolution. Since the turn of the twentieth century, numerous workers in the field of flowering plant taxonomy have attempted to produce systems of classification reflection the phylogeny of the flowering plans. This is very difficult task, owing party to the poorness of the fossil record and partly to the complexity and diversity of the plants themselves. The difficulty of developing a phylogenetic system is also complicated by the fact that some workers in this field have based their systems on characters that they believe to be primitive rather than on factual evidence.

Speculative information does not lend itself to the development of phylogenetic systems having high predictive value. By the 1980s most phylogenetic system differ only in detail, and the authors of such attempt to avoid the pitfalls of earlier system. In recent years, two school of workers, the pheneticists and the cladists, have developed methodology in an attempt to bring about greater objectivity in the construction of classification.

Numerical Phenetics

Phenetics involves the organizing of data on the basis of similarity for the purpose of obtaining a classification. *Pheneticists*, the workers in this field of taxonomy, emphasize the need for objectivity in classification. They attempt to develop schemes having high predictive values that allow the maximum number of generalizations to be made from predictive values that allow the maximum number of generalization to be made form the schemes. They stress the importance of using a large number of characters, at least 60 but preferably 80 to 100, that can be correlated on the basis of similarity. Phenetic systems are based on data derived from the phenotypes of organisms-hence the designation phanetic systems. The idea of phenetic system classification was proposed by Michel Adanson in the latter part of the eighteenth century. With the availability of computers in the 1950s, workers such as C. D. Michener, R. R. Sokal, P. H. A. Sneath, A. J. Cain, and G. A. Harrison began to propose computer-based methods for quantifying similarity and for the grouping of taxa using quantitative methods.

One premise was that it should theoretically be possible to develop methodology that would quantitatively allow an experienced or inexperienced person to arrive at basically the same classification. Sneath and Sokal summarized their philosophy in 1963 and in a revised edition in 1973. They believed in equal weighting of characters and made sweeping claims about the value of numerical phenetics. However, their refusal to weight characters and their tendency to ignore phyletic information had an initial negative impact on the validity and acceptance of their method. Modifications of these extreme viewpoints by later pheneticists has brought about wider acceptance of the methodology. In an attempt to be objective, pheneticists avoid referring to taxa by the species or generic names. Instead, they replace these groupings with *Operational Taxonomic Units* (OTU'S) the term given to the lowest-ranking taxon studied in the investigaion. However, the establishment of OTU'S require subjective judgments and considerable knowledge and experience.

Another common problem is character definition and description; for instance, can leaf shape be defined as a single character or is it actually a summarization of several independent variables that determine the shape? Simply stated, phenetic methodology requires that character states be determined for each OUT. A data matrix (OTU's versus character states) is prepared and the data are codified for computer processing. In cluster analysis, the computer sorts out (clusters) the OTU's according to their overall similarity, that is according to the number of attributes or characters in common. The available computer programs produce a *phenogram* that is a dendrogram of phenetic relations in which less and less similar OTU's are successively linked together with the contourlike lines encircling different numbers of OTU's or by linking OTU's with straight lines of varying lengths. These computer methods attempt to demonstrate a classification graphically that can be comprehanded by the taxonomist. Some workers have attempted to delimit subjectively taxa by phenetic method especially by equating different levels of similarity on the phenogram but, thus far, this has not been successful or practical.

The overall purpose of phenetics is to lend objectively to what have previously been the somewhat subjective methods employed by traditional taxonomist. While traditional taxonomists have not been enthusiastic about the methodology, the younger generation of numerical taxonomists seem to appreciated better the difference in information content of characters and is willing to weight characters

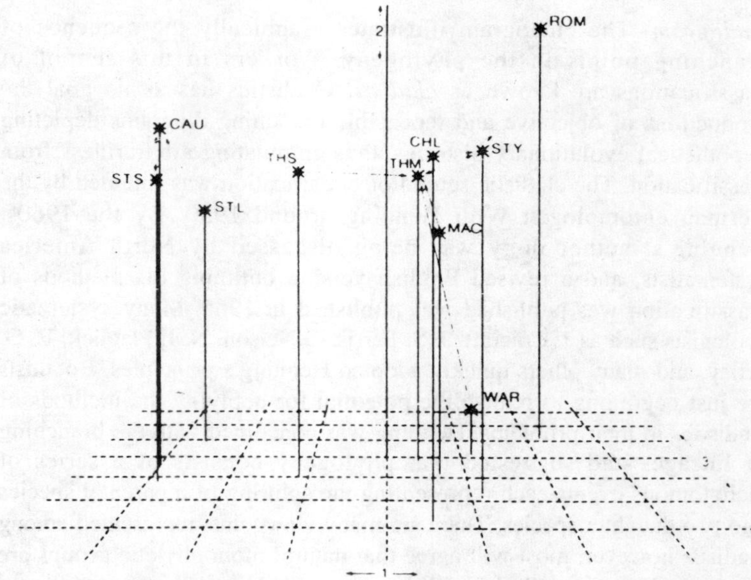

Fig. 3.2. First three dimensions of a principal components analysis of the genera of the tribe Thelypodieae (Cruciferae). A nearest neighbour network, measured by Euclidean distance is superimposed on the taxa; arrows point to nearest genus. Generic names are abbreviated: CAU—Caulanthus, CHL—Chlorocrambe, MAC—Macropodium, ROM—Romanschulzia, STY—Stanleya, STL—Streptanthella, STS—Streptanthus, THS—Thelypodiopsis, THM—Thelypodium, WAR—Warea.

based on a *posteriori* evidence. The basic philosophy of pheneticists has contributed to the development of logical and objective methodology and the formulation of better definition of characters in plant systematics. Numerical method appear to be useful in difficult groups where other methods have not proved satisfactory. Most workers today agree that classifications developed by numerical methods are probably no better or no worse tha:. classifications produced by the traditional taxonomists. One major advantage of the numerical approach is that it forces the taxonomist to provide definitions of the characters, as well as data that may be used by other in reassessing classification. Computer techniques allow the taxonomist to incorporate many character analyses than could be done by traditional "brain power."

Cladistics

Cladistic analysis attempts to summarize knowledge about the similarities among organisms in terms of a branching diagram called a *cladogram*. The cladogram illustrates graphically the sequence of branching points in the phylogeny. Workers in this school of classifications are known as *cladists*. Cladistics has as its goal the productions of objective and repeatable branching diagrams depicting hypothetical evolutionary histories, thus eliminating arbitrariness from classification. The cladistic school of classification was founded by the German entomologist Willi Henning around 1950. By the 1960s, Henning's methodology was being discussed by North America systematists, and a revised English version outlining his methods of classification was published was published in 1966. Many systematic zoologists such as J. Cracraft, J. S. Farris, G Nelson, N. I. Platnick, E. O. Wiley, and many others quickly adopted Henning's principles. Botanists are just beginning to realize the potential for applying the methods of cladistics to their problems. Henning was concerned with the branching of lineages and suggested that phylogeny consists of a series of dichotomous events each representing the splitting of a parental species into two daughter species. There are many viewpoints represented among cladists; however, most will agree that natural monophyletic groups are recognized by uniquely derived characters termed *synapomorphies*, and that only natural groups defined by synapomorphies are to be included

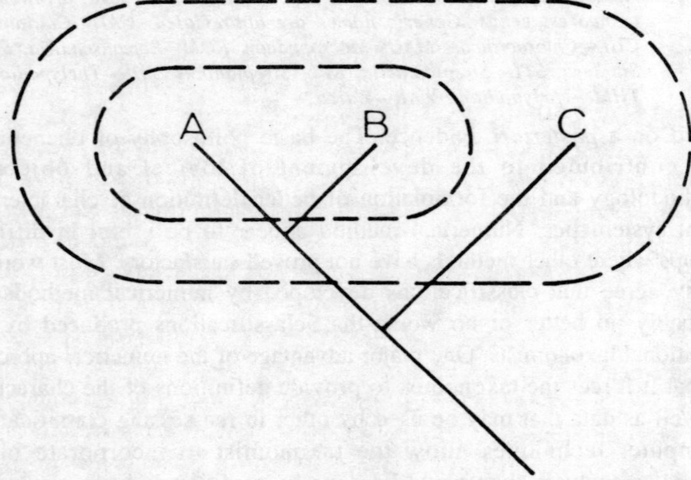

Fig. 3.3. A directed or rooted cladogram illustrating hypotheses about synapormophic statements. Taxon A is more closely related to taxon B than either are to taxon C; i.e., A and B share synapomorphies nested within a larger synapomorphic pattern which includes C. A+B is a sister group of C; A is a sister group of B.

in classification schemes. A shared primitive character-state is known as a *sympleisomorephy*.

Cladists attempt to apportion similarities on a branching diagram according to the hierarchical level at which the similarities characterized groups, that is, the level at which they are postulated to be a synapomorphy. Cladists reason that branching points are determined by backward tracing of uniquely derived characters since derived characters are to be found only among the descendants of the ancestor. In general, two comparative methods are used for evaluating characters-ontogenetic criterion and outgroup criterion. Both methods attempt to establish the partitioning of characters into ancestral or derived. A primitive or general character is one that is more inclusive, whereas a derived character is less inclusive.

An *ontogenetic criterion* uses data from comparative embryology to assess the direction of character transformation. Although potentially useful, developmental information is not usually available for plants. For this reason, most botanical cladists use *outgroup comparison*. By this method, if a shared character within the group being studied is also found outside the group in closely related taxa, it is reasoned that the character is too general, or primitive, to define a subgroup. This requires that the systematist know the relationship of one group to other related groups in order to analyze critically the relationship within the group. Basically, cladists attempt to maximize the agreement of characters while attempting to account for their distribution. In theory, cladistic analysis has value as a method for delimiting monophyletic groups. It forces the careful analysis of all characters; it introduces a new concept of character weighting, that is, the possession of uniquely derived characters; and it helps to clarify the relationship betwreen phylogeny and classification by permitting the testing of phylogeny in the absence of a fossil record. Problems in communication exits because of the excessively complicated terminology of the cladists.

There are further difficulties in determining primitive characters and the determination of the direction of evolutionary change. Some groups resist analysis because of the absence of easily detectable synapomorphies. Continuous characters and variation within a taxon present difficulties when using cladistic methods. Also, OTU's with hybrid origin are usually omitted from cladistic analyses or require special techniques because of their reticulate evolutionary history. Hybridization, while not of great importance in the origin of animal taxa, must be considered when examining plant taxa, as many appear to have been derived in this way. At times there may be two cladistic hypotheses that are equally good at explaining character distributions. In such a case it is necessary for the systematist to judge the relative merits of the branching patterns and specify the hypothesis exhibiting the most agreement in the synapomorphy pattern.

To this problem, cladists apply the principle of *parsimony* by selecting the hypothesis that can best be defended by the investigator as having the fewest character state changes. One application of cladistic principles to flowering plant classification can be seen in the following example. Since (1) monophyletic groups are established by the use of unique, derived features inherited from the immediate ancestor of the group, and (2) species splitting has been mainly dichotomous, than the search for sister species or sister groups (Subsets of a set defined by a synapomorphy) is the main procedure in the reconstruction hypothesis of the phylogeny. Using this criterion, Bremer and Wanntorp (1978) suggest that the division of the angiosperms into the dicotyledons and monocotyledons at equal rank is untenable.

It is usually agreed that minocotyledons are derived from a group within the dicotyledons. If true, the monocotyledons cannot be a sister group of the dicotyledons and, therefore, the monocotyledons cannot have the same rank in classification as the dicotyledons. Using this reasoning, the monocotyledons should have rank as their sister group within the dicotyledons. From this and other examples in the literature, it is clear that the application of cladistics to botanical classification will have an impact on the practice of developing classification. For a bibliography of papers on cladistics see Funk and Wagner (1982).

Biogeography

Classification depends on the evolutionary developments of the taxon, the present day relationships to the environment, and the analyses of the taxon's distribution pattern in relation to paleoclimatic and paleogeographic data. The study of the patterns of distribution of plants and animals is known as *biogeography*. The analysis of the distributional patterns of organisms is essential to the classification of all taxa at various ranks. Distribution patterns may be shown graphically by means of a single line enclosing the area of distribution or may be rapresented by dot distribution maps. The dots may represent the exact collection sites or may simply indicate the presence of the taxon in a political region such as a county (or parish), state, or province. Taxa coexisting in the same general geographic region are said to be *sympatric*, while taxa whose ranges do not overlap are referred to as *allopatric*.

Many taxa are partially sympatric. When the area occupied by taxon includes sizable regions in which the taxon is not found, that taxon is said to have *discontinuous* or *disjunctive* distribution and is discribed as a disjunct. Disjunction are usually explained on thc basis of disruptions of former wide distribution by paleoclimatic changes, by present day environmental or topographic factors, or even by the possibility of random long-distance mechanisms. Major discontinuities in distributional patterns may occur within a species, genus, or family when the range includes widely separated parts of a continent or location on separate continents. For example, taxa with disjunctions may be found among the floristic elements of the eastern Asia-eastern North America flora, and among the southern South America-Australasia-South African flora.

The geophysical theory that lands masses of the earth have been steadily moving during the course of geological time has been called *continental drift*, or more properly, *plate tectonics*. The concept of plate tectonics provides a geophysical explanation for the isolation of land areas by sea floor spreading, the uplift of mountain ranges, the formation or disappearance of islands, and the shifting position of continents. Latitudinal changes, rafting of floras, and environmental stresses due to plate tectonics surely played a major role in the evolution of the angiosperms. For these reasons, *biogeographers*, who study the distribution of plants and animals, have recently reexamined distribution patterns fossil records, and evolutionary history of the flowering plants in respect to plate tectonic theory. The earliest known angiosperm pollen has been found in lower Cretaceous strata. During this geologic period, the northern group of continents (North America and Eurasia), which geologists call *Laurasia*, had largely separated from the southern group of continents (South America, Africa, Australia, Antarctica, and the Indian subcontinent), called *Gondwanaland*.

The components of Gondwanaland and laurasia were adjacent to each other or in close proximity when the angioperms first appeared in the earth near the beginning of the Cretaceous Period. The pollen record strongly suggests that much of the radiation of the angiosperms occurred when the continents of Gondwanaland were essentially close together, allowing direct interchange of floristic elements. Taxa have subsequently radiated in various directions, occupying suitable habitats and forming their present ranges.

Endemism refers to a distribution restricted to one region or area. An endemic is a taxon whose distribution is limited to a special or very restricted geographic region. Certain habitats or geographic regions have a high proportion of endemics and are described as areas of endemism. For example, the granitic outcrops of the Piedmont of Georgia and the Carolinas have a number endemic such a number of endemic species. The bluffs of the Appalachicola River in Florida have a number of endemics such as the gymnosperms *Torreya taxifolia* and *Taxus floridana*. The native angiosperm flora of Hawaii is more than 90 percent endemic. It is known from fossil evidence that *Ginkgo biloba* was once widespread but now occurs naturally only in a small area in China.

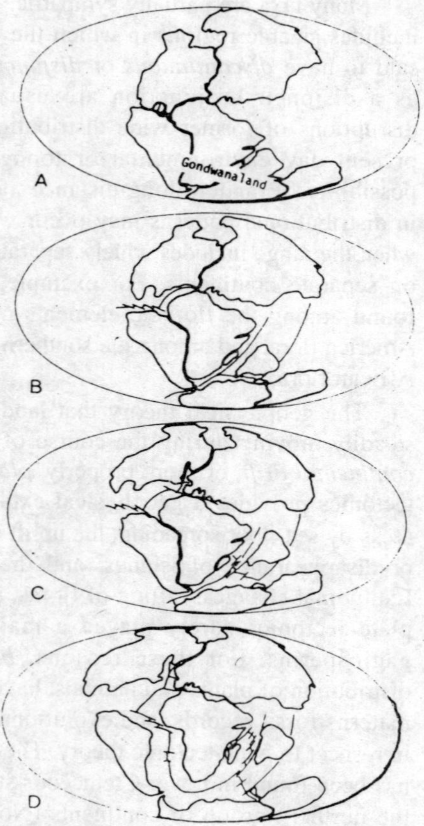

Fig. 3.4. Continental drift. The maps show the relative position of the continents—(A) 200, (B) 180, (C) 135, and (D) 65 million years ago.

The explanation of endemism is related to the past geographic and climatic history of the region and the subsequent isolation of the area involved as well as the present day environment of the area. A given genus or family may show a marked concentration of related species or genera in one or more geographic regions. An area with such concentrations of related taxa is called the *center of genetic diversity* for that genus or family. For example, one center or genetic diversity fro the genus *Vernonia* is south central Brazil. At one time, the center of genetic diversity was regarded also as the *center of origin* for the taxon. However, present evidence indicates that this is not necessarily true. The geographic and climatic histories of the area and their influence on the evolutionary history of the taxa may have shifted the center of genetic diversity away from the center of origin.

Vicariance Biogeography

In the past 25 years, a method of study has been formalized to analyze the relationships and histories of biotas by searching for common patterns of distributions of organisms. These common distributional patterns are then coupled with hypotheses relating to the evolutionary histories found among the organisms inhabiting these biotas. One a number of distributional patterns reflecting some common pattern have been examined, it is possible to compare the distributions to events that might have influenced the evolution of the taxa. Since by definition each monophyletic taxon has a single center of origin, it may be possible to determine pathways of migration and evolution.

The recurrence of replicted geographic patterns among unrelated taxa allows the taxonomist to sort out a series of distributional patterns that have a general explanation. This has led to the development of an approach called *Vicariance biogeography*. When one species is replaced by a similar, closely related speceis in separate geographic or ecological areas, the phenomenon is called *vicariance*, and the taxa involved are known as *vicariance*. The recurrence of replicated geographic patterns in unrelated taxa represent one or more vicariance events that fragmented a continuous ancestral biota. Such vicariance events include continental drift, mountain building, climatic change disruption of drainage patterns by stream capture, etc.

Vicariance biogeographers recognize that disjunctions may also result from dispersal of taxa from their center of origin, followed by evolution into new taxa. As part of their methodology, vicariance biogeographers assess the frequency of agreement in the distribution of unrelated taxa. They attempt to relate these facts to the occurrence of common paleogeographic and paleoclimatic events that affected the evolution and distribution of two or more groups of organism. Phylogenetic hypotheses and coinciding areas that the species inhabit can be depicted graphically to form a branching *area cladogram* in which the species name is replaced by the area in which the species occurs. Vicariance events are reflected in the branching pattern of the cladogram.

Vicariacne biogeographers suggest that by developing a synthesis of distributional patterns and evolutionary events, much new information can be provided as well as uncovering questions for future study. Not all systematists, however, would agree with the promises made by the proponents of vicariance biogeography. Briggs (1983) and other argue that the concept of vicariance brings nothing new to evolutionary

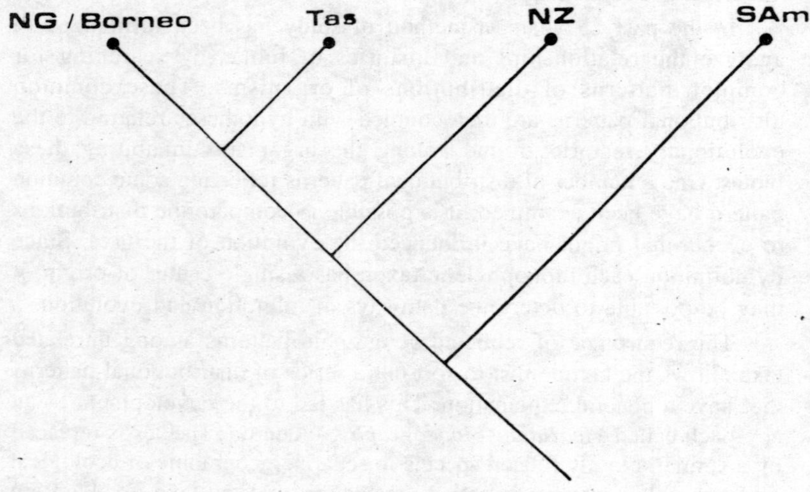

Fig. 3.5. A combined area-cladogram of Negria, Depanthus, Rhabobthammus, Corathera, and Fieldia (Gesneriaceae). The geographic areas represented include New Guinea (NG), Borneo, Tasmania (Tas), New Zealand (NZ) and South America (SAm).

biology, that it downplays long-distance dispersal and that its application is limited group, such as fishes, that have limited means of dispersal. Thus far, the concept and techniques of vicariance have been applied largely to zoological systematists. Certainly, it does seem that general patterns of relationships among certain areas of endemism are best explained by vicariance of once widespread biota rather than solely by independent, random dispersal events such as long-distance migration. Some areas of endemism of flowering plants are interrelated among themselves, and it seems that such relationships can be broken down also into species relationships.

Parallelism and Convergence

If two organisms of different lineage are alike in their feature, the resemblance is usually attributed to either parallelism or convergence. *Convergence* is the resemblance between two or more distinct phyletic lines brought about through evolution by adaptation to similar environments or to similar reproductive biologies. Adaptation to similar environments is most often apparent in the vegetative characters of plants, but it may occur in reproductive parts. Convergence in vegetative features of unrelated taxa occurring in the Mediterranean vegetation of

California and Chile show the similarities in morphology caused by similar climates. Many desert species of Euphorbiaceae from Africa and Cactaceae form North and South America demonstrate convergence with their "cactus-like" appearance, although their flowers immediately suggest the proper family. This convergence in vegetative features presumably reflects adaptive responses to aridity. Adaptation of reproductive biology in pollination or seed-dispersal mechanism causes convergence in flowers, fruit, or seed.

The weed *Camelina sativa* (Cruciferae) has an ecotype that infests flax fields. The flax field ecotype mimics the flax plants in size and weight of the seeds. This adaptation results when *Camelina* seed resembling flax seeds are selected in the harvest and seed separation process. Another example of convergence, seemingly related to their pollination biology, is the development of pollinia in both the orchids and milkweed. With *parallelism*, the resemblance is due to a common ancestry and genetic background. Two phyletic lines may diverge up to some stage and cease to diverge and then run parallel. Parallel evolution is apparently a common feature of flowering plants but is difficult to document with certainty because of the poor fossil record.

Parallel evolution may take place in related lineages at any stage in their separation , or at the level of infraspecific taxa, species, genera, families, or so on. Parallelism and convergence are difficult concept at best. If the angiosperms are monophyletic, the dividing line between parallelism and convergence is purely arbitrary. Given the sparse fossil record of the angiosperms, it may be impossible to distinguish between parallelism and convergence with closely related order.

Guiding Principles of Angiosperm Phylogeny

The angiosperm phylogenist is face with major problems of an inadequate fossil record, the prevalence of convergent evolution, and the extreme structural modification of many flowering plants. The relatively soft tissues of most angiosperm flowers and certain other plant parts provide nothing analogous to bones, exoskeletons, or shells for fossilizaion. Flowering plants do have resistant pollen, and some have hard woody tissue or waxy leaves suitable for fossilization. Even though the fossil record of flowering plants is meager, it is sufficient to indicate some general trends in the angiosperms. In addition, ancestral characteristics and the direction of trends in the angiosperms. In addition, ancestral characteristics and the direction of trends of specialization can often be recognized in so called "living fossil."

The classification of angiosperms is primarily based on the examination of contemporary flowering plants and establishing correlations of characters. The necessary use of contemporary plants to establish hypothetical lineages and inferred evolutionary relationships has inherent problems. The prevalence of convergent evolution in flowering plant has made it difficult to develop monophyletic groups and the taxonomist is faced with the problem of deciding whether certain characters are due to common ancestry or to convergence. To overcome the problem of convergence, information from all possible sources regarding the plant must be assembled and evaluated. A voluminous literature of comprehensive comparative studies has been developed on groups thought to contain primitive characters. In 1915 Bessey presented a set of guiding principles of angiosperm phylogeny, or what he called "dicta," to facilitate piecing together data into phylogenetic system of angiosperm classification.

The information reflected in these dicta has largely been determined from studies of comparative morphology and paleobotany. These "Besseyan principles" have been modified through the years by numerous phylogenists as new information appeared in the literature. Although clearly Besseyan in origin, the present principles have been refined and restated, and most modern classification conform to them. Give the diversity of the angiosperms, there are exceptions to these principles. They are not dogma and should be used only as guides.

Possible Evolutionary Trends in Angiosperms

1. Flowering plants with collateral vascular bundles arranged in a cylinder (dicotyledons) are more primitive than those with scattered vascular bundles arranged in a cylinder (monocotyledons). Unstoried wood with scalariform endplates, scalariform endplates, scalariform side wall, and apotracheal wood parenchyma are regarded as primitive.
2. The many parted, spirally imbricate flower is primitive, while those with few parts, which are whorled or valvate, generally are more advanced. Regular (actinomorphic) flowers preceded irregular (Zygomorphic) flowers.
3. In most groups of flowering plants, woody plants usually have preceded the harbs, vines, and climbers. Perennials gave rise to biennials, and annuals have been derived from both. Terrestrial seed plants usually preceded closely related aquatics or epiphytes, saprophytes, and parasites.

4. Alternate leaves secretory cells and stipules are primitive; opposite of whorled leaves without secretory cell or stipules are regarded as advanced. In nearly all instances, simple, pinnately vained, evergreen leaves preceded compound leaves.
5. Perianth parts that are separate and poorly differentiated into sepals and petals are more primitive than laterally fused perianth parts that are sharply differentiated. Usually, flowers with petals preceded apetalous ones.
6. Bisexual flowers have traditionally been regarded as primitive, but evidence is accumulating that unisexual flowers may also represent a primitive condition. Most agree, however, that flowers borne separately in the axils of subtending leaflike bracts are primitive.
7. Those plants with leuco-anthocyanins and ellegitannins are regarded as primitive; recently these compounds have been referred to as pro-anthocyanins.
8. The primitive seed is arillate, has a small embryo, ad contains abundant nuclear endosperm. Nonedospermic seeds are considered to be derived. Primitive families have ovules that are crassinucellate, while those in more advanced groups tend to have tenuinucellate ovules.
9. Many separete stamens are more primitive than few or united stamens.
10. Axile placentation preceded free central placentation, and latter is regarded as derived.
11. Hypogyny (superior ovary) is the primitive condition, and perigyny and epigyny (inferior ovary) are derived.
12. Single fruits preceded aggregate fruits formed from several ovaries. The capsule preceded the drupe or berry.
13. Generally, Numerous, separate, carpels represent a more primitive condition than fewer fused carpels.
14. Pollen grains with one pore preceded those with 2 or more pores.

The trends in the evolution of the flower have come about by reduction in number, fusion, specialization of parts, and changes of symmetry. Simple structures are not necessarily primitive but have become simple as a result of reduction from more complex part. The rate of evolution is not always the same for all structures of the plant. Some parts of a plant may become more specialized than others, and some taxa may have both advanced and primitive features.

Following are some principles of taxonomy.

1. Vicariance biogeography examines the relationships and histories of biotas by searching for common pattern of distributions among various organisms. These common distributional patterns are then coupled with hypotheses related to their evolutionary histories.
2. Careful consideration should be given to the facts regarding the early angiosperms as they are known from the fossil record.
3. The goal of taxonomy is to develop a workable classification that reflects evolutionary relationships and provides for nomenclature and identification.
4. Cladistic taxonomy involves summarizing knowledge about the similarities among taxa in terms of a branching diagram called a cladogram depicting the hypothetical evolutionary histories of the organism.
5. Categories such as species, genera, families, and orders are not rigid but are flexible and individually delimited fore each group. Their sequence in the hierarchy is established by he International Code of Botanical Nomenclature. Categories are subjective and are definable only with respect to their position in hierarchy.
6. Whenever possible in the development of classifications, taxa should be sampled throughout their range, and all taxa at lower ranks should be examined.
7. Species represent lineages produced by evolution, and branching genetic relationships exit among the taxa of each group.
8. Morphological indicators of phylogeny may provide guidence to primitive versus advanced features and aid in developing phylogenies. Thinking should remain flexible so that modifications in classification may be made as new evidence become available.
9. Biogeography analyzes the patterns of distribution of organisms and relates these to the systematics of the organisms.
10. Taxa are based on the correlation of characters and discontinuities in the variation pattern. Characters may be selected from any attribute of the plant and do not have a fixed value at all ranks.
11. Ancestral features and trends of diversity may often be recognized in the structure of living angiosperms.
12. For delimiting taxa, characters should be constant and show little environmental variation.
13. Taxa may resemble one another because of either convergence of parallelism.

14. Taxa should be monophyletic. On a practical basis, the monophyletic requirements for higher ranking taxa must be broadly interpreted.
15. Phenetic taxonomy organizes and classifies taxa on the basis of similarities of the phenotypes of the organisms.
16. Evolution may result in reduction or loss of part. It can also proceed toward greater structural complexity.
17. Tadonomic treatments should be practical and consistent in their use of the various categories.

Sources of Taxonomic Evidence

Taxonomic evidence for the establishment of classifications and phylogenies is gathered from a variety of sources. Because all parts of a plant of all stages of its development can provide taxonomic characters, data must be assembled from many diverse disciplines. The use of information from studies on comparative anatomy, embryology, phlynology, cytogenetics, chemistry, and so on has greatly improved the modern classification of plant.

Morphology

The features of floral morphology are the most important characters in the classification of flowering plants. These features are easily observed, and they are practical for use in keys and descriptions. Morphology currently provides most of the characters used in constructing taxonomic system. When considering the evolutionary lines of the flowering plants, each represented by perhaps thousand of species, it should not be expected that all morphological characters should occur uniformly among all species. The occasional absence of a synthetic or diagnostic character, or its rare occurrence in a group from which it is normally absent, may be expected. Terms used in identification of angiosperms are morphological, and these are illustrated or defined. *Morphology of the Angiosperms* provides a comprehensive review for the flowering plants.

Natural selection, associated with successful reproduction, maintains a basic similarity of the reproductive features of flowers, fruits, and seeds within the various species, genera, and families. This general constancy makes these structure ideal for characterizing taxonomic groups. In addition to being more constant than vagetative features, reproductive characters are generally more numerous and therefore provide more feature to differentiate taxa. Floral features are often fundamental in defining natural group.

Variations in ever conceivable part of the flower, fruit, and seed have proved to be useful in the classification of one or more groups of flowering plants. Modifications in floral morphology usually can be related to the mode of pollination or specialized reproduction. Wind-pollinated taxa frequently have unisexual, reduced flowers that are individually inconspicuous. Inset-pollinated plants (which typically have large, colourful, bisexual flowers) are valuable to pollinators because of their nectar or pollen, and the showiness of he flowers helps attract pollinators. Corolla colour, pollinator guides, and other modification of the corolla are related to insect pollination. Other modifications of flowers (in stamen number, anther position, ovary position, style length, stigma shape, number of corpels, number and fusion of perianth parts, inflorescence type, fruit and seed types) contribute to the reproductive success of the species. Although these characters play important functional roles in the plant's reproductive biology, the adaptive advantages are not always obvious.

It is quite possible that some floral features have no adaptive advantage in themselves, but that either they had such advantages historically, or their elimination may require too much genetic manipulation, when the nonadaptive trait is linked to some essential feature. When one become familiar with a number of families of flowering plants, it is possible to see that generalizations can be made about the use of certain reproductive features in the classification of one or more groups For example, the mature fruits of the mustards (Cruciferae) and carrots and their relatives (Umbelliferae) are significant as diagnostic characters distinguishing the genera. For this reason, specimens of these families should include mature fruits to insure identification. Certainly, fruit and flowers types are no less important to insure identification. Certainly, fruit and flower types are no less important in the classification of the Rosaceae.

Likewise, the stamens and corollas of the figworts (Scrophulariaceae) and mints (Labiatae) are important floral features. Seed structure of taxa within the pink family (Caryophyllaceae) provides much information for classification at various taxonomic ranks within the family. The value or characters drawn from reproductive morphology varies from group to group within the angiosperms. The growth habit (herbaceous or woody) of plants may be of primary usefulness in classification. The condition may be constant within a genus or family, or it may be variable. All members of the family Cruciferae are herbaceous. Some families, such as the Compositae, have both woody and herbaceous members.

Genera or families that contain both tropical and temperate taxa are more likely to have both woody and herbaceous growth habits. Some semitropical or tropical herbaceous perennial species very in their tendency to woodiness. With some taxa, vegetative features are of little taxonomic value, but in other taxa they can be of major importance. For example, flowers and fruits are of little value in the elms (*Ulmus*), but leaf shape is an important taxonomic feature.

Leaf characters are significant features also in the identification and classification of the oaks (*Quercus*) and birches (*Betula*). The diagnostically important basal rosette leaves of the Cruciferae may wither and die before flowering and are often lacking on fruiting specimens. Stem types, buds, thorns, spines, and so forth on woody plants are often useful in identification. Leaf arrangement, type form, duration, and venation are frequently used in both classification and identification. In trees with small reduced flowers of short duration, vegetative features often assume great importance in plant taxonomy and should be carefully observed. Vegetative underground structures such as rhizomes, corms, and bulbs may sometimes characterize a group.

Comparative Plant Anatomy

For over a century, taxonomy has used comparative plant anatomy to aid in classification, and several principles on the use of anatomical data have become established. These principles are the following: (1) anatomical features have the same inherent problems of other characters, that is, sampling, reliability, parallelism and convergence; (2) anatomical characters must be used in combination with other features; and (3) anatomical characters tend to be most useful in classification the higher categories and less useful below the rank of genus. Further information on comparative plant anatomy may be found by consulting the following sources; Esau (1965) for a general review of plant anatomy; Metcalfe and Chalk (1979-1983) and Metcalfa (1960, plus several additional books in this series) for anatomical information arranged by taxa; and Radford et al. (1974) for a list of pertinent citations and a summary of anatomical techniques for systematics. Other useful references include Metcalf (1954, 1963), Robson, Cutler and Gregory (1970), and Fahn, (1982). Since the 1930s the value of the evolutionary trends of specialization of the secondary xylem has been clearly established.

A progressive series from tracheids (commonly found in the gymnosperms) to specialized vessel elements occurs in the secondary xylem of angiosperms. All stages of specialization, from vesselless wood to highly specialized vessel elements, may be found in contemporary flowering plants. Those angiosperms with vesselless wood often have other features regarded as primitive,. This evolutionary series of vessel elements has been used in combinations with other morphological features to develop hypotheses about the phylogeny of the angiosperms. Little information for classification has been obtained from anatomical studies of the phloem because of its lack of apparent variation. However, types of sieve-element plastids have recently proved useful in classification. Some comparative plant anatomists suggest that correlations of nodal anatomy with other feature might provide significant information about angiosperm phylogeny. Other fine this of limited value because of convergence and reduction. Recently, nodal anatomy was used to help define the broad outline of dicot relationships among the primitive dicots. It is often possible to interpret vestigial or modified parts of flowers by studying their vascularization. This has been valueble in clarifying the relationships of some taxa.

Anatomical features of leaves frequently provide characters. For example, investigation of the anatomy of leaves, associated with C_4-photosynthetic pathways, has resulted in a revised classification for several genera in the grass family. C_4 species have prominent chlorenchymatous vascular bundle sheaths. Venation patterns of fossil leaves are beginning to yield significant information when examined with reference to the Cronquist-takhtajan system of classification. Petiolar vascularizaton has been helpful in classification of certain genera such as *Rhododendorn* but has not proved applicable to classification problems of the higher categories. Variation patterns of epidermal hairs or trichomes may provide characters for classification at species-genus-family levels. For example, trichomes are diagnostic characters for certain species of *Vernonia*. Emphasis is placed on the structure of trichome; that is, the size, shape, and arrangement of cell making up the hair. The presence and structure of trichomes, as well as their distribution patterns among taxa, are taxonomically important. In the Compositae and several other families, tichomes are of value in analysis of suspected hybrids. A glossary of epidermal hair terminology can be found in Payne (1978).

Stomatal types, produced by characteristic arrangements of guard cells and subsidiary cells, can be of taxonomic use. Some 31 patterns of stomata and subsidiary cells are known to exist in the vascular plants. The patterns seem to be most valuable at the higher taxonomic levels. Perhaps one of the best examples is their use to characterize the subclasses of monocots. Some workers regard anatomical features as conservative characters that are not easily modified by growing conditions. Closely related species or genera will often have the same anatomy, but there are exceptions. Leaves grown in sun may differanatomically from those grown in shade. Anatomical features, like other characters, can be greatly influenced by selection. The stem of the hemp cultivar of marijuana (*Cannabis sativa*) have more fiber bundles than the stems from plants high in intoxicating resins. In most instances, variation such as this would not be expected in stem anatomy within a species.

Anatomical features of vegetative structures have importance in separating gymnosperms from angiosperms and monocotyledons from dicotyledons. However, such features have proven of relatively minor value for the delimitatin of subdivisions of these groups. Anatomical features tend to be most useful at the higher ranks or in determining relationship where there have been great structural modifications or reductions. The concept of anatomical characters as conservative has been challenged by Carlquist (1969), who considers that anatomy is always related to adaptation and function. He stresses that the anatomy of floral structures should be viewed as part of a dynamic functioning system related to pollination biology and dispersal mechanisms.

EMBRYOLOGY

Embryology includes micro- and megasporogenesis, development of gamatophytes, fertilization, and development of endosperm, embruo, and seed coats (Palser, 1975). For general, systematic, and comparative information on plant embryology, the following reference book may be consulted: Bhojwani and Bhatanagar (1979), Davis (1965), and Maheshwari (1950). Although there are a few minor differences in embryology among the flowering plants, there is strong embryology unity throughout the angiosperms, as exemplified by double fertilization.

The major embryological separation the dicotyledons from the monocotyledons is the number of cotyledons. Cotyledon characters have been shown to be useful in developing classifications in the family Gesneriaceae. The embryological features most common to all flowering plants are (1) four microsporangia per anther; (2) two-called pollen grains; (3) eight-nucleate embryo sac; and (4) nuclear endosperm (Palser, 1975). Dahlgern (1975) has compared the distribution of some embryological characters of the flowering plants within the framework of his system of classification. The number of nuclei in the pollen grains at the time of pollen dispersal is either two or three. It is generally regarded that binucleate pollen is primitive and the trinucleate pollen is derived. Embryological features are ordinarily constant at the family level in the angiosperms. In those families where variation has been found, the features are usually constant at the generic level. Embryolocy has had systematic significance in the grass family, where with several other characters, it has been used to revise the family's classification. Embryological features tend to be less useful as taxonomic characters at the rank of order, subclass or class. When used judicioulsy and in combination with other characters, they may be useful in determining relationships within families, genera, and species. The technical work and time required to obtain sufficient embryological information for comparative purposes has limited its overall value in taxonomy.

Cytology

Although cytology refers to the study of the cell, only information about the chromosomes-that is, chromosome number, shaper, or pairing at meiosis-is used for classification purpose. *Cytotaxonomy* refers to the use of chromosome number and morphology as data for classification. *Cytogenetice* includes those studies dealing with observation of chromosome pairing or behaviour at meiosis. The haploid number of chromosome in angiosperms ranges from n = 2 in *Haplopappus gracillis* (Compositae) to around n = 132 in *Poa littoroa* (Gramineae). Most angiosperms have chromosome number ranging between n = 7 and n 12. About 35 to 40 percent of the flowering plants are polyploids. *Polyploids* are organisms that have higher chromosome number because of multiplication of chromosome sets (i.e. , genomes). In an extensive review article, Raven (1975b) suggests that the original basic chromosome number for most angiosperms was probably 7. Raven speculates that an initial burst of polypoidy occurred during the early evolution of flowering plants, since many of the families with primitive feature have high chromosome numbers. There are several kinds of polyploid number relationships in flowering plants. Some genera such a *Pinus* with n = 12 have no deviation among the species and the species are termed *homoploids*. In other genera, a polyploid series is present, for example, *Aster* (Compositae) with different species having n = 9, 18, or 27.

There are still other genera with number that show no simple numerical relationship to one another (*aneuploids*)- i.e. , *Brassica* (Cruciferae), with n = 6, 7, 8, 9, or 10. *Vernonia* in the Old World has n = 9 or 10, with polyploids of n = 18, 20, or 30, whereas in the New World it has n = 17, 34, 51, or 68. As with many other characters, the value of cytotaxonomic data depends on the group or the category under consideration. A combination of infromation on chromosome number and chromosome morphology has proved useful or improving the classification of families, such as Agavaceae, or genera in the family Ranunculaceae. Base numbers and chromosome size have been very useful in understanding relationships in the grass family.

In the family Onagraceae, information on chromosome number and morphology, in combination with experimental hybridizations and analysis of the chromosome set, provide clues for tracing evolution. Because relatedness of taxa is often reflected n homology (similarity) of the chromosomes, determination of the amount of chromosome pairing at meiosis in hybrids of two species may aid in understanding relationships of closely related species. It is generally assumed that the more completely two chromosomes pair at meiosis, the more homologous they are. Or, if dealing with entire sets of chromosomes, the higher the percentage of pairing between the individual chromosomes of the two, sets, the more closely the plants are presumed to be related.

Electron Microscopy

Unlike its application to the study of lower plants, electron microscopy is a relatively new approach to the study of flowering plants. Feature observable only with the electron microscope are providing useful characters relevant to the phylogeny of the angiosperms. *Ultrastructure* refers to those features observable with transmission electron microscopy (TEM), or with scanning electron microscopy (SEM). SEM provides an image of unequaled depth of field, which is ideal for comparative studies of pollen grains and plant surfaces. SEM has a wide range of magnification and is relatively easy to use. Used to supplement light microscopy, SEM has been widely and readily adopted by plant systematists for pollen grains, small seeds, trichomes, and other surface features of plants. Relatively few new characters have been uncovered by SEM, but the technique has enabled rapid and recordable, comparative study of numerous features (Barthlott, 1981; Heslop-Harrison, 1981).

Recently, using SEM micrographs of starch grains, the laticifer cell of *Euphorbia* were demonstrated to be useful in the interpretation of evolution within the genus (Biesbor and Mahlberg, 1981). Using TEM, Behnke (1972, 1977, 1981; Behnke and Dehlgren, 1976) demonstrated the taxonomic value of sieve-element plastids in flowering plant classification. Behnke found two distinct classes of plastids; the s-type which accumulates starch and the p-type that accumulates protein or protein and starch. This character was then successfully applied to taxonomic problems in the Caryophyllidae and the angiosperms.

TEM studies of thin section of pollen grain walls can yield reliable taxonomic information of certain group. For example, some member of the family Compositae have differing types of wall structure and these have proved useful as characters. Dilated differing types of wall structure and these have proved useful as characters. Dilated cisternae of endoplasmic reticulum are a characteristic feature of members of Cruciferae and Capparaceae of the order Capparales. Membrane bound protein bodies found in seeds may contain crystalloids that very in size, shape, and number. The cotyledon mesophyll cell bodies in the Cucurbitaceae generally contain crystalloids, where those of the Composite and Cruciferae do not.

Although seed protein bodies of relatively few species have been studied with TEM, there are indications that protein bodies could be a useful character for studies in plant systematics. In general, TEM has not been as successful as SEM in providing useful ultrastructural characters, for three reasons; (1) the general uniformity of cell organelles within each type of tissue; (2) the small number of taxa and tissues sampled; and (3) the greaster difficulty and time required for specimen preparation. Ultrastructural characters may be found with TEM once tissues are compared and surveyed in groups that appear to have classification problems.

Palynology

Palynology is the study of pollen and spores. As a result of the stimulus and availability of SEM, taxonomists no longer overlook pollen as a source of characters. The development and ready availability of SEM has revolutionized the study of the surface features of pollen grains by providing a depth of focus never possible with light microscopy. The availability of countless pollen samples from herbarium sheets and relatively rapid techniques for preparation allow a palynological survey of many taxa in a relatively short period of time. The taxonomic characters provided by pollen grains include pollen wall structure, polarity, symmetry, shape, and grain size. Pollen grains do not differ within most genera; however, pollen has been very useful in determining patterns of spices relationships in Vernonia. The order Caryophyllales is a large natural association of families diagnosed in part by a curved embryo surrounded by perisperm storage tissue. Two unique features found among certain families in the order, namely the betalain pigments and protein sieve-tube plastids, have brought the order under scrutiny.

Using SEM, Nowicke and Skvarla (1979) have shown that pollen morphology reinforces the close relationships accorded the families and supports their distinctness from other nearby families. Two basic kind of pollen grains are found in the angiosperms : monosulcate and tricolpate. *Monosulcate pollen grains* are boat-shaped and have one long germinal furrow and one germinal aperture. Pollen of the monosulcate type is characteristic of the primitive dicotyledons, the majority of the monocotyledons the cycads, and the pteridosperms. Palynologists agree that the first flowering plants probably had monosulcate pollen grains. *Tricoplate grains* are globosesymmetrical, typically have three germinal apertures, and are characteristic of the advanced dicotyledons. Walker and Doyle (1975) concluded that with some exceptions; pollen morphology is consistent with systems of classification.

PALEOBOTANY

Paleobotany uses microfossils, such as pollen, or macrofossils of leaves, stems and other plant parts as sources of data. Paleobotanists attempt; (1) to elucidate the composition and the evolution of the floras of the past; (2) to trace these evolutionary development through stratigraphic sequences; (3) to integrate paleobotanical data with comparative morphology; and (4) to determine past ecological conditions. Until recently, most systematists generally agreed that paleobotany could provide little evidence on the origin and diversification of the flowering plants. However, evidence is rapidly accumulating that paleobotany can provide significant information about important aspects of the early history of angiosperms.

New technique and approaches to the study of fossil flowering plants are helping taxonomists discover exciting infromation. The most encouraging effect of the recent upsurge in research on angiosperm origins is the recognition that the fossil record can furnish critical evidence for the origin and diversification of the flowering plants as it has for the vertebrates. In recent years, the discovery by Dilcher and his associates of several angiosperm reproductive structures has greatly modified a number of early ideas about primitive flower types. There have been many marvelous discoveries of new fossils in recent years that have contributed a wealth of information about flowering plants.

The application of the technique of SEM, has provided a clearer picture of the complex micromorphology of fossil pollen grains, and the application of TEM is now providing information on the details of pollen ultrastructure. The new approaches have been especially rewarding. Recent collections of well-preserved angiosperm flowers from Eocene sediments of the Southeastern United States and from Kansas include an interesting array of morphological types. Diversified representatives of the present-day "Amentiferae-line" were common by Middle Eocene time. They were well adapted for wind pollinations, and it appears form the fossil evidences that adaptation to wind pollinations had, by that point, reached a high level.

Flowers and inflorescences with structural features allowing pollination by beetles, flies, bees, and butterfies were also extant by Middle eocene. Fossil evidence recently uncovered by Crepet, Dilcher and other, coupled with observations of modern angiosperms, suggests that coevolution of flowers with insects was important in the radiation of the flowering plants. Their work clearly demonstrates that the Middle Eocene flora Can help provide comprehensive information about the evolution of the flower. An extraodinary data source has been provided by the organic chemical profiles of Miocene leaf specimens of five genera (*Acer*, *Celtis*, *Quercus*, *Ulmus* and *Zelkova*), Which were compared with those of present-day counterparts. A high degree of correlation between chemical compounds of the fossils and those of the modern species was found. This new information has allowed paleobotanists to formulate new concepts of early flowering plant evolution. These differ some what from the traditional view that was largely derived by circular reasoning from the morphology of so-called primitive group. Data provided by paleobotany allow taxonomists to deal with the facts to the fossil record about angiosperms and their possible progenitors. Clearly, the best source of evidence is actual data from the fossil record.

CHOMOSYSTEMATICS

For centuries, chemosystematics (or biochemical systematics, or chemotaxonomy) has been practied by human cultures as they recognized and classified plants by tastes, smell, colour, and whether or not they were poisonous or of medicinal value. For example, the Greek philospher Socrastes was forced to take his own lie by dirinking a broth os poison hemlock (*Conium maculatum*). By the first century after Christ, Dioscorides had recognized the aromatioc mints and grouped them together. Later, the externsive explorations to the far corners of the world by Europen adventurers and merchents were fueled by the desires of Europeans for exotic speces and condiments. The value of the spices was for food preservation and maskihg odors and tastes of spoiling meat.

Taxonomists of the nineteenth century often recognized chemical characters and used them in their work before the compounds were completely identified.

Plant *chemosystematics* is the application of chemical data to systematic problems. It is rapidly expanding interdisciplinary field concerned with using chemical constitutents for explaining relationships plants and inferring phylogeny. The development of several types of chromatography and various types of spectroscopic and instumentation techniques since the 1940s has simplified survey procedures and considerably eased compound indentification. With the development of modern methods, information on natural plant products for phylogenetic comparisons is becoming increasingly available. Plants produce many types of natural products, and quite often the biosynthetic pathways producing these compounds differ from one taxonomic group to another.

In many instances, the distributions of these compounds and their biosynthetic pathways correspond well with existing taxonomic arrangements based on the traditional morphological features. In other cases, chemical data have contradicted existing classifications, thus requiring a reexamination of the classification, or have provided decisive infromation where other forms of data were insufficient. Since many kinds of molecules are produced by plants, the discussion in this brief summary is neccessarily selective. Chemosystematistis often divide the vast array of natural plant products into two major groups according to molecule size.

Compounds of relatively low molecular weight, that is, those compounds with a molecular weight of 1,000 of less—such as alkaloids, amino acids, glucosinolates (mustard oils), flavonoid pigments and other phenolic compounds, fatty acids, and terpenoids-are termed *micromolecules*. Compounds of high molecular weight (over 1,000), such as proteins, DNA, RNA, cytochrome c, ferredoxin, and complex polysaccharides, are referred to as *macromolecules*. The information-carrying proteins, such as DNA, RNA, and other proteinaceous materials are sometimes referred to as *semantides*.

Flavonoids

Among micromolecules, the flavonoids have most widely and effectively used in chemosystematics. The reasons for this are several: their ubiquitous occurrence in nearly all plants; their ease of isolation and identification even from small amounts of plant material; their great structural variation (over 2,000 different flavonoids have been isolated from plants); the demonstrated genetic basis for this variations; their chemical stability, which allows analysis years after the material was collected; and the fact that they can be used at all taxonomic levels in most groups of plants. The simple but elegent use of the chromatographic patterns of flavonoids in the 1960s by R. E. Alston and B. L Turner to document hybridization between species of *Baptisia* set the stage fot the effective use of flavonoid data in plant systematics.

Occasionally, when two species of Baptisia occur together, hybridizaton occurs. The hybrids resulting from this accidental crossing of two species often look different from either parent and sometimes confused plant taxonomists. Alston and Turner examined the flavonoid profiles of the two parents and found them to be matually exclusive in all or at least in several compounds. They demonstrated that interspecific hybrids could be identified on the basis of *additive profiles*; that is, certain species-specific compounds from each parent were found together in the hybrids. Indeed the *Baptisia-type* work is so well documented that it is oftern assumed to be standared for flavonoid work. In the classical flavonoid situation, when chemically different parents are hybridized, the F_1 generation exhibitis an additive profile of both parental types. However, the early enthusiasm generated by the classical *Baptisia-type* studies has been tempered somewhat by later ones showing that hybrids do not always exhibit additive flavonoid profiles and the compounds not found in either parent may be found in the hybrids.

In a flavonoid survey of the deciduous Rhododendrons of Eastern North America, it was found that complementation did not occur in the classical sense. Instead, variable, nonadditive profiles were present in both natural and artificial hybrids. Giannasi and co-workers have shown both complementation and noncomplementation in hybrids between two species of *Solidago* often with the production of unique nonparental compounds. Additional, detailed Studies of flavonoid inheritance and expression in diploid and polyploid species are needed. There has been gradual shift in flavonoid work in the past 10 years from intensive studies of infraspecific variation in the mannar of Alston and Turner toward comprehensive detailed flavonoid surveys from the interspecific to the subclass showed the flavonoids can be useful in clarifying certain relationships. Also, there data suggest that there is a need for circumscribing a number of families using their overall flavonoid compounds.

The controversial placement of *Psilotum* as a living representative of the most primitive ferns has not been supported by recent flavonoid evidence. The possession of biflavonyls seems to place *Psilotum* closer to some lycopods and gymnosperms than to the leptosporangiate ferns as had been proposed. Flavonoid data used in connection with information on betalain distribution have also been helpful in another problem area. In the subclass caryophyllidae, the distribution of the betalain-producing families and those that produce only anthocyanins has led to reclassification of that subclass. The presence of betalains in the family Cactaceae is an example of chemical data aiding in determining relationships at higher taxonomic levels, the Cactaceas (once placed elsewhere in classification schemes) should be placed in the Carophyllidae, the only subclass with betalains. Further progress in flavonoid systematics may depend on work in comparative enzymology of a better understanding of biosynthetic pathways. It is likely that the result will provide a more biologically sophisticated and genetically meaningful interpretation of flavonoid chemistry.

Terpenoids

Terpenoids have used extensively in the chemosystematics of groups where they are distributed; for example, mints, unbellifers, citrus plants, gymnosperms, and so on. However, they have been used less than the flavonoids in systematics because of their limited distribution among vascular plants and because of the complex instrumentation and experience need for their analysis. The technology of gas chromatography used with terpenoids provides both a quantitative and qualitative measure of chemical differences between plants. Terpenoids can be used for distinguishing specific and subspecific entities, Geographical races, and for documenting hybridization. In the flowering plants, interesting studies have been carried out using rind and leaf terpenoid patterns in *Citrus* in attempts to determine the origin of certain *Citrus* cultivars.

In the gymnoperms, terpenoids of the cortical oleoresins have been used to distinguish geographic races of Douglas fir (*Pseudotsuga menziesii*). Extensive studies of the terpenoids of *Juniperus virginiana* and *J. ashei* refuted previous hypotheses about extensive hybridization and introgression between the two species. A major contribution of terpenoid chemosystematics has been the use of sesquiterpene lactones, which for the most part are characteristic of the Compositae family. Many of the tribe of the Compositae are characterized to a greater or lesser extent by the distinct type of sesquiterpene lactones they produce. Within the Compositae tribes, sesquiterpene lactone chemistry are characterized to a greater or lesser extent by the distinct types of sesquiterpene lactones they produce.

Within the Compositae tribes, sesquiterpene lactone chemistry can be used effecitvely both phylogenetically and phytogeographically. Sesquiterpene lactones have been used in the tribe Vernonieae to show that the largest genus in the tribe, *Vernonia*, has two major centers of distribution, one in the Neotropics and the other in Africa. Future studies of terpenoids will likely continue with surveys of taxa at higher and lower ranks and with investigations related to geographic factors and to hybridization. Additional work is needed to determine the genetic control of terpenoid expression.

Alkaloids

From ethnobotanical, medicinal, poisonous, chemical, and systematic viewpoints alkaloids have elicited more intense interest than any other group of compound. Relative to their value in plant systematics, alkaloids are a difficult and heterogenous group of compounds, both structurally and biosynthetically. Further, the published literature on alkaloids is overwhelming. Despite these difficulties, alkaloid distribution has been useful in taxonomic studies in *Veratrum* (Liliaceae) and related genera, in *Papaver* and *Argemone* (Papaveraceae), as well as *Lycopodium*, *Lupinus*, and taxa in the order Caryophyllales.

Glucosinolates

Glucosinolates (mustard oil glucosides) are useful in the classification of the order Capparacles, including Cruciferae (mustards) and the Capparaceae (capers). At one time the Cruciferae, Capparaceae, Papaveraceae, and Fumariaceae were included in the single-order Rhoeadales, but chemical and other evidence supports the placement of the Cruciferae and capparaceae in the order Capparales (which produce glucosinolates) and the Papaveraceae and Fumariaceae in the order Papaverales (which are alkaloid-containing). Two families of uncertain affinities, Bataceae and Gyrostemonaceae, were once placed in the Caryophyllales, but the presence of glucosinolates suggests that the two families be removed from this order. Detailed studies in the genus *Cakile* (Cruciferae) have shown that glucosinolates are also effective infrageneric characters and that glucosinolate patterns may be used to document hybridization.

Iridoids

Iridoids represent a class of compounds (mostly monoterpene lactones) that have increased in taxonomic significance. The iridoids have shown promise in clarifying interfamilial relationships in the orders

Rubiales, Caprifoliales, Scrophulariales, and Cornales, and they bear on the debate over the probable ancestral progenitor of the Compositae. Several iridoid families have been suggested as the putative progenitor of the Compositae, including the Rabiaceae, Dipsacaceape, Calyceraceae, Capeifoliaceae, Scrophulariaceae, and Cornaceae, all or which produce iridoids, which are not found in the Composite. Only the Campanulaceae, Araliaceae, and Umbelliferae, which do not produce iridoids, remain as possible progenitor lines. The latter two families are also linked to the Compositae by sesquiterpene lactones and polyacetylenes, raising the possibility of an ancient relationship.

New Approaches in Micromolecular Chemosystematics

A new and intriguing approach in plant chemosystematics in comparative phytoalexin inducion studies on legumes. This technique uses fungi such as Helminthosporium to induce chemical "antibiotic" interactions between tissues of different plant species and the fungus. Also of interest is the fact the qualitative and qualitative differences in nectar amino acids have been found among the flowers of different plant species. This technique can be used to document hybridization between plant species and to study pollinator relationships. The biological and evolutionary significance of micromolecules as allelopaths, herbivore-feeding deterrent or attractants, visual or olfactory stimulants, phytoalexins, and phytotoxins point to the critical functions of micromolecules in the evolutions and survival strategies of plants. Thus, the ecological role or micromolecules also plays a part in the phyletic distribution of micromolecules.

Macromolecules

Analysis of macromolecules (i.e. DNA, proteins) provides promise for helping to discover relationships among major groups of angiosperms, and more attentions will be directed to their future. Comparative protein analyses can be used as fundamental taxonomic characters, since proteins represent direct products of the DNA code. The genetic material (DNA) can also be analyzed and compared between plant taxa. Thus, certain direct questions of plant relationships can be addressed in systematic investigations of macromolecules that would not be possible with many micromolecules, which often involve complex biosynthetic pathways. There are, however, significant problems associated with the time-consuming techniques of macromlecular analysis requiring much technical laboratory expertise, as well as numrical problems associated with data handling and the proper interpretation of generated figures and graphs. Because of the specialization in the field, it is likely that teamwork among researchers sharing their expertise will yield the most promising new insights into plant relationships. Macromolecular data will be considered under the following general heading: electrophoresis, amino acid sequencing, systematic serology serology, and nucleic acid sequencing.

Electrophoresis

Profiles produced by the electrophoretic separation and subsequent staining of seed and pollen proteins have been employed in a number of systematic studies (Gottleib, 1977). The technique is based on the assumption that proteins from different plants will migrate in a slab of agar gel under an electric current. Those proteins that migrate to the same place in the gels and yield similar bands upon staining with a market dye represent homologous proteins. This approach of using electrophoretic band patterns has been used in the investigation of the origin of polyploid taxa from diploids, for instance, *Solanum*, *Phlox*, *Brassica*, and so on. It has also been used at the interspecific level to determine whether difference exist among plants that are treated as distinct species.

As with any technique, there are a number of technical and interpreitive restrictions that require care in the protein profiles. The electrophoresis of specific enzyme types has been widely used animal systematics (less so in plants), allowing the products of individual genes to be studied. In this method, the different alleles (Variants) of a locus (gene) coding for molecular variants of a gene can also be made visible as coloured bands in gels. From a systematic standpoint the method is significant because a large number of different enzyme (gene) systems can be examined among a large number of plant species. The data obtained are helpful at the level of the population, subspecies, species, of sometimes genus. Minor variation in allozymes is compared between individuals within populations, or between populations, or among closely related taxa at low rank. Such data are not generally useful at higher ranks because of the wide distribution of similar allozymes in unrelated groups. They are ideally suited detect interspecific gene glow and percentage of polyploids.

Amino Acid Sequencing

The phylogenies that have been generated from sequencing the amino acids making up a protein have produced data that require the use of complicated computer programs. Such computer-derived data and their systematic implications are sometimes more controvesial than simpler types of distributional chemical information because of the uncertain acceptability of the numerical approached themselves. The evaluation of these complex numerical data has not been easily comprehended, and thus, information from amino acid sequences has not been readily or universally accepted by the practicing plant systematist for solving systematic problems.

Yet helping to understand phyogenetic relationships among the flowering plants as the technology and understanding of the statistical techniques become more comprehensible to the nonspecialist. Amino acid sequence data for cytochrome *c*, plastocyanin, and ferredoxin have been determined for about 30 species of vascular plants. With this approach, it is possible to compare amino acid sequences of the same protein among various plants. From the similarities and differences in the amino acid chain, it is also possible to infer the DNA nucleotide similarities for the same gene in various plants since specific units of DNA are coded for each amino acid of protein. Thus, the amount of genetic (DNA) divergence may also estimated. This has provided some phylogenetic information, but it is complicated by the likelihood that different classes of proteins (and hence their DNA) evolve at different rates.

Some workers, however, have assumed that nucleotide substitutions occur at a generally constant rate for a given protein; that is, there is a molecular "clock." While this has been convincingly illustrated in animals, workers in plant evolution have questioned whether there is a "molecular evolutionary clock" and just how accurate is it since the clear evolutionary evidence seen in animals is not observed in amino acid sequeness in plants. While the paucity of data and their interpretation have been questioned, amino acid sequence studies have produced excitement about the possibilities of such an approach, even though they have had as yet little actual impact on plant taxonomy.

Systematic Serology

Systematic serology had its origin in the early 1900s with the discovery of serological reactions and the development of the discipline of immunology. The technique was applied to comparative problems involving birds by J. Bordet, who noted that the immune reaction was relatively specific, and that the degree of cross-reactivity was proportional to the degree of relationships between organisms. The systematic implications of the immune reaction resides in the fact that the antigen-antibody responses are very specific. An animal will produce precise types of antibodies in response to the interaction with particular kinds of antigens. It is then possible to determine the degree of similarity between organisms (Species, genera, families) by comparing the reaction of antigens from various plant taxa with antisera (antibodies) raised against the antigen of a particular taxon.

Crude protein extracts from seed or pollen are commonly used for antigen production. However, recent developments of techniques in serology have led to the comparison of single proteins from different taxa. Plant systematic serology has provided valuable data for the classification of flowering plants. Serological data have contributed to the classification of orders and the placement of families in the Apiales, Capparales, Caryophyllales, Cornales, Dilleniales, Dipsacales, Fagales, Juglandales, Magnoliales, Myricales, Nymphaeales, Papaverales, Primulales, Ranunculales, Restionales, Rubiales, Sapindales, Scrophulariales, Solanales, Typhales and many other (Fairbrothers, 1983). A definite and substantial rise in the use and incorporation of serological data in angiosperm classification has taken place in the past few years.

Nucleic Acids

The total number of plant taxa compared by using nucleic acid techniques has been few, largely because the necessary expertise and equipment are unavailable to most systematists. In addition, much of he methodology has been developed only recently, and there has been little time for the accumulation of data. Basically, nucleic acid sequencing involves several technique. Initially, the natural double-stranded DNA is separated into single strands. The single strands of DNA from one taxon are allowed to reassociate with similarly treated DNA from another taxon. Comparison is made with controls consisting of reassociation of strands from a single taxon. The percentage of reassociation of DNA between the two species. Observable changes in sequences of DNA can provide information on speciation events that will aid in the reconstruction of phylogenies. Such techniques of DNA hybridization have also been used in studies of ribosomal RNA (rRNA)/ DNA homolgies among plants.

Comparisons of plant organellar DNA (chloroplasts, mitochondria) also may offer some advantages in systematic studies. Plant organellar DNA represents of smaller amount of genetic code and thus allows sequencing of a less complicated molecule than that of nuclear DNA. Using the approach of digestion of organellar DNA with restriction endonucleases, the molecule can be broken up into smaller, more easily identifiable fragments whose specific gene function may also be identified. Comparison of precise DNA nucleotide sequences (and specific gene function) among species provides a powerful taxonomic tool. Indeed, recent studies comparing eight species of *Atriplex* (Chenopodiaceae) have provided estimates on time origins for the lineages as well as information useful in developing a new fromal classification.

Ecological Evidence

Information on the ecology of flowering plants is basic to systematics in providing an understanding of (1) the distribution of taxa, (2) the variation within taxa, and (3) the adaptations of plants. In delimiting taxa, it is necessary for taxonomist to comprehend developmental responses of plants as distinguished from genetically fixed characters. Ecological studies have demonstrated that the characters states of many morphological features are correlated with environmental factors such of many morphological features are correlated with environmental factors such as light, moisture, and soil fertility. Ecology contributes to systematic interpretation of the evolutionary process by seeking environmental explanations for discontinuities in the structure, function, and distribution of plants.

Plant ecologists examine ecotypic variation, edaphic specializations, pollination mechanisms, the effect of habitat on hybridization, plant-herbivore interactions, seed-dispersal mechanisms, ecology of seedling establishment, function of plant structures, and reproductive isolating mechanisms. These features are important in the adaptation of population to their environments and represent areas where ecologists make substantial contributions to systematics. Much of the information derived from ecology has implications for classification below the level of genus.

Yet ecological research has provided generalizations that may be applicable to the evolution of the flowering plants. For example, Stebbins (1974) presents the hypothesis that adaptive rediation of the angiosperms occurred in ecotones, or marginal or transitional habitats, under stresses of seasonal drought or cold, or in a mosaic of local edaphic conditions. Consequently, there may have been strong interactions between environmental factors and natural selection. He also favours a hypothesis of angiosperm origin in a climate having marked seasonal drought with a short season favourable for the formation of flowers. Under these conditions, natural selction might have favoured the reduced angiosperm reproductive cycle. Although these hypotheses can be neither proved nor completely rejected, they do reflect the role that environmental factors play in the development of concepts about the origin of flowering plants. Furthermore, they express the idea that ecological data must be considered in the classification of plants.

PHYSIOLOGICAL EVIDENCE

Physiological and biochemical evidence is providing data of increasing importance to plant systematics. Of particular significance are data dealing with metabolic systems and biochemical pathway. Recently, it has become apparent that a syndrome of anatomical and physiological features, related to a high-efficiency carbon fixation process, occurs in a large number of plants from tropical or warm temperate regions. This syndrome has been called the *Kranz syndrome*, C_4-photosynthesis, or the Hatch-Slack pathway. The most common pathway of carbon fixation is the *Calvin-Benson cycle*, or C_3-photosynthesis. In the algae, mosses most ferns, gymnosperms, and many families of flowering plants, C_3 photosynthesis is the only know carbon fixation cycle. C_4 Photosynthesis occurs in approximately 10 unrelated families of monocotyledons and dicotyledons.

Crassulacean acid metabolism (CAM) has been found in both monocotylen families and in dicotyledon families, and in ferns. CAM and C_4 both represent adaptations to arid climates. It was originally thought that the distribution of CAM would provide a useful taxonomic character; however, the random distribution of CAM has prevented it from being used in classification. The kranz syndrome has proved useful in characterizing *Panicum* and other taxa in the grass family. In the dicotyledon genus *Euphorbia* (Euphorbiaceae), C_3, C_4, and CAM species are known. Both C_3 and C_4 carboxylation occurs in *Zygophyllum* (Zygophyllaceae) and *Atriplex* (Cheonopodiacea). Comparative studies of unique physiological processes of flowering plants might provide data of taxonomic significance. For example, the ability of halophytes to grow in highly saline soils seems to be restricted to relatively few families, but the phenomenon has not been well explained. There is much interest in the field of physiological plant ecology, but the major contributions to data have been in the physiology of ecotypes. As more plants are examined with biochemical and physiological technique, useful evidence will undoubtedly be found and applied to systematic problems.

4

ORIGIN OF ANGIOSPERMS

The Angiosperms are the flowering plants and have the greatest number of species and occupy more types of habitats than any other group. They include trees, shrubs, perennials annuals and lianas. They range in size from tiny duckweeds to giant trees. Adaptive radiation of angiosperms has produced parasites, saprophytes without chlorophyll, and epiphytes. Insectivorous species such as pitcher plants, Venus flytrap, and sundew represent unusual leaf adaptations. Diversity in flower structure is another remarkable feature of flowering plants. The term *flower* usually refers to a structure containing sepals, petals, stamens, and carpels. However, flowers may consist only of a single ovary or a single stamen. Charles Darwin once called the origin of angiosperms an "abominable mystery" because little evidence was available to him to help trace the angiosperms back in time.

The origin and evolution of angiosperms continues to present an intriguing challenge to those palobotanists who specializa in the study of fossil angiosperms. The collections of fossils, the accumulated literature, and the experience available today provide a much better understanding of angiosperm origin. With the accessibility of increasing information from many sources, modern angiosperm systematics allows a synthesis of data never before, possible. Angiosperm classification research is in a new phase; basic and far-reaching results will produce the best classification possible.

CHARACTERISTICS OF FLOWERING PLANTS

The angiosperms and gymnosperms constitute the seed plants. Generally, the difference between them is the extent to which the ovules are exposed at the time of pollination. The charateristics of angiosperms are the following: (i) vessels in the xylem, (ii) sieve elements and companion cells in the phloem, (iii) an embryo sac of eight of eight nuclei (one egg, two synergids, three antipodal, and two polar), (iv) double fertilization, and (v) closed carpels. There are numerous exceptions to these characteristics. Xylem vessels are not peculiar to angiosperms because they occur among some gymnosperms and there are certain angiosperms that have vesselless wood. Many exceptions are known to the eight-nucleate embryo sac.

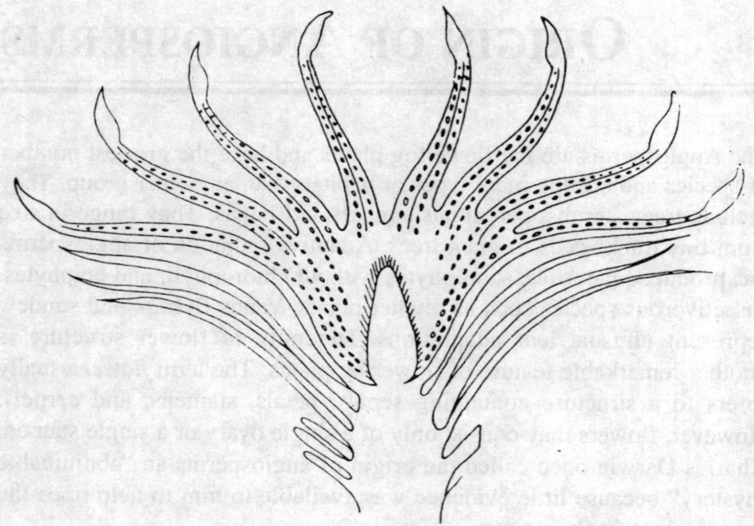

Fig. 4.1. Bisexual strobilus of Cycadeoidea.

The gametophyte generation or flowering plants is generally more simplified and reduced than that found in the gymnosperms. The male and female gametophytes of flowering plants are small and function efficiently with relatively few cell division. Angiosperms have *double fertilization* with one sperm fusing with the egg nucleus and the other sperm fusing with two polar nuclei in the embryo sac. The latter results in initiation of *endosperm*, the nutritive tissue of the seed. In angiosperms, the embryo is formed directly from the zygote. Double fertilization is not known to occur in other groups of plants. In the angiosperms, the specialized megasporophylls (or carpels) that enclose the ovules are unique and provide the basis for the name *angiosperm*, "covered seed." The carpels cover the ovules and facilitate adaptations for seed dispersal. Pollen collects and germinates on a stigmatic surface,

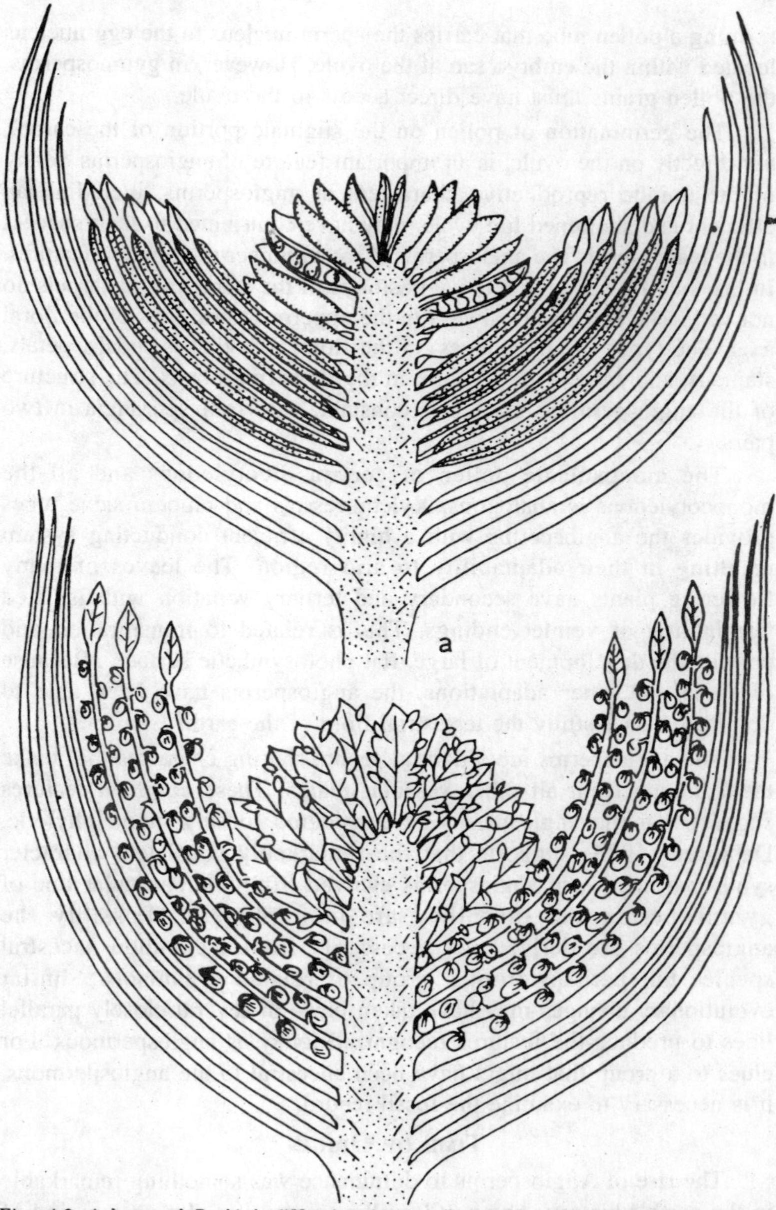

Fig. 4.2. Arber and Parkin's "Hemiangiosperm". (a) Hypothetical angiospermous strobilus in the primitive conditions. (b) The proanthostrobilus of hypothetical Hemiangiosperme.

forming a pollen tube that carries the sperm nucleus to the egg nucleus located within the embryo sac of the ovule. However, in gymnosperms, the pollen grains must have direct access to the ovule.

The germination of pollen on the stigmaic portion of the carpel, *not* directly on the ovule, is an important feature of angiosperms. Many of the unique reproductive characters of angiosperms, including an efficient and shortened life cycle, may have contributed to their success in the plant world. The angiosperms share a number of common features. In bisexual flowers, the relative positions of the stamens and carpels do not very. The carpels are always in the uppermost positions of the floral axis. The types of appendages on the floral axis-that is sepals, petals, stamens, and carpels-are uniform in the flowering plants. The structure of the angiospermous stamen is constant, with four sporangia in two pairs.

The monosulcate pollen of certain dicotyledons and all the monocotyledons is analogous. Xylem vessels and phloem sieve tubes provides the angiosperms with a highly efficient conducting system resulting in their adaptability to arid region. The leaves of many flowering plants have secondary and tertiary venation with isolated termination of veinlet endings. This is related to transpiration and permits the development of large, flat photosynthetic surface . Because of these and other adaptations, the angiosperms have been able to dominate successfully the terrestrial flora of the earth.

The angiosperms are a natural group, sharing charactes that make them unique from all other vascular plants. These common features suggest that the angiosperms be considered loosely mono phyletic. Dahlgren (1983) suggests that the remarkebly consistent character syndrome in present angiosperms evolved from one particular line of gymnosperms and thus they are monophyletic. Possibly the angiosperms are derived not necessarily from a common ancestral species but perhaps from a group of related organisms. Similar evolutionary changes may have taken place in several closely parallel lines to produce the features recognized today as angiospermous. For clues to a group that might have been ancestral to the angiospermous, it is necessary to examine the fossil record.

Time of Origin

The rise of Angiosperms to dominance was something remarkable in the earth's history. About 100 million years ago, during a period of only 10 million years, the flowering plants spread over the earth from the equator to the poles. This distribution was apparently undeterred by ocean barriers or competition from the eatablished vagetation that was dominated by gymnosperms and ferns. How the angiosperms managed this remarkable dispersal has long been a puzzle, but the most encouraging effect of the recent upsurge in research in angiosperm origins is the recognition that the fossil record can furnish critical evidence for the origin and diversification of the flowering plants. In the past, paleobotanists, having been unsuccessful in illuminating the origin or the angiosperms, formulated three hypotheses to explain this failure: (i) incompleteness of the fossil record and the lack of fossil flowers; (ii) a long and slow evolutionary history of flowering plants; and (iii) evolution of early angiosperms in upland areas away from the necessary basins where fossil could have been preserved. These so-called "escape" hypothese led some botanists to suggest that the angiosperms originated in the early Mesozoic or even the late Paleozoic and that they underwent extensive diversification by the Aptian-Albain stages of the lower Cretaceous. However, after reviewing the supporting evidence for this hypothesis, Wolfe, Doyle, and Page (1975) found ".... no unequivocal evidence that indicates a pre-cretaceous origin for the angiosperms."

The best available evidence form the fossil record indicates that the angiosperms originated during the early Cretaceous about 130 to 135 million years ago. Taylor (1981) suggest that pollen and megafossil data indicate the origin of the angiosperms during the very early Cretaceous or the latter stages of the Jurassic. It is likely that these early angiosperms were represented by a relatively small group of plants. In the older Cretaceous sediments, the angiosperm fossil are overshadowed by those of ferns and gymnosperms, and it is not until the late part of the Cretaceous that the angiosperms become dominant. The first monosulcate angiosperm pollen grains, characteristic of the primitive dicotyledons and the monocotyledons, appear in the Barremian stage of the lower Cretaceous.

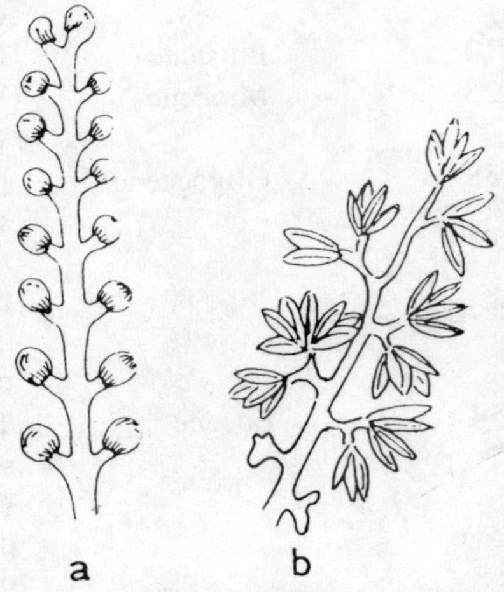

Fig. 4.3. The Caytoniales: (a) Pinnately arranged cupules; (b) branched stamens.

Tricolpate pollen grains, which are found in the more advanced dicotyledons, are first reported from slightly younger Aptian rocks. Pollen sequences have clearly shown a pattern of increasing diversity in the Cretaceous. The pollen of primitive angiosperms, however, is almost indistinguishable from gymnospermous pollen, creating uncertainty about the identification of the older fossil pollen. At an early date the angiosperms split into the monocotyledons and dicotyledons and very slowly gained prominence in the terrestrial flora that was largely dominated by ferns and gymnosperms. The progression of pollen from primitive types in older strata to more derived types in younger sediments indicate to paleobotanists that angiosperms underwent much diversification during the Cretaceous. By the Turonian and Coniancian stages, angiosperms pollen had become more abundant then fern spores and gymnosperm pollen.

Table 4.1. Geologic record of events.

10^6 Years ago since start	*Epoch or period*	*Geologic events, climatic trends, plant and animal development*
2.5	Pleistocene	Cascades, Sierra and coast Ranges rise, Rockies, Andes and Himalayas renewed; successive glaciations and warmings; development of modern floras, man evolves, large mammals develop and then die out.
12	Pilocene	Climate similar to present
25	Miocene	Himalayas rise; cool at first, warming later.
38	Oligoccene	Northern and southern continents widely separated, Alps rise, Rockies and Andes erode; warm early, cooler later; rapid apporach to modern angiosperm families, Compositae pollen first appears.
54	Eocene	Temperature rise near end; angiosperm infloresences and flowers well adapted to wind and insect pollination appear; modern lineages of vertebrates and insects appear; radiation of higher insects.
65	Paleocene	Lowland swamps decrease, Rockies and Andes complete major rise;

		cilmate cooler and more varied; groups of higher insects appear, primite arise; the Alismatidae appear.
110	Upper Cretaceous	Andes and Rockies arise, especially later much of North America covered by seas; climate cooling at end; angiosperms become dominat; ca. 90-100 million years ago modern lineages of angiosperms appear, mammals and birds begins to diversify, dinosaurs become extinct at end; the Commelinidae, Liliidae, Arecidae, Dilleniidae, Caryophyllidae and Rosidae appear.
135	Lower Cretaceous	Climate warm and moist; oscilation of epicontinental seas; first documented angiosperms appear, monosulcate pollen ca. 125 million years ago, tricolpate pollen ca. 122 million years ago; the Hamamelidae and Magnoliidae appear.
180	Jurassic	Widesprad seas in North America but with continental emergence toward end Atlantic Ocean forming as continents drift; dinosaurs flourish, first birds, mammals and modern amphiblan types; perhaps first angiosperms in a world dominated by gymnosperms and ferns.
225	Triassic	Continental drift, present continents from; climates varied first dinosurs; conifers highly diversified with most modern families present.
270	Permian	Appalachian Mountains arise; ginkgo-like plants appear, cycads appear.
325	Carboniferous (Pennsylvanian)	Climate warm, humid; coal deposited; major development of terrestrial vegetation, conifers appear, seed

		ferns abudant, bryophytes appear, lycopods dominat vegetation vegetation, equisetum-like plants reach maximum diversity.
350	Carboniferous (Misissippian)	Marine limetones deposited; marine evolution; ferns appear.
405	Devonian	Climate seasonal; fern-like plants appear, locods appear early, equisetum-like plants appear, progymnosperms appear, terrestrial plants and animals arise, tree-like land plants appear amphibians appear, seed-ferns appear.
440	Silurian	Vascular plants appear.
500	Ordovician	Radiation of marine animals, first fishes.
600	Cambrian	Seas expand; first abundant fossil.
4500	Pre-Cambrian	Earliest life arises.

Paleobotanists such as David Dilcher working with fossils of the Middle Cretaceous have provided important new ideas about angiosperm evolution. Angiosperm reproductive structures found by Dilcher and colleagues indicate that some early angiosperms bore separate male and female flowers and other were bisexual. They also suggest that some of the plants were wind pollinated or were generalists, that is pollinated by both wind and insects. Also, the seeds lacked adaptations for animal dispersal, suggesting that these plants were generalists in seed dispersal as well. It also appears that early Cretaceous fossil pollen is predominantly "generalist" in size and structure. Dilcher has found bisexual flower that is 95-93 million years old, the oldest bisexual flower thus far found. His work has shown that lineages of bisexual and unisequal flowers thus far found. His work has shown that lineages of bisexual and unisexual flowers coexisted in the early history of the angiosperms.

The fossil record indicates an extremely rapid radiation of types of angiosperms reproduction during the early Crataceous. It appears possible that the extinct ancestors of today's major angiosperms groups may have been derived in the early phases of angiosperms evolution. By the Maestrichtian stage at the close of the Cretaceous, a number of modern orders are represented by fossil pollen and leaves. Members of such groups as the Magnoliales, Hammamelidales, Ranunculales, and

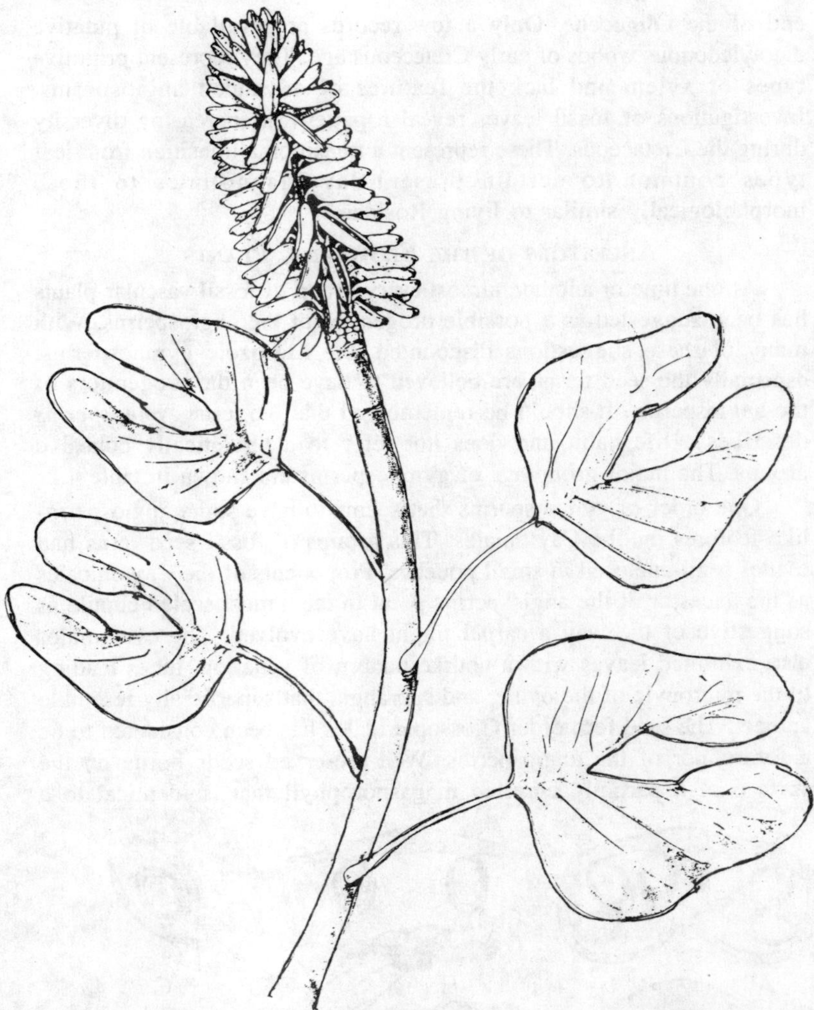

Fig. 4.4. Archaeanthus linnbergeri. Reconstruction of a leafy branch bearing a multifollicular axis from the mid-Cretaceous (uppermost Albian/mid-Cenomanian). Clusters of follicles were borne terminally on the axis; the scars at the base are interpreted as representing the places where stamens and perianth parts were attached.

Theales and some monocotyledons had appeared by this time. Evolution diversification of the angiosperms continued into the Cenozoic. Fossil flowers are known from Eocene strata. Although there is some evidence that the group may be older, pollen of the Compositae appears at the end of the Oligocene. Only a few records are available of putative dicotyledonous woods of early Creteceous age. They represent primitive types of xylem and lack the features of advanced angiosperms. Investigations of fossil leaves reveal a pattern of increasing diversity during the Cretaceous. These represent a transition a transition from leaf types common to certain present-day Magnoliales to those morphologically similar to living Rosidae.

Ancestors of the Flowering Plants

At one time or another, almost every group of fossil vascular plants has been suggested as a possible progenitor of the angiosperms. With many of these suggestions discounted, the Mesozoic gymnosperms, especially the seed ferns, are believed to have been the progenitors of the angiosperms. It should be remembered that the term *gymnosperms* describes a life habit and does not refer to a phyletically cohesive groups. The major groupings of gymnosperms are shown in table 4.2.

One order of gymnosperms that seems to have some angiosperm-like features ins the Caytoniales. This groups of fossil seed ferns had ovules semi-enclosed in small pouches. Proponents of the Caytonicales as the ancestor of the angiosperms point to the almost sealed cupule as suggestive of the way a carpel might have evolved. The cyatoniales also exhibited leaves with a netlike pattern of venation, tubes leading to the micropyle of the ovule, and sporangia that superficially resemble anthers. The seed fern order Glossopteridales has been considered to be a forerunner of the angiosperms. Well-preserved seeds borne on the surface of a partially enrolled magasporophyll that is identical to a

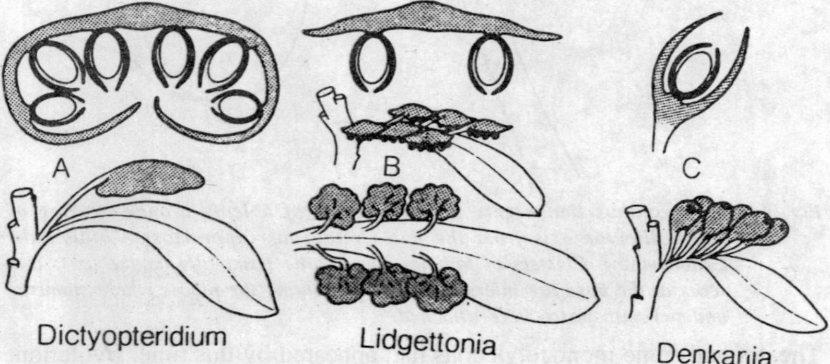

Fig. 4.5. Ovuliferous structures in glossopterids (A-C). It is suggested that the "cupules" in the glossopterids correspond to the outer integument of an angiosperm ovule.

typical leaf of the group support this assumption. However, the pollen of the Glossopteridales is unlike that of early angiosperms.

Table 4.2 Classification of the Gymnosperms

Division—Pinophyta (Gymnosperms)
- Subdivision—Cycadophytina
  - Class—Lyginopteridopsida* (seed ferns)
    - Order—Lyginopteridales
    - Order—Medullosales
    - Order—Callistophyales
    - Order—Calamopityles
    - Order—Caytonianles
    - Order—Corystospermales
    - Order—Paltaspermales
    - Order—Glossopteridales
  - Class—Bennettitopsida* (Cycadeoids)
  - Class—Cycadopsida (Cycads)
- Subdivision—Pinophytina
  - Class—Ginkgoopsida (ginkgo)
  - Class—Cordaitiopsida*
  - Class—Coniferopsida
    - Order—Voltziales* (transitioon conifers)
    - Order—Coniferales (conifers)
    - Order—Taxales (taxads)
- Subdivision—Gnetophytina
    - Order—Ephedrales
    - Order—Welwithchiales
    - Order—Gnetales

*known only from the fossil record

The Czekanowskiales, an order of Jurassic seed plants belonging to the class Gink-goopsida, are regarded by some as early angiosperms. As in the Caytoniales, the ovule was surrounded by a cupule having a flange that somewhat resembled a stigmatic surface; however, the pollen apparently landed directly on the micropyle of the ovule.

The Bennettitopsida (cycadeoids) also have often been suggested as a progenitor of the angiosperms. This hypothesis is based largely on the nature and organization of the reproductive structures of the cycadeoids, which were superficially similar to a "flower." This structure consisied of a whorl of microsporophllys surrounding numerous stalked ovules. Among the ovules were scales that possibly fused to form a carpel-like structure. The cycad lines has also been proposed as a progenitor of the angiospermous condition. The recently described fossil family Dirhopalostachyaceae of the late Jurassic and early Cretaceous had paired capsules each with a flattened seed. Again, the suggestion is based largely on the ovule bearing unit.

The search for angiosperm features among fossil of the late Jurassic and early Cretaceous focuses on the enclosure of the ovules to form a carpel-like structure. However, other unique features of angiosperms, such as double fertilization and endosperm, will not likely be detected in fossil specimens due to their microscopic aspect. This it is extremely difficult in pre-Cretaceous fossils to determine precisely what is and what is not angiosperm. Taylor (1981) suggests that, "Perphaps a more realistic and rewarding avenue of investigation would be to view fossil angiosperms as representing a particular level of evolution, with the earliest forms not demonstrating the total complement of modern angiosperm characters". Althought some groups have been eliminated as likely ancestors of angiosperms, Darwin's "abominable myster" is not completely solved. However, it appears likely that plants having some angiosperm features could have evolved from the seed fern line in the late Jurassic or very early Cretaceous. Certainly a vast gene pool was available at the time to give rise to the first angiosperms, which might be called the *proangiosperms*.

Early Flowering Plants

The early ideas of angiosperm evolution were based on analysis of modern, tropical plants. These models accounts for the paucity of fossils by hypothesizing that the first angiosperms were upland plants that diversified in arid or tropical regions in the Permian about 250 million years ago. The lack of fossils was explained by the fact that the plants from upland areas are not usually preserved as fossils. Migrations of the angiosperms to coastal and alluvial areas about 110 million years ago would have allowed deposition of plant remains and fossilization. One early popular idea suggested that the first angiosperms were tropical, mesophytic trees with pinnate leaves and clusters of large arillate follicles. Another hypothesis based on living angiosperms suggested that the first angiosperms were *Magnolia*-like evergreen trees in tropical upland and were pollinated by insects. This classical theory holds that the primitive angiosperm flower was solitary, terminal, bisexual, actinomorphic, with numerous sepales and petals (or tepals), stamens,

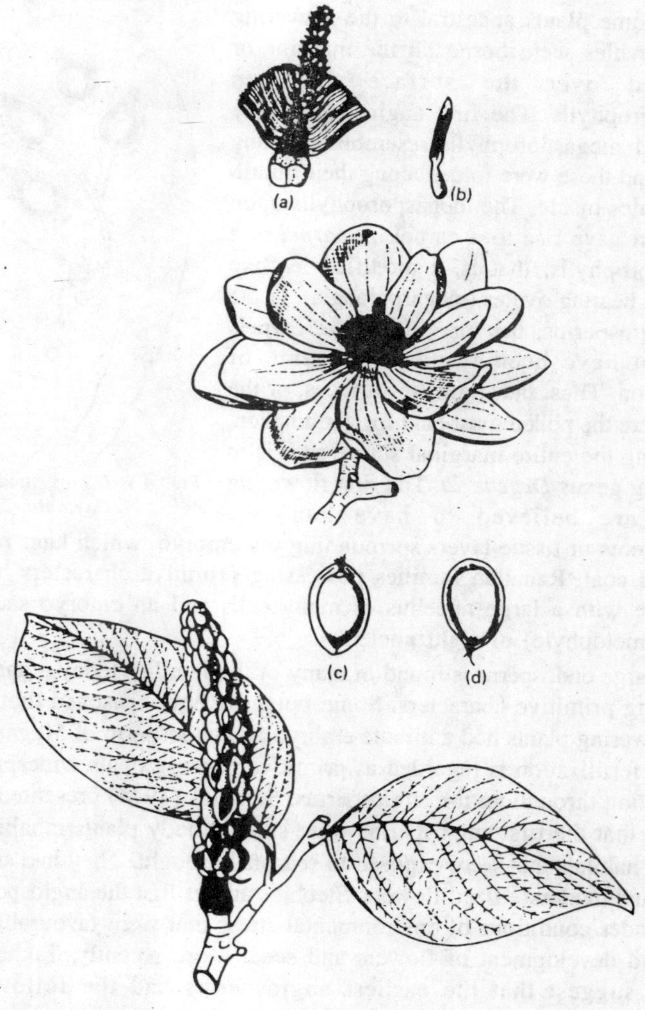

Fig. 4.6. Magnolia campbell (Magnoliaceae). (a) Section showing arrangement of stamens and carpels. (b) Stamen. (c) Seed. (d) Section of seed.

and carpels all spirally arranged on an elongate axis. The stamens were broad and leaflike. The carpels were large lacking both well-developed styles and distinctive terminal stigmas similar to those found in present-day Magnoliaceae. Flower structures with spirally arranged carpels similar to follicles (not unlike some of the magnoliales) are common among fossil remains in the Lower Cretaceous.

In some plants ancestral to the flowering plants, ovules were borne on the margins or scattered over the surface of open megasporophylls. The first angiosperms may have had megasporophylls resembling young leaves, and these were folded along their midrib with ovules inside. The megasporophylls upon reduction gave rise to a carpel. A *carpel* is a megasporophylls, that is, a modified leaflike structure bearing ovules (megasporangia). In the first angiosperms, the margins of the carpels may not have been fused at the time of pollination. Thus, the stigmatic surfaces, or the area where the pollen must land for germination, was along the entire marginal suture as it is in the living genus *Degeneria*. The first flowering plants are believed to have had two integuments or tissue layers surrounding the embryo, which later form the seed coat. Ranalian families possessing primitive characters have an ovule with a large nucellus of many cells and an embryo sac (= megagametophyte) of eight nuclei.

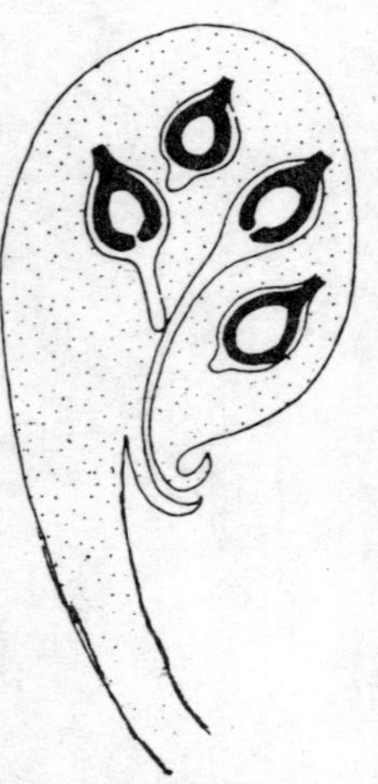

Fig. 4.7. L.S. of cupule of Caytonia.

A large endosperm is found in many of the families today regarded as having primitive characters. Some botanists have reasoned that the first flowering plants had a minute embryo and endosperm of *Magnolia*. Double fertilization is regarded as primitive because of its widespread distribution throughout the angiosperms. Stebbins (1974) presented the opinion that the first angiosperms were small woody plants inhabiting pioneer habitats that were exposed to seasonal drought. They had small leaves and moderate sized flowers. Stebbins argues that the angiosperms arose under conditions of environmental stress that were favourable to the rapid development of flowers and seed. More recently, Takhtajan (1980) suggest that the earliest angiosperms had the following characteristics: small woody plants; leaves simple, entire, pinnately veined; flowers solitary, both terminal and axillary, of moderate size, with an elongated recaptacle, perianth of modified bracts; stamens leaflike; pollen monosulcate; and carpels unsealed, conduplicate with the stigmatic surface along the margin.

Alternative view of the Primitive Angiosperms Flower

The prevailing ideas regarding the primitiveness of angiosperm flowers have changed in recent years. Morphologists and paleobotanists

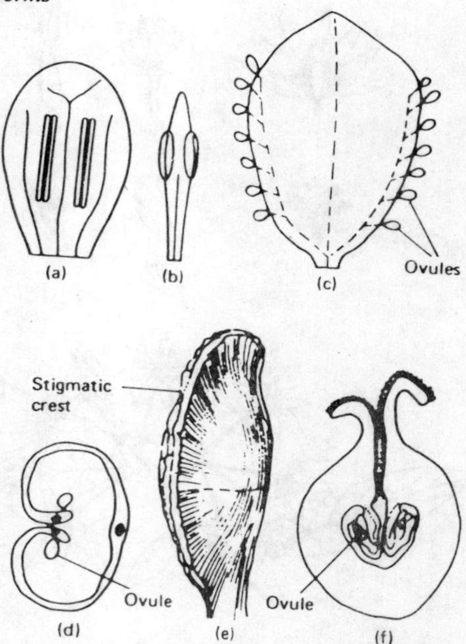

Fig. 4.8. Stamens and carpels. (a) Leaf-like microsporophyll (stamen) believed by Takhtajan to be primitive. (b) Stamen from Winteraceae. (c) Hypothetical megasporophyll with ovules and three vascular strands. (d) Hypothetical meagasporophyll (cross section) closed to form a carpel. (e) Primitive conduplicate carpel of Drimys (Winteraceae); note stigmatic crest along suture. (f) Cross section of a conduplicate carpel of Degeneria (Degeneriaceae); cross section cut in about the position of the dashed line shown in c.

are questioning some of the concepts of the classical "*Magnolia* is primitive" school. They reject the idea that the long floral axis of *Magnolia*, with its many micro- and megasporophylls, is the prototype of flowers in the angiosperms. They object to the concept of leafy stamens and conduplicate (folded lengthwise) carpels as being primitive. However, Dilcher's fossils point out that flowers that are clearly Magnoliidae appear early in the history of the angiosperms. As an alternative hypothesis, some morphologists suggest that the most primitive flowers were probably middle-sized and grouped together into lateral clusters or inflorescences much like those in family Winteraceae.

The Winteraceae have small flowers in lateral clusters, with several stamens and one to several carpels. The flowers of *Drimys* (Winteraceae) are shown in. The Winteraceae are regarded as primitive because they have : (i) great similarity between their micro-and megasporophylls; (ii)

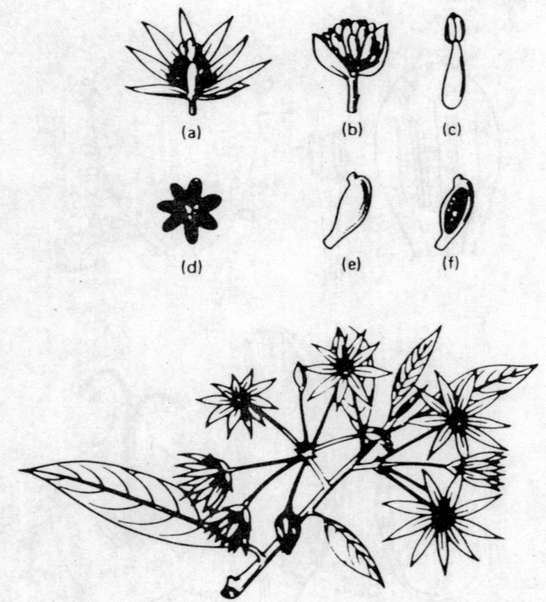

Fig. 4.9. Drimys winter (Winteraceae). (a) Vertical section of flower. (b) Flower with petals removed. (c) Stamen. (d) Cross section of carpels. (e) One carpel. (f) Carpel in vertical section.

unifacial (one-front surface) carpels and stamens; (iii) high chromosome numbers suggesting a long evolutionary history; (iv) morphology similar to the pteridosperms; and (v) vesselless wood similar to that of the gymnosoperms. Additionally, the similar pollination of *Drimys* is or a less specializad type than the of *Magnolia*. The latter has many functional modifications for attracting beetles (Gottsberger, 1974).

Fossil Record

Paleobotanists such as Dilcher (1979), reason that the Magnoliales should no longer be thought to epitomize the only primeval angiosperms flower. Dilcher and his colleagues suggest that the floral structures of the Ranalian complex (the subclass Magnoliidae) may represent only line of early angiosperm evolution and not necessarily the most primitive. The point out that the fossil record indicates that the very earliest angiosperms were probably woody, small-leaved plants occurring in the rift valley system adjoining Africa and South America. Retallack and Dilcher (1981) suggest that some of these angiosperms adapted to the disturbed coastal environments and became widespread during changing sea levels during the early Cretaceous.

In North America angiosperms soon extended from the present Gulf of Mexico as far north as Alberta, Canada, around 110 million years ago. As they flourished the angiosperms replaced the previous dominant plants of the coastal swamps, marshes, and levees along streams. However, in the uplands, the conifer-dominated vegetation was not yet affected. Retallack and Dilcher further hypothesize that by the late Cretaceous these ancient angiosperms had become increasingly isolated because of both the continental drift and the retreat of the seas.

The angiosperms may have migrated into a variety of disturbed and depositional environments and moved into the interior along the alluvial flood plains of streams. Migrations of these pioneering plants allowed further contact with insects and other animals and favoured co-evolution with animals in relation to pollination and seed dispersal. Eventually, the angiosperms moved into the uplands and became dominant around 100 to 90 million years ago. Dilcher (1979) argues that the first angiosperms were probably generalists, pollinated by both wind and a variety of insects. As their numbers increased, adaptive soon produced species pollinated exclusively by wind or by insects as well as species whose pollination was that of a generalist. Fossils thus far studied suggest that the early angiosperms had unisexual and bisexual flowers with simple floral parts clustered on elongate axes or in catkinlike or haedlike structures. This suggest that some of the features of the subclass Hamamelidae may be primitive rather than derived. The fossil record suggest that early angiosperm reproductive structures had few (3-10), free, floral parts, few to many carpels, radial symmetry, and symmetry, and small seeds dispersed by wind and water.

Fig. 4.10. Ligettonia: fertile branch with subtending bract.

Some paleobotanists, such as D. I. Axelrod, suggest that the angiosperms evolved from some primitive gymnosperms, probably a shrub, in the Mesozoic and Paleozoic periods. This is earlier than can be documented from the fossil record. Axelrod suggests that early evolution of the angiosperms took place away from the lowland areas where fossils are normally found and that the first angiosperms had pollen that could not be distinguished from gymnosperm pollen. In this scenario, the angiosperms evolved into a variety of forms in the uplands of the tropics without the opportunity for their remains to have been part of the fossil record. These tropical uplands had a climate with seasonal droughts, and under ecological stresses a number of adaptations developed that made angiosperms successful, for example, a shortened life cycle, small tough leaves, vessel elements, and a small seed that protects the embryo,. This scenario by Axelrod differs primarily from that by Dilcher in the time and place of origin.

Interrelationships of Angiosperms and Animals

Biologist now realize that the flowering plants are closely interrelated with animals in the evolutionary processes. Noting the amazing correlation between host plants and certain animals (particularly insects), biologists have proposed that patterns of adaptations are generated by a process of reciprocal interactions called co-evolution. Krassilov (1977) believes angiosperms differentiation was directly related to mammalization and other processes in the development of Cenozoic flowering plants. One of the important areas of interaction and adaptation has been that of pollination and flower developments.

Pollination

The early seed plants, Various gymnosperms, were pollinated by wind. Ovules borne within cones on the edges of leaves exuded drops sticky sap from their microphyles, trapping cones on the edges of leaves exuded drops of sticky sap from their micropyles, trapping pollen grains as in present-day gymnosperms. It is likely that insects, probably beetles, very early were attracted by both the sticky droplets and pollen, and inadvertently began to carry pollen from plant to plant. According to Dilcher, the fossil record indicates that the first angiosperms were wind pollinated.

In the case of the insect pollinated lines of angiosperms, the more attractive the plants were to beetls the more frequently they would visited by the pollen-laden beetls and the more seeds they would produce. They activities of insects were encourages by the evolutionary development of edible flower parts, protein rich pollen, and nectaries. Beetles were plentiful when the angiosperms appeared and may have been the first animals pollinators. Today they assist with the pollination of *Magnolia*, *Drimys*, and numerous other flowering plants with primitive characters. The accidental beetle pollination provided a more efficient method of pollination than chance pollination by wind. The insects-bees, butterflies, and so on—we usually think of as pollinators had not yet appeared when the first angiosperms evolved. Pollen distri-bution was more effective, and therefore seed production increased when plants efficiently attracted insects. Through natural selection, the ancestral angiosperms developed adaptations to encourage beetle pollination.

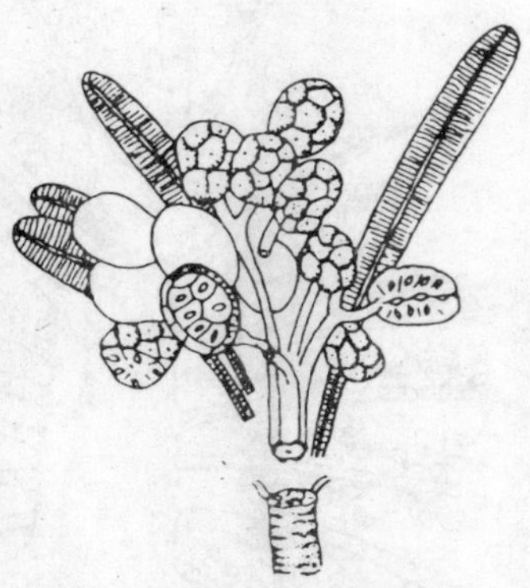

Fig. 4.11. The Pentoxylates: female branch.

The development of a bisexual flower with adjacent carpels and stamens made beetle visits more efficient. Plants that provided food in the form of nectar, pollen, sap, and succulent tidbits of tissue attracted more beetles, giving these plants an advantage over plants that attracted fewer insects. This successful attraction of insects created a problem. The ovules of many early flowers were probably exposed on megaosporophylls and required protection to prevent them from being eaten. The hollow closed carpel functioned to protect the ovules from insect predation. The protection afforded by the closed carpels allowed the ovules to become smaller and capable of faster development. The development of shorter life cycle and reduced energy requirements was advantageous under conditions of stress and seasonal drought-conditions under which the angiosperms had an advantage over the gymnosperms and true ferns. Before the evolution of the closed carpel, each ovule collected its own pollen by way of the micropyle droplet. After the carpel closed, the pollen-collecting function had to shift from each individual ovule to a centralized stigmatic area.

Eventually, during the evolution of the flowering plants, a style and stigma developed that required the germination of pollen some distance from the ovule. By the beginning of the Cenozoic era, the fossil record shows that the higher insects (including moths, butterflies, bees, wasps, and flies) were undergoing diversification. The rise of the higher insect orders was coincidental with the radiation of flowering plants, which provided energy sources for the insects. The higher insects profoundly influenced the evolutionary diversificaton of the angiosperms. The angiosperms underwent a transition from promiscuous pollination by wind and unspecialized beetles to restricted pollination by insects and other animals specialized for flower pollination.

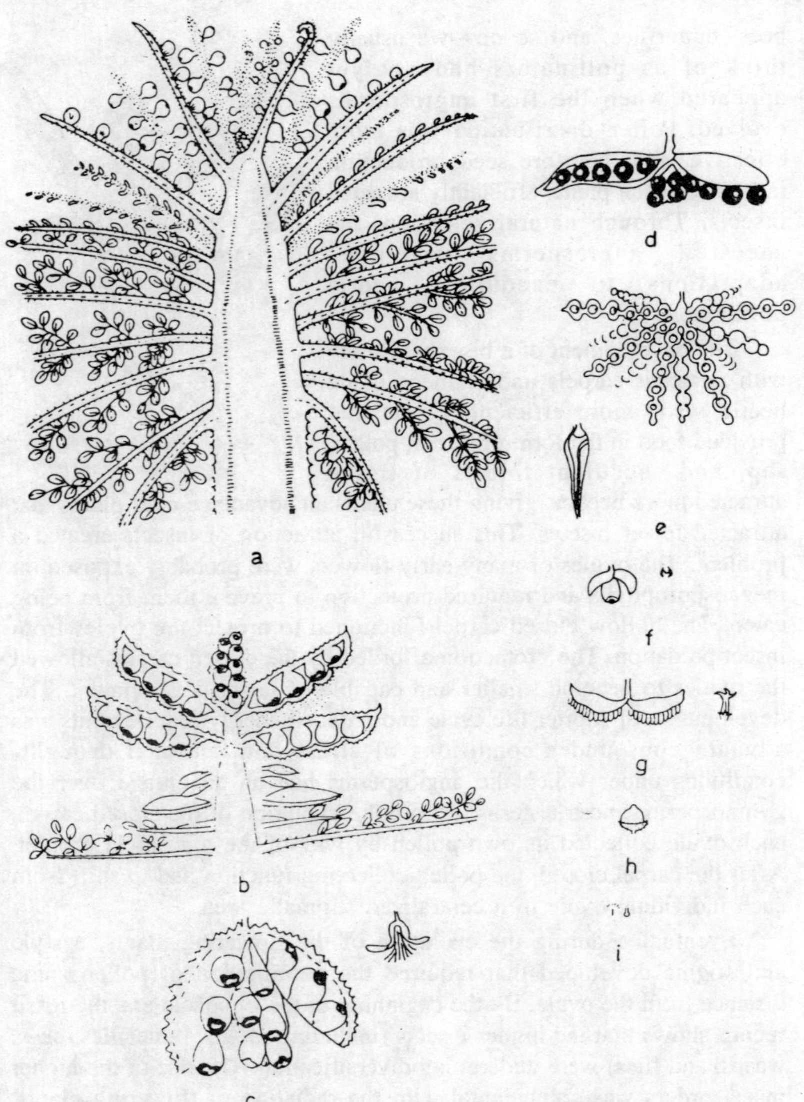

Fig. 4.12. Corner's reconstructions of ancestral angiosperms according to Durian theory: (a) Ancestral pachycaulous spinous flowering plant terminated by flowers and fruits. (b) Primitive fruit with spinous arillate pods. (c) Fruit and flower of Durio. (d) Fruit and flower of Sterculia. (e) Fruits of Desmos. (f) Fruit and flower of Horsfieldia. (g) Pods and flower of Tabernaemontana. (h) Acorn and flower of Quercus.

By means of natural selection, many plant species developed floral feature to accommodate their insect visitors. The evolution of structural change in flowers permitted more effective pollination, resulting in the production of more of its kind. Diversification of the flowering plants also caused evolution in insects of features adapted for flower visitation. Thus there was a mutual influence of plants. Transfer of pollen by bees, wasps, moths, butterflies, flies, birds, and bats is generally considered advanced as compared with the primitive but effective pollination by beetles. Of course, there have been reversals in this trend. Secondary wind pollination has reappeared a number of different times in families, e.g., regweed (*Ambrosia*) compositae and the grasses (Gramineae).

Beetle pollination

Plants pollinated by beetle typically have dull or white flowers, spicy, fermenting or fruity odors, edible petals or food bodies, and ovules heavily protected against the chewing mouthparts typical of beetles. In the members of the 16 families of beetles known to visit flowers, the sense of small appears better developed than sight, thus explaining the strong odors and dull colours of beetle pollinated flowers.

Bee pollination

Most of the 20,000 species of bees pollinate flowers and feed on nectar and pollen. Bee see in the ultraviolet part of the spectrum, thus shades of red appear black to a bee. Bees- pollinated flowers are typically brightly coloured, usually blue or yellow (never pure red), often with a distinctive pattern or honey guides indicating the positions of the nectaries. Since the bee lands and enters the flower, a landing platform is typical for bee-pollinated species. Observations of bee behaviour suggest that in some cases bees exhibit a high degree of constancy to particular species of flowers. This flower constant behaviour is an important contributing to the evolutionary development of angiosperms.

Moth and Butterfly pollination

Plants that have co-evolved with butterflies and with diurnal (day active) moths often have flowers similar to bee-pollinated flowers. However, there is some evidence that butterflies may be able to see red as well as blue and yellow. Most moth, however, are night active and pollinate white or yellow flowers that are open at night or those that stand out against a dark background. Many moth- pollinated flowers have heavy fragrances that assist the moths is locating them. Typical flowers have long slender corolla tubes of hovering moths and butterflies.

Since moths and butterflies hover over the flower and do not land, moth- and butterfly-pollinated flowers do not have landing platforms typical of bee-pollinated flowers.

Bird pollination

In the New World, hummingbirds visit numerous angiosperms, and in other parts of the world, specialized families of birds are adapted for flower visitation. Typical bird- pollinated flowers have large amount of thin nectar, little fragrance, and bright red or yellow colouration. Birds have a poor ability to detect odors but a well-developed sense of visual acuity and often show flower constant behaviour.

Bat pollination

Flower-visiting bats occur in the tropical areas of both the Old and New World. Typical bat-pollinated flower open at night and are dull in colour. Flower visiting bats have long tongues, often with brushy tips, and feed at night, attracted by fermenting or fruity odors. Bats fly from tree to tree, lapping nectar, eating flower part and protein rich pollen, and in the process, they carry pollen from plant to plant.

Seed Dispersal

According to the fossil record, the seeds of the early angiosperms were probably dispersed by water and by wind. At the time angiosperms first appeared, reptiles were dominant, and there is little evidence of adaptations of seeds for animal dispersal. It is conceivable that angiosperm seed dispersal occurred as the abundant reptile populations moved about their habitats. Then, in the late Mesozoic, birds and mammals replaced reptiles as the primary dispersers of seeds. Today flowering plants have many adaptations that promote seed dispersal.

Coevolution

Some groups of flowering plants have evolved various natural products such as alkaloids, tannins, flavonoids, terpenes, and even insect hormones. Many of these chemicals are regarded as defensive compounds that tend to protect the plants form selected herbivores. Still other herbivores may be found associated with these "protected" plants. Ehrlich and Raven (1964) coined the term *coevolution* to refer to any situation in which a pair of organisms act as selective agents for each other. According to their theory, plant and phytophagous insects have undergone adaptive rediation in a stepwise mannar, with plant groups evolving new and highly effective chemical defenses against herbivores and the herbivores continually evolving means of overcoming these defenses.

Probably part of the evolutionary success of angiosperms is linked to the chemical defenses they have erected. Ehrlich and Raven suggest that "...the plant-herbivore 'interface' may be the major zone of interaction responsible for generating organic diversity." Mustard oils, characteristic of the family Cruciferae, are toxic to many animal. Yet the volatile mustard oils attract other herbivores to plants of that family. The white adult of the imported cabbage worm (Lepidoptera), often seen flying around cabbage and broccoli, is guided to the cabbage by mustard oils. The larvae cause major damage to cultivated cabbages. The imported cabbage by mustard oils. The larvae cause major damage to cultivated cabbages. The imported cabbage worm has exploited the chemical defenses of cabbage and uses the mustard oils to locate a host plant for laying its eggs. Present in the genus *Hypericum*, the chemical hypericin repels almost all herbivores.

An exception is the beetle genus *Chrysolina*, which can detoxify hypericin and use it to locate the host plant. According to coevolutionary theory, not all herbivores will respond to a chemical defense in the same way. For example, laboratory feeding tests with five species of insects and two species of mammals demonstrated that a bitter sesquiterpne lactone from *Vernonia* is an effective deterrent to some *but not all* herbivores. Defensive compounds act as filters, screening only a portion of potential herbivores. Some sex attractants in insects are derived from plants on which the insects feed. The insects of one sex concentrate these substances and use them to attract the opposite sex.

Many complex systems involving plant-herbivore interactions occur in the tropic. In Mexico and Central America certain shrubby species of *Acacia* and the ants that inhabit them provide a remarkable example of this complexity. The Acacias occur in full sun in secondary shrubby thickets where competition with other plants is keen. The ants live in the swollen thorns of these Acacias and feed on nectaries located on the petioles of the leaves and Beltian bodies located at the tip of each leaflet. The ants defend the Acasicas against other insect herbivores and girdle the shoots of competing vegetation. If the ants are removed from the Acacias, growth slows, and the Acacias often die. This same type of interaction has been observed between a species of cactus (*opuntia acanthocarpa*) and species of ants.

The cactus possesses extrafloral nectaries embedded in the aureoles of new reoproductive and vegetative growth. The nectar secreted by these glands attracts ants and is a food source for them. Members of one attracted ant species are aggressive and efficient defenders of the plants against other cactus-feeding insects. Some natural plant ingested by insects aid in protection the insects against predators or microbial pathogens. For example, the monarch butterfly ingests cardiac glucosides form milkweed plants (Asclepias). These poisons accumulate in the larval stages of the monarchs and are retained by the adults. The ingestion of monarch butterflies by blue jays causes the jays to become violently sick. Blue jays learn to recognize the toxic, brightly coloured monarch. Thus, the monarch gain protection form predators using the cardiac glucosides of milkweed.

Classification of the Angiosperms

Truly phylogenetic classifications are the ultimate aim of taxonomy. Such classifications are highly useful to systematists, other biologists, and lay people. Numerous systems have been developed, such as those of de Jussieu, de Candolle, Bentham and Hooker, Engler and Prantle, Bessey, Hallier, Hutchinson, and more recently Takhtajan, Sporne, Cronquist, Dahligren, Stebbins, and Thorne. Once widely accepted and still widely used in published floras and herbaria, the Englerian system has received considerable criticism from a phylogenetic viewpoint. This criticism is especially directed to the question of relative primitiveness of various groups and the linear sequence of families. One fault is the positioning of the monocotyledons before the dicotyledons, a placement reversed in the most recent edition of the *Syllabus der Pflanzenfamilien*. The major problem concerns the position of the Amentiferae, or catkin bearers, and other apetalous groups. The families making up the so-called "Amentiferae" were perhaps regarded by Engler and his followers as the most primitive.

The Amentiferae (which are now considered to be an advanced, heterogenous group and not a phylogenetic unit) were therefore placed before the petaliferous families such as Ranunculaceae and Magnoliaceae and were regarded as evolutionarily primitive. The flowers and inflorescences of the Amentiferae are now regarded as the products of phenomena similar to those associated with the miniaturization of flowers and inflorescences. The critical weakness of the Englerian system is the failure to recognize the significance of reduction, and for this reason "simple" was equated with "primitive". Since all families were treated in the Englerian system, it became the standard way to arrange herbaria and manuals in the United States. The practice of using this outmoded linear sequence continues mainly because it is a convenient and well-known filing system. Although largely original, the Besseyan system of classification is rooted in de Jussieu's work (1789), further expanded by de Candolle in 1813, and again later expended by Bentham and Hooker (1862-1883). Bessey also drew information from the Engler and Prantl classification scheme. The Besseyan school regards the "ranalian complex" (those plants having flowers with many, free, equal, and spirally arranged parts) as primitive.

Hence, the phylogenetic system developed by Bessey, often called the *renalian concept of evolution*, is essentially the same as the "*Magnolia* is primitive" school. Diagramed, it took on the form of a cactus plant and came to be called "Bessey's cactus". Bessey considered the angiosperms monophyletic and derived form a cycadeoid ancestor with bisexual strobili. Bessey began his scheme with the group considered the least modified from the ancestral prototype. Believing the primitive angiosperms were insect-pollinated, Bessey concluded that the wind-pollinated Amentiferae had resulted form reduction and much evolutionary change. He disagreed with Engler and Prantl, who regarded the Amentiferae as primitive.

Table. 4.3. Comparison of the Englerian and Besseyan Concents

Feature	*Englerian school*	*Besseyen school*
Primitive flower;	Apetalous; unisexual	Polypetalous, perianth of many, free, equal parts; bisexual
Primitive pollination mechanism:	Wind pollination	Insect pollination
Dicotyledons began with:	Amentiferae	Ranales
Monocotyledons derived from:	Gymnosperous stock	Primitive dicotyledons
Ancestors:	Coniferoid or gentoid gymnosperms	Cycadicae Gymnosperms, most likely, seed ferns
Philosophy:	Simpleflowers are primitive	Many parted flowers are primitive

Bessey's major contribution was first intentionally phylogenetic classification scheme; he attempted to create a taxonomy that would reflect evolution. In the process, Bessey proposed "his" to provide the scientific guideline for understanding the construction of his system. In Bessey's cactus the implication of direct descent of the orders from one another is open to criticism. Likewise, certain evolutionary lines proposed by Bessey are not accepted today. Ragardless of these criticisms, the most recent taxonomic schemes of Cronquist, Takhtanjan,

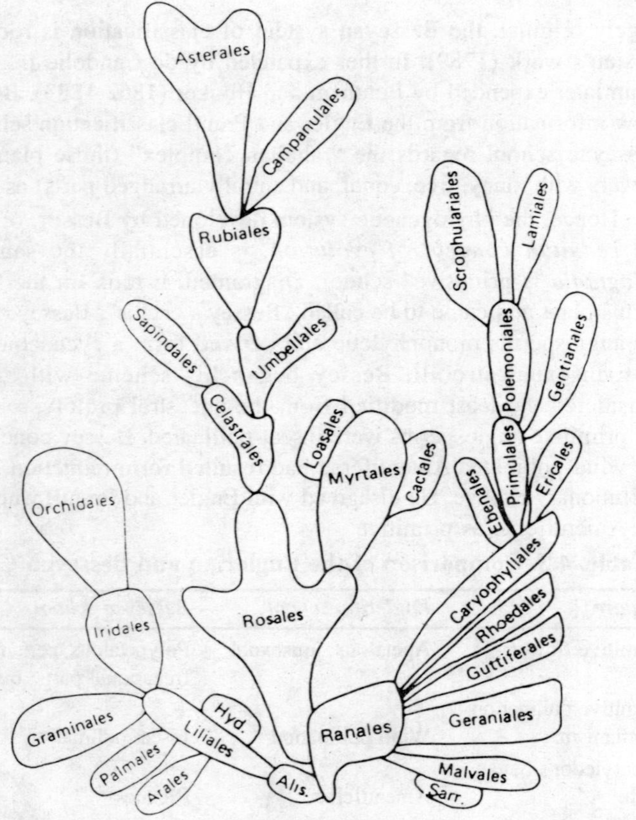

Fig. 4.13. Bessey's cactus. The chart shows the relationship of orders recognized by Bessey.

and thorne remain close to Bessey's system. Hutchinson's classification was a modification of the Bentham and Hooker-Bessey tradition.

Like Bessey, Hutchinson (1973) considered the flowering plants monophyletic, derived from a cycadeoid ancestor. Hurchinson, however, made the unfortunate error of separating the dicotyledons into two lines; one line primarily herbaceous and the second line primarily woody. This unnatural division caused families with close affinities to be separated. For example the family Labiatae was separated from the closely related Verbenaceae and the Araliaceae from the related family Umbelliferae. Evidence indicates that these pairs of families share many common features. Hutchinson also failed to provide evidence for his taxonomic decisions. However, he was extremely knowledgeable about flowering plant families on a worldwide basis, and his family treatments provide a major contribution to the understanding of angiosperms. Takhtajan is an adherent of the Besseyan philosophy and has been strongly influenced by some of the more progressive German workers. Published in Russian, the earlier versions of Takhtajan's work were not readily appreciated in the West. This disadvantage was largely overcome with publication in 1969 of an English translation and summary prepared by Charles Jeffrey.

Takhtajan has used much recent information and many diverse characters in the development of his classification scheme. One criticism of his work is that his taxa are too are too narrowly defined and that as a result unnecessary splitting of related group has occurred. Takhtajan (1980) classifications the angiosperms as Division Magnoliophyta and subdivides them into the dicotyledons (Class Magnoliopsida) and monocotyledons (Class Lilliputian). The flowering plants are further divided into 10 subclasses, three of which are monocotyledons and the other seven dicotyledons. Takhtajan has used the rank of superorder (With a *-anae* ending) as a supplementary rank placed between subclass and order. The superordinal names, as well as all the other names of taxa above the rank of genus, are derived from generic names. Cronquist published a classification conceptually similar to Takhtajan's system, but differing in detail. A modified diagram of Cronquists's classification is shown in. Since it has been widely distributed in book format and offers detailed explanations, the system by Cronquist is widely used and referred to in the United States.

A comparison of the placement of families in Takhtajan, Thorne, Engler, Hutchinson, and Bentham and Hooker in relation to the scheme of Cronquist 1968 is provided by Becker (1973). A brief survey of the arrangement of Cronquist is presented in what follows. Cronquist considers the angiosperms to merit the rank of division (Magnoliophyta). In his 1981 treatment, Cronquist provides a synoptical arrangement of the flowering plant families within the orders and of the orders within the subclasses, etc., a classification similar to Takhtajan's. Cronquist classifies the angiosperms at divisional rank and treat the dicotyledons and monocotyledons as classes that in turn are broken into subclass.

Unlike Takhtajan, thorne, and Dahlgren, cronquist does not use the category of superorder. Cronquist's system has five subclasses of monocotyledons, versus three for Takhtajan. Cronquist includes Takhtajan's subclass Ranunculidae in the subclass Magnoliidae accounting for the discrepancy in number of dicotyledon subclasses. In the monocotyledons, Cronquist segregates the subclasses Commelinindae and Zingiberidae from Takhtajan's huge subclass Liliidae. Among the angiosperms, Cronquist recognizes 83 orders and 383 families versus Takhtajan's 92 orders and 410 families. Robert Thorne (1968, 1976, 1983) has proposed a phylogenetic classification of the angiosperms. Thorne refers to the diagrammatic representation of the evolutionary interrelationships of the extant angiosperms as a "phyletic shrub" viewed from above. Following the Besseyan tradition, Thorne divides the angiosperms into the dicotyledons and monocotyledons at subclass rank. The two subclasses are further divided into superorder (With the ending of *-iflorae*), orders, and families. An outline of his classification is provided in Appendix 4.

K. R. Sporne (1976) has attempted by means of simple statistical tests to identify primitive and advanced characters among the angiosperms. Using the evidence available to him and especially by examining the fossil record, sporne as assembled 26 pairs of characters representing the primitive and advanced states for each character pair. With these data, Sporne constructed an advancement index for each family based upon the proportion of advanced characters possessed by the family. For example, among the dicotyledons, the order Magnoliales has the lowest advancement index and the subclass Asteridae has on the average the highest.

Sporne has presented his conclusions in the form of a circular diagram. The degree of advancement is indicated by succeedingly larger concentric circles, with the distance of a taxon from the center of the diagram proportional to its degree of divergence. In sporne has arranged the dicotyledon orders of Cronquist (1968) to reflect Cronquist's views of affinity and to show also the range of the advancement indexes of the families within each order. It can be seen that the orders belonging to the subclass Magnoliidae are grouped together within an area enclosed by a broken line and that the Magnoliales extend on the advancement index from 20 to 567. It should be noted that the three major subclasses (Magnoliidae, Dilleniidae, and Rosidae) all contain very primitive families and that some orders have a wide range of primitive to advanced characters. Sporne has constructed a similar advancement index for the monocotyledons; however, he feels less confident of this arrangement.

Following chiefly the system of Cronquist (1968), but with modification according to the system of Takhtajan, Thorne, and

Hutchinson, and using many original concepts, Stebbins (1974) has presented a classification of the orders families of angiosperms. It is presented diagramatically in a circular format suggestive of Sporne's advancement index. Stebbins favours this circular format to depict relative degrees of specialization of the orders of angiosperms rather than a conventional phylogenetic tree. He reasons that the common ancestor of any two or more orders was probabl·· so different from any of the extant orders that the common ancestor, if known, would be placed in a separate extinct order. Stebbins arranges the angiosperms into subclass representing the monoctyledons and the dicotyledons, which are in turn broken down into superorders, for instance, Magnoliidae, and orders, for example, Magnoliales. Rolf Dahlgren (1983) of the University of Copenhagen has developed an informative classification of the angiosperms based on the distribution of a wide range of phenetic characters.

Among plant taxonomists, Dahlgren's classification has an increasing following and many consider it the best yet developed. In his classification the angiosperms (class Magnoliopsida) are equivalent to the main groups of gymnosperms. The angiosperms are further divided into two subclasses, representing the dicotyledons and monocotyledons, the Magnoliideae and Liliidae repectively. Thus, Dahlgren's subclasses have a much broader circumscription than the subclass used by Cronquist and Takhtajan. He notes that the division of the flowering plants into the dicotyledons and monoctyledons would not be allowed it one followed rigid cladistic approaches. However, Dahlgren believes the monocotyledons to be unique group worth of subclass rank. Dahlgren further divides the subclasses into superorders similar to Thorne's. These bear the termination *-iflorae*. Dahlgren indicates that the use of superorders and orders allows increased flexibility in indicating affinities that might otherwise be restricted by the subclasses of Cronquist and Takhtajan. Further, he suggests that this approach allows better placement of superorders of unclear affinities.

Two Classes of Magnoliophyta: Magnoliopsida (Dicotyledons) and Liliopsida (Monocotyledons)

Division Magnoliophyta includes all the angiosperms. The natural group consists of two major subgroups: the monocotyledons and dicotyledons. To conform to the International Code of Botanical Nomenclature, formal names with Latinized endings have been applied to each of the categories, Magnoliopsida for the dicotyledons and Liliopsida for the monocotyledons. The English names are useful and continue to be employed. In general, the dicotyledons are much more diverse in habit than monocotyledons. About 50 percent of the present-day species of dicotyledons are woody. Few of the monocotyledons are woody, and most of these are found in family Palmae.

The Palmae are regarded as relatively advanced in the monocotyledons. Most woody dicotyledons are highly branched, but the woody monocotyledons are generally unbranched. Most systematists agree that monocotyledons were phyletically derived from the dicotyledons early in the evolutionary history of the angiosperms. Cronquist (1968) believes the monocotyledons arose from a primitive vesselless ancestor resembling present-day Nymphaeales. If this viewpoint is correct, the monocotyledons had an aquatic origin profoundly channelizing the evolutionary pathways available to them. If aquatic in origin, the first monocotyledons probably had apocarpous flowers with undifferentiated tepals and monosulacate pollen, and they lacked a functional cambium and xylem vessels.

Cheadle (1953), using evidence from xylem vessels, rejected the idea of an aquatic origin. Instead, he suggested that monocotyledons arose from terrestrial woody dicotyledons that lacked xylem vessels. The characters separating the monocotyledons from the dicotyledons are usually regarded as secondarily derived with but one exception. The monosulacate type of pollen of monocotyledons is shared with certain living dicotyledons, as well as with their gymnospermous ancestors. This one exception really poses no problem, since a lineage of pollen types can be extended from the monocotyledons back through the primitive dicotyledons into gymnospermous stock.

In addition to the morphological evidence, data from comparative biochemistry point to an origin of monocotyledons from dicotyledons. In recent years the classical circumscription of the monocotyledons has been challenged with some workers suggesting the placement of the Nymphaeales among the monocotyledons. The similarities in certain features between the dicotyledonous Piperales and some monocotyledons make the border line between the dicotyledons and monocotyledons appear less distinct than once believed. Other workers have noted the resemblance between the monocotyledons Dioscoreals and certain orders of the Magnoliales. Dahlgren (1983), although noting the lack of sharp discontinuities, feels that the monocotyledons can be clearly circumscribed on the combination of (i) an embryo with a single cotyledon and (ii) sieve-tube plastids with cuneate protein bodies. There are even unusal hypotheses, such as by Burger (1981), who argues that the ancestral angiosperms were monocotyledons. Whether the monocotyledons are monophyletic from one dicotyledonous ancestor of polyphyletic from two or more dicotyledonous ancestors requires further study.

Characteristics of the Subclasses of the Cronquist System of Classification

Division: Magnoliophyta

Angiosperms or flowering plants (two calsses, 11 Sub classes, 83 orders, 383 families, and about 219,300 species).

Class: Magnoliopisida

Dicotyledons (Six subclasses, 64 order, 318 families, and about 169,400 species).

Subclass: Magnoliidae

This subclass includes those dicotyledons that have retained one or, more often several primitive characters. The subclass originated about 122 million years ago in the Lower Cretaceous. Magnoliidae generally have a many-parted, developed perianth of tepals, often differentiated into sepals or petals, but sometimes apetalous. The stamens are numerous and mature in a centripetal manner. The pollen typically is binucleate and monosulcate. The gynoecium is typically apocarpous with bitegmic and crassinucellate ovule. All families is the largest order. The Magnoliidae have their own set of chemical defenses, with a great many of the taxa producing isoquinoline alkaloids.

Subclass: Hamamelidae

Hamamelidae is the smallest subclass of dicots. The subclass originated about 100 million years ago in the Lower Cretaceous as a group characterized by wind pollination and floral reduction. They were well established by 80 million years ago. With the exception of some taxa in the order Urticales, the families are typically woody and often contain relatively few species. The subclass is characterized by reduced, often unisexual flowers, which in the advanced group are arranged in catkins. The perianth is either absent or poorly developed. The mature fruit contains a single ovule. At some stage in their evolution the Hamamelidae began to use tannins as their chemical defense against herbivores. Evidence is accumulating that this subclass, held together mainly by the bonds of reduced floral structure, probably should be regarded as only loosely monophyletic. Many familiar deciduous trees of eastern North America are members of subclass Hamamelidae.

Subclass: Caryophyllidae

Several of the families in this mostly herbaceous subclass contain succulents or halophytes. The fossil record extends the Caryophyllidae back about 70 million years ago. The perianth of the group is morphologically complex and diverse. The primitive members have only a single perianth whorl, and from this, various modified perianths developed with apparent sepals and petals. The stamens are centrifugal in maturation sequence and produce trinucleate pollen. Placentation is free central to basal. The ovules are bitegmic and crassinucellate. They are either campylotropous or amphitropous and when mature the embryo are often surrounded by perisperm. Betalains, a distinctive class of pigments are found in many of the families in order Caryophyllales. Often called the Centrospermae, order Caryophyllales contains around 10,000 species representing the bulk of the subclass. The Caryophyllales have a unique P-type sieve tube plastid.

Subclass: Dilleniidae

With the notable exception of the apocarpous order Dilleniales, subclass Dilleniidae is distinguished from typical members of subclass Magnoliidae by syncarpy. Members of subclass dilleniidae have a centrifugal maturation of stamens and binucleate pollen, except for the family Cruciferae, which is trinucleate. Ovules are unitegmic or bitegmic and have crassinucellate to tenuinucellate endosperm. This subclass contains many woody species. Pollen that may represent the Dilleniidae has been found in the fossil record about 100 million years ago at the beginning of the Upper Cretaceous.

Subclass: Rosidae

This is the largest subclass, in terms of numbers of families, but about the same size as the Asteridae in terms of the numbers of species. The flowers have numerous stamens that mature in centripetal sequence. The ovules are bitegmic or unitegmic and crassinucellate or tenuinucellate. The flowers have a polypetalous corolla, although a few apetalous and sympetalous flowers occur. The fossil record suggests that the Rosidae began to appear about 110 million years ago in the Lower Cretaceous.

Subclass: Asteridae

Asteridae is the second largest subclass of dicotyledons. About one-third of the species belong to the family Compositae (Asteraceae), the largest family of dicotyledons. Flowers of this subclass are sympetalous and only rarely apetalous or polypetalous. The stamens are few and alternate with the petals. Members of subclass Asteridae usually have two carpels with unitegmic and tenuinucellate ovules. Of the dicotyledons, subclass Asteridae is probably the best defined. Present evidence strongly suggests that the Asteridae were likely derived from the evolutionary Rosidae line. The Asteridae are the most advanced subclass of the dicotyledons and are the most recently evolved, appearing in the fossil record around 65 million years ago, but not becoming abundant until about 30 million years ago.

Class Liliopsida

Monocotyledons (five subclasses, 19 orders, 65 families, and about 49,900 species.)

Subclass: Alismatidae

Alismatidae characteristically inhabit aquatic or wetland habitats and are herbaceous. Most are apocarpous and have trinucleate pollen. When mature, the seeds lack endosperm. Two subsidiary cells surround the stomates. Regarded as a relic side branch of the monocotyledons, subclass Alismatidae has retained a number of primitive characters. The fossil record extends the subclass back about 60 million years ago.

Subclass: Arecidae

In habit, the members of subclass Arecidae vary from tiny macroscopic duckweeds to huge woody palms. About 50 percent of the species are arborescent. The flowers tend to be numerous, small, and often aggregated in a spadix subtended by a spathe. The stomatal subsidiary cells are mostly four but may be two or three. Many species have features that are atypical of monocotyledons, such as broad, petiolate, net-veined leaves. All but the order Arales contain vessels. Over half of the specie belong to the order Arecales, which includes only the palm family (Arecaceae). The fossil record of the palms extend back to the Upper Cretaceous, about 80 million years ago.

Subclass: Commelinidae

The vast majority of the species subclass Commelinidae are herbaceous. They inhabit sites ranging from aquatic to terrestrial or even herbaceous. They inhabit sites renging from aquatic to terrestrial or even epiphytic. The flowers may have either well-defined sepals and petals or a perianth which is cheffy, bristly, cr lacking, Primitive Commelinidae have showy insect-pollinated flowers, whereas advanced members with reduced perianths are wind-pollinated. Commelinidae pollen is either trinucleate or, less commonly, binucleate. About 50 percent of the species belong to the family Gramineae (Poaceae), and another 30 percent to the Cyperaceae. The subclass extends back in the fossil record about 85 million years, with the Gramineae (one of the more advanced families) appearing about 60 million years ago.

Subclass: Zingiberidae

The vast majority of the members of Zingiberidae are tropical and are either terrestrial or epiphytes with regular to irregular flowers, septal nectaries, and an inferior ovary. The two orders are distinctive, and although associated here, many other of their features appear to have developed independently. They differ from the many other monocotyledons in having retained the floral nectaries and in having achieved epigyny.

Subclass: Liliidae

The members of the monocotyledonous subclass Liliidae are syncarpous, with petaloid sepals and petals. They are highly developed for insect pollination. The majority are terrestrial or epiphytes herbs. Leaves are linear and parallel-veined to broader with net venation. The ovary in subclass Liliidae is frequently inferior. Stomatal subsidiary cells are usually absent, but sometimes two or more are present. More than 80 percent of the species belong to the families Liliaceae and Orchidaceae. Pollen of the order Liliales appears in the fossil record at the close of the Upper Cretaceous, 70 million years before present.

5

PHYTOGRAPHY

Because the identification and classification of plants is based chiefly on the details of their external features, a knowledge of the terminology of plant morphology is essential. The descriptions of plant structures is called *phytography*, and it includes the descriptive terminology of whole plants and their component part. The major components of a plant are referred to as *organs*. The *vagetative organs* are the roots, stems, and leaves. The *reproductive organs* are the flower, fruit, and seed. Adjective referring to each of these organs are used to designate their various states or forms. The commonly encountered terminology for descriptive plant taxonomy is illustrated and defined in this chapter and is arranged under broad heading for each organ. Mastery of these selected terms will allow one to use most manuals. The best way to learn the terms is to studey them concurrently with selected plant material. The learning experience is much pleasant and rewarding if critical features are seen in actual plant material.

VEGETATIVE MORPHOLOGY

Duration

Duration is the length of time an individual plant exists. Note the following definitions having to do with duration:

Annual. A plant that lives for one growing season and then dies. A *winter annual* germinates in the fall, overwinters, produces seeds in the spring, and then dies.

Biennial. A plant that requires two growing seasons to complete the life cycle of two years's duration.

Herbaceous perennial. A nonwoody plant that lives for several years; the shoot system dies back each winter.

Woody perennial. A tree or shrub. The shoot system remains above ground winter; the stems and the roots becomes woody live for a number of years. Some woody perennials are *monocarpic*, living for several years then flowering and dying-for example, *Agave*, the century plant.

Roots

Unlike stems, roots do not have regularly spaced buds and leaves. They branch irregularly and may produce adventitious buds. In general, roots provide relatively few features for identification and classification; however, they are often essential to determine whether the plant is an annual or a perennial. In some families, such as grasses, it is essential that roots be collected if the specimens are to be correctly identified. Terminology essential to the description of roots includes:

Adventitions. Developing from something other than the hypocotyl or another root.

Aerial. Growing in the air.

Fibrous. Threadlike and usually tough.

Fleshy. Relatively thick and soft, with storage tissue.

Haustorial. Specialized, penetrating other plants and absorbing nutrients form them.

Primary. Developing from the radicle of the embryo; the root the first appears from the seed.

Secondary. Developing from the primary root; branch roots.

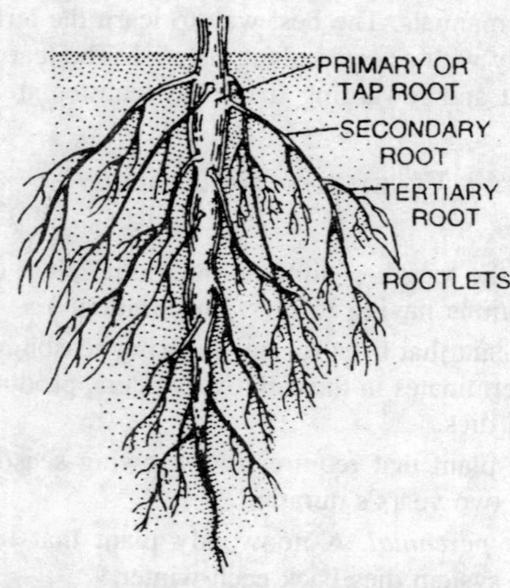

Fig. 5.1. Typical or racemose tap root system.

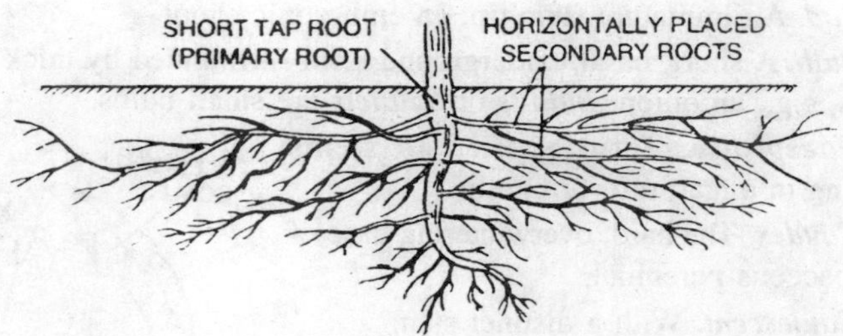

Fig. 5.2. Fibrous tap root system.

Tap. A main primary root that more or less enlarges and grows downward.

Tuberous. Term applied to a relatively thick and soft root: enlarged and with storage tiss ue.

Stems

Stems are the main axes of plants. As opposed to roots, they have nodes and internodes and usually leaves with a bud in the leaf axil. Stems may bear flowers. Many stems are highly specialized, particularly those underground stems which function as storage or overwintering mechanisms. Variation in stems is of importance in classification, and stems provide many useful taxonomic characters. Many genera of native trees may be easily recognized in their winter condition by examination of their stem features. Essential stem terminology include the following:

Acaulescent. Stemless, or with the stem inconspicuous.

Arborescent. Becoming treelike and woody, usually with a single main trunk.

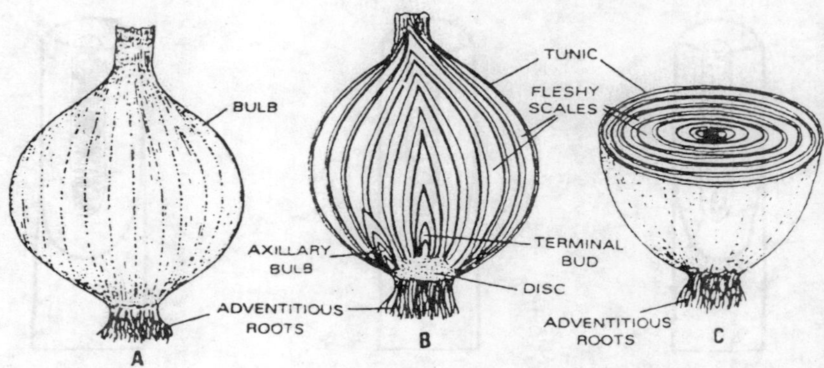

Fig. 5.3. Tunicated bulb of Onion. A—external view. B—V.S. Bulb. C—T.S. Bulb.

Bud. An immature stem tip; an embryonic shoot.

Bulb. A short, basal, underground stem surrounded by thick fleshy leaves, e.g., an onion. *Bulbils* or *bulblets* are small bulbs.

Caespitose. Term applied to stems growing in a mat, tuft, or clump.

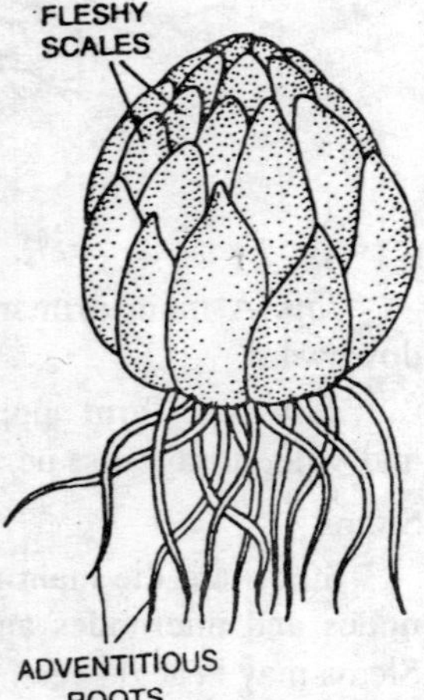

Fig. 5.4. Scaly bulb of Lily.

Caudex. The hard, overwintering base of a herbaceous perennial.

Caulescent. With a distinct stem.

Cladophyll. A flattened, leaflike, green stem.

Climbing. Clinging to other objects, e.g., English ivy.

Corm. A short, upright, hard or fleshy bulblike stem, usually covered with papery, thin, dry leaves, e.g., *Gladiouls*.

Decumbent. Term applied to stems lying flat on the ground but turning up at the ends.

Erect. Growing upright.

Fruticose. Shrublike and woody.

Herbaceous. Not woody; dying to the ground at the end of the growing season.

Internode. The part of the stem between two successive nodes.

Lenticels. Lens-shaped or warlike patches of parenchymatous tissue on the surface of a stem.

Node. The area of a stem where a leaf and bud borne.

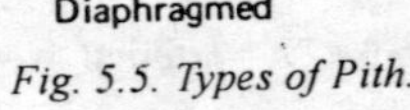
Fig. 5.5. Types of Pith.

Pith. The spongy tissue in the center of a stem. The pith may be (1) *continuous*, or solid throughout; (2) *diaphragmed*, or continuous, but with firmer cross plates at intervals; or (3) *chambered* in which the tissue between plates has disappeared *prickle*. A sharp pointed extension of the cortex and epidermis.

Prostrate. Growing flat on the ground.

Rhizome. A horizontal, prostrate, or underground stem with reduced scalelike leaves, as in German iris.

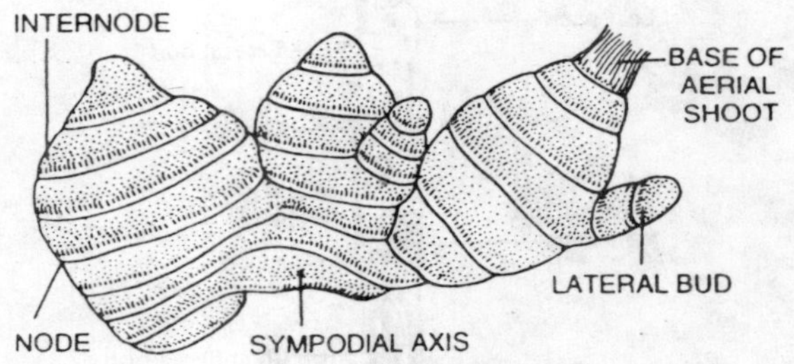

Fig. 5.6. Sympodial rhizome of Ginger.

Rootstock. The underground portion of a plant; often referring to a caudex or a rhizome.

Runner. A horizontal aboveground stem, usually rooting and producing plants at the nodes, e.g., strawberry. *See stolon.*

Scandent. Climbing.

Scape. Leafless flowering stem, arising from the ground, sometimes with small scale-like leaves.

Scapose. Having a scape.

Scars. The remains of a point of attachment. These include: (1) *leafscars*, where a leaf was attached: (2) *stipule scars*, where stipules were attached (in some taxa, e.g., Magnoliaceae, they form a ring around the stem called a *stipular ring*); and (3) *scale scars*, where the bud scales are sloughed off. In the temperate zone, the age of twigs may be determined by counting each set of *terminal bud scale scars*.

Shrub. Plant having several main stems, more or less woody throughout, and usually less than 4 to 5 meters in height.

Spine. A stiff pointed outgrowth.

Stolon. Horizontal stem rooting at the nodes; a runner.

Suffrutescent. Tending to be woody, especially at the base.

Tendril. A small, twisting appendage that attaches a scandent plant to other plants or objects.

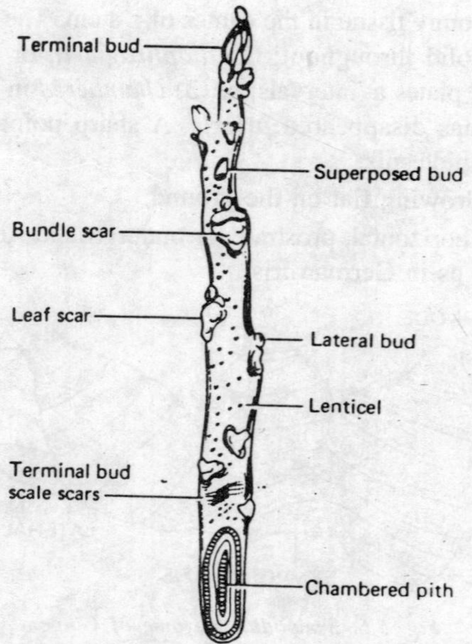

Fig. 5.7. A stem tip in the winter condition. The terminal bud scale scars indicate the location of last year's terminal bud scales.

Thorn. A reduced, sharp, pointed stem, See Prickle.

Tree. A woody plant, usually with one main trunk and normally 4 to 5 meters or more in height.

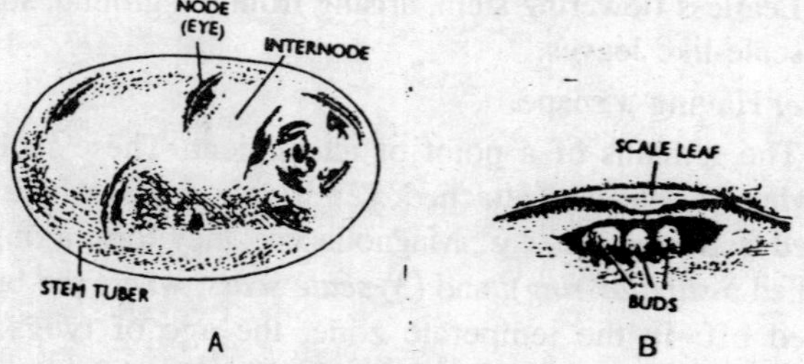

Fig. 5.8. A—external view of the stem tuber of Potato showing many eyes; B—an enlarged eye.

Tuber. An enlarged fleshy tip of an underground stem; e.g., Irish potato.

Woody. Hard in texture and containing secondary xylem.

Buds

Buds are short, embryonic stem tips bearing leaves or flowers or both. They may be quite useful in the identification or woody plants in the winter condition. Some of the common descriptive terms for buds include the following:

Accessory. An extra bud produced to either side of an axillary bud.

Adventitious. Term applied to a bud developing from some place other than a node of a stem.

Axillary. Term applied to a bud located in the axil of a leaf.

Collateral. Bud(s) lateral to the axillary bud.

Dormant. Inactive; may be due to dry weather or winter condition.

Flower bud. A stem tip containing or producing embryonic flowers.

Infrapetiolar. An axillary bud surrounded by base of petiole, e.g., sycamore.

Lateral. Produced on the side of a stem, as opposed to *terminal*.

Leaf bud. A stem tip containing leaves.

Mixed. Term applied to a bud containing both embryonic leaves and flowers.

Naked. Not covered by bud scales.

Pseudoterminal. Term applied to a lateral bud which takes over the function of the terminal bud, e.g., persimmon.

Reproductive. Containing embryonic flowers.

Scaly. Covered by bud scales; a scaly bud is sometimes called a *covered* bud.

Superposed. Bud (s) located immediately above the axillary bud.

Terminal. Located at the stem tip.

Vegetative. Containing embryonic leaves.

Leaves

Leaves are generally broad flattened and photosynthetic and are borne at the nodes of a stem. Just above the point of attachment of the leaf base of petiole, there is an axillary bud. Leaves are of considerable value in identification and classification. Some flowering plants have a particular form of leaf and may be recognized by their leaves alone. Some angiosperms are leafless and a few have leaves which are variable in form.

Leaf parts

A complete leaf is composed of a blade, petiole, and stipules, and one of which may be lacking or highly modified. The expanded, flattened part of the leaf is the *blade* (or *lamina*). The supporting stalk is called the *petiole*. The base of the petiole is variously shaped, or is may be rather enlarged and termed a *pulvinus*. In some taxa, the petiole is lacking and the leaf is said to be *sessile*. *Stipules* are a pair of appendages which, when present, either are located at the base of the petiole or are attached to the stem at the node on either side of the petiole. Stipules are usually green and leaflike or scalelike, but they may be modified into tendrils (e.g., *Smilax*) or prickles (e.g., *Robinia*), or they may form a papery sheath around the stem called an *ocrea* (e.g., *polygonum* and *Rumex*). A plant lacking stipules is said to be *exstipulate*. In some plants, such as the grasses and the sedges, the basal part of the leaf surrounds the stem and is called the *sheath*. In grasses, there often is a *ligule*, which is a small portion of the sheath extending beyond the junction of the sheath and the blade.

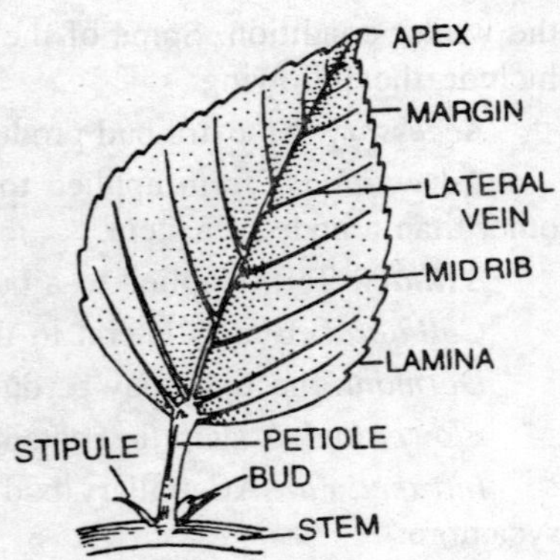

Fig. 5.9. Parts of a typical leaf.

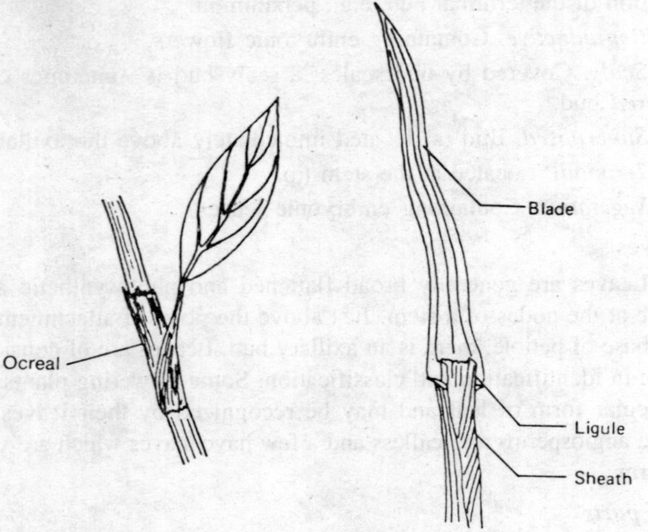

Fig. 5.10. Specialized parts of leaves. Left—An ocrea of Polygonum. Right—A grass leaf with a ligule and sheath.

Venation

The positioning of the veins (vascular bundles) in the leaf blade is called *venation*. Some leaves, especially those of monocots, have

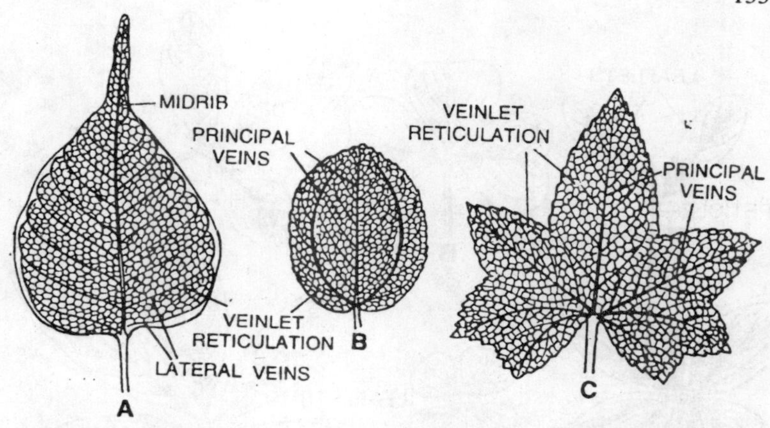

Fig. 5.11. Types of reticulate venation. A—unicostate of Peepul; B—multicostate convergent of Zizyphus; C—multicostate divergent of Luffa.

parallel venation, that is, the primary veins occur side by side without intersecting. In contrast, most dicots have *net* or *reticulate* venation, with primary veins that branch and intersect. There are two main patterns of net venation: (1) *palmate*, with the main veins radiating from the point where they join the petiole; and (2) *pinnate*, with one central vein or *midrib* that has lateral veins arising along its length and at angles from it. In dichotomous venation, the veins fork into equal-sized branch

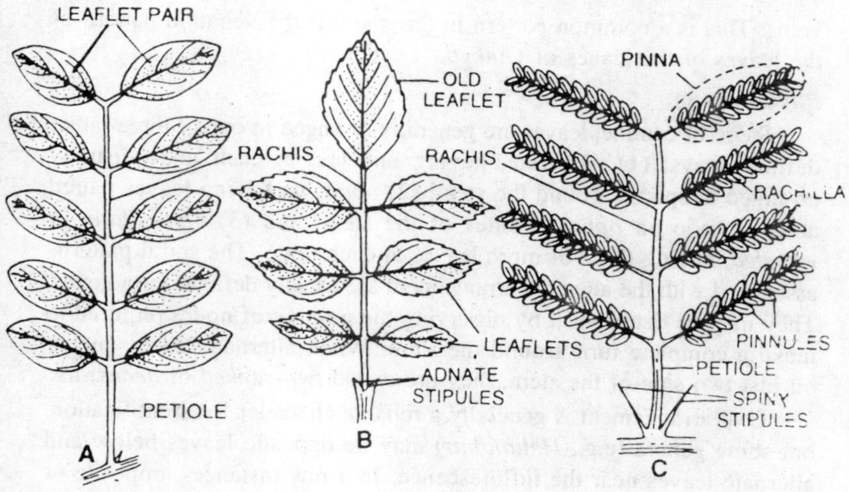

Fig. 5.12. Pinnate compound leaves. A—paripinnate; B—imparipinnate; C—bipinnate

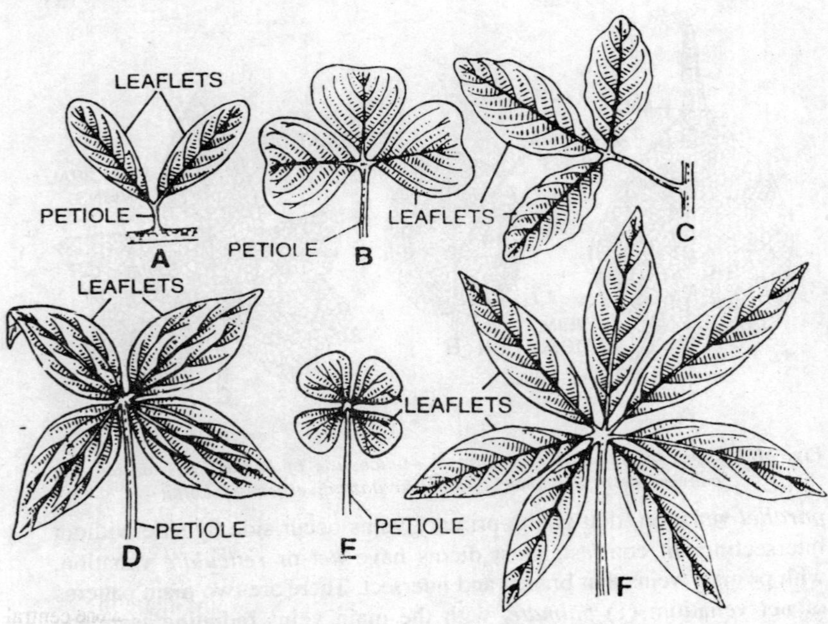

Fig. 5.13. Palmate compound leaves. A—Bifoliolate of Balanites; B—trifoliolate of Oxalis; C—trifoliolate of Aegle (Bael); D—quadrifoliolate of Paris quadrifolia; E—quadrifoliolate of Marsilea; F—multifoliolate of Bombax (Simbal).

veins. This is a common pattern in ferns and is the venation pattern of the leaves of the leaves of *Ginkgo*.

Phyllotaxy

Stem, or cauline, leaves are generally arranged in one of three rather definite ways: (1) *alternate*, having one leaf at each node, usually arranged in spirals around the stem; (2) *opposite*, having leaves paired at each node on *opposite* sides of the stem; and (3) *verticillate*, or *whorled*, having three of more leaves at each node. The spiral patterns associated with the alternate arrangement are usually definite for a genus. They may be determined by observing the number of nodes required to make a complete turn around the stem. When alternate leaves appear on just two side of the stem, they are called *two-ranked* or *decussate*.

Leaf arrangement is generally a reliable character for identification, but some genera (e.g., *Helianthus*) may be opposite leaves below and alternate leaves near the inflorescence. In a few instances, opposite or whorled may intergrade as a result of either extremely fast or slow growth.

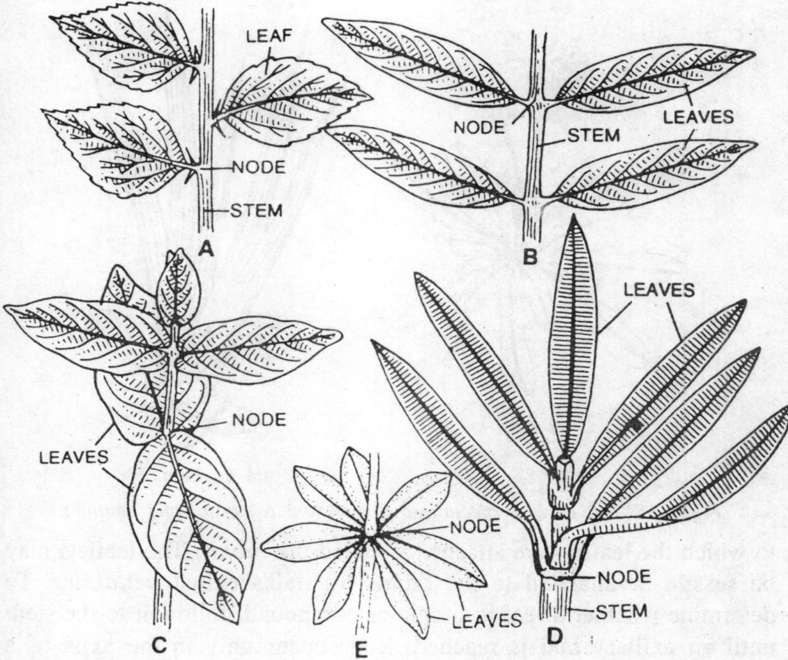

Fig. 5.14. Phyllotaxy. *A—spiral of Shoe flower; B—opposite and superposed of Jamun; C—opposite and decussate of Calotropis (Ak); D—whorled or verticillate of Oleander (Kaner); E—whorled of Alstonia.*

With woody plants, leaf arrangement should be determined by examining normal, terminal, long shoot rather than by observing lateral, dwarf, or spur shoots; accurate determination of the leaf arrangement may be difficult because of the very short internodes on spur shoots. In *acaulescent* species (without conspicous leafy stems), the leaves often form a cluster at ground level and are said to be *basal* or *radicle*. An acaulescent plant, such as dendelion, has a stem, but the internodes are exremely short, thereby producing a stacked spiral of leaves that forms a rosette on the ground. *Iris* has two-ranked leaves with overlapping leaf bases and the leaves folded together lengthwise; this arrangement is called *equitant*.

Leaf type

A leaf with the blade in a single part, although it may be varioulsy divide, is a *simple* leaf, while one with the blade divided into smaller, bladelike parts is termed *compound*. The individual bladelike parts of a compound leaf are called *leaflets* or *pinnae*. The main axis of the leaf

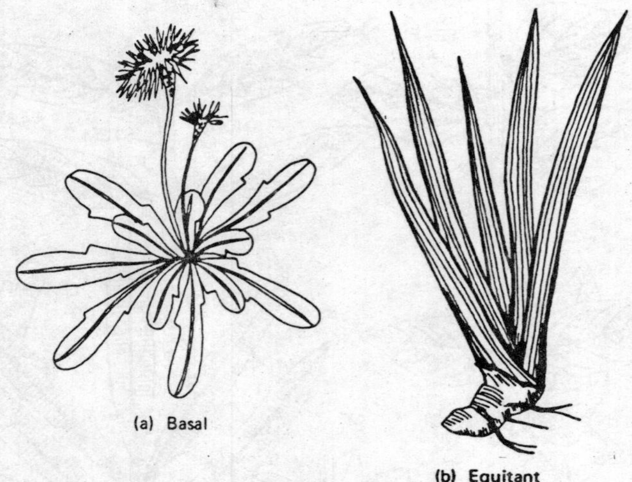

Fig. 5.15. Leaves with specialized disposition: (a) basal and equitant.

to which the leaflets are attached is called the *rachis*. The leaflets may be sessile or attached to the rachis by stalks called *petiolules*. To determine whether a leaf is simple or compound, follow it to the stem until an axillary bud is reached. Buds occur only in the axils of a complete leaf, not in the axils of the leaflets. In some plants, especially if immature, it may be difficult to find the axillary bud. It may be necessary to determine whether a leaf is a varioulsy divided simple leaf or if it is a compound leaf.

Generally, if the leaflets are well-developed and separate from the rachis, the leaf is regarded as compound. If the leaflets are attached to both sides of one central rachis, the is *once-pinnately compound*. If they diverge from a common point at the end of the petiole, much like fingers from the palm of the hand, the leaf is *palmately compound*. Pinnately compound leaves may have the leaflets dived. This type of leaf is said to be *twice-pinnately compound* or *bipinnate*. If the leaflets are divided twice, the leaf is *thrice-pinnately compound* or *tripinnate*. A once-pinnately compound leaf with paired leaflets and one leaflet at the tip is *odd-pinnate*, because it has an odd number of leaflets. If it does not have a leaflet at the tip, it is *even-pinnate*. A leaf with three leaflets is called *trifoliolate* or *ternate*.

Duration of leaves

Leaves are often temporary organs, and *duration* refers to the length of time they function. In the temperate zone, duration is usually related to the onset of lower temperatures, but in regions with seasonal rainfall, it may be regulated by drought. The following definitions are important:

Deciduous. Falling at the end of the growing season; not evergreen; referring to leaves falling in the autumn.

Evergreen. Persistent, not deciduous; often used to describe the plant.

Fugacious. Falling shortly after development; soon falling.

Marcescent. Withering, but not falling off.

Persistent. Evergreen, not deciduous; functioning over two or more growing seasons; often used to describe the leaves.

Shape of blade

Although the overall size of the leaf blade may be environmentally influenced, the shape of a leaf or leaflet is often characteristic for the species. Some common leaf shapes are listed below;

Cordate. Heart-shaped, with a basal notch.

Cuneate. Wedge-shaped, tapering toward point of attachment.

Deltoid. Triangular.

Elliptical. Having the shape of a flattened circle, usually more than twice as long as broad.

Hastate. Arrowhead-shaped.

Lanceolate. Lance-shaped, tapering from a broad base to the apex; much longer than wide.

Linear. Long and narrow with nearly parallel sides.

Oblanceolate. Lanceolate, but with the broadest part near the apex.

Obovate. Ovate, but with broadest part near the apex.

Ovate. Egg-shaped, with the broadest part toward the base.

Peltate. With the petiole attached to the lower surface of the blade rather than on the margin; shield-shaped.

Perfoliate. Term used to describe opposite leaves having their bases united around the stem and the stem apparently passing through the leaves.

Reniform. Kidney-shaped.

Subulate. Tapering from a broad base to a sharp point, awl shaped.

Apex of leaf blade

The tip of a leaf farthest removed from the petiole is termed the *apex*. Some common types are listed below:

Acuminate. Tapering gradually to a prolonged point.

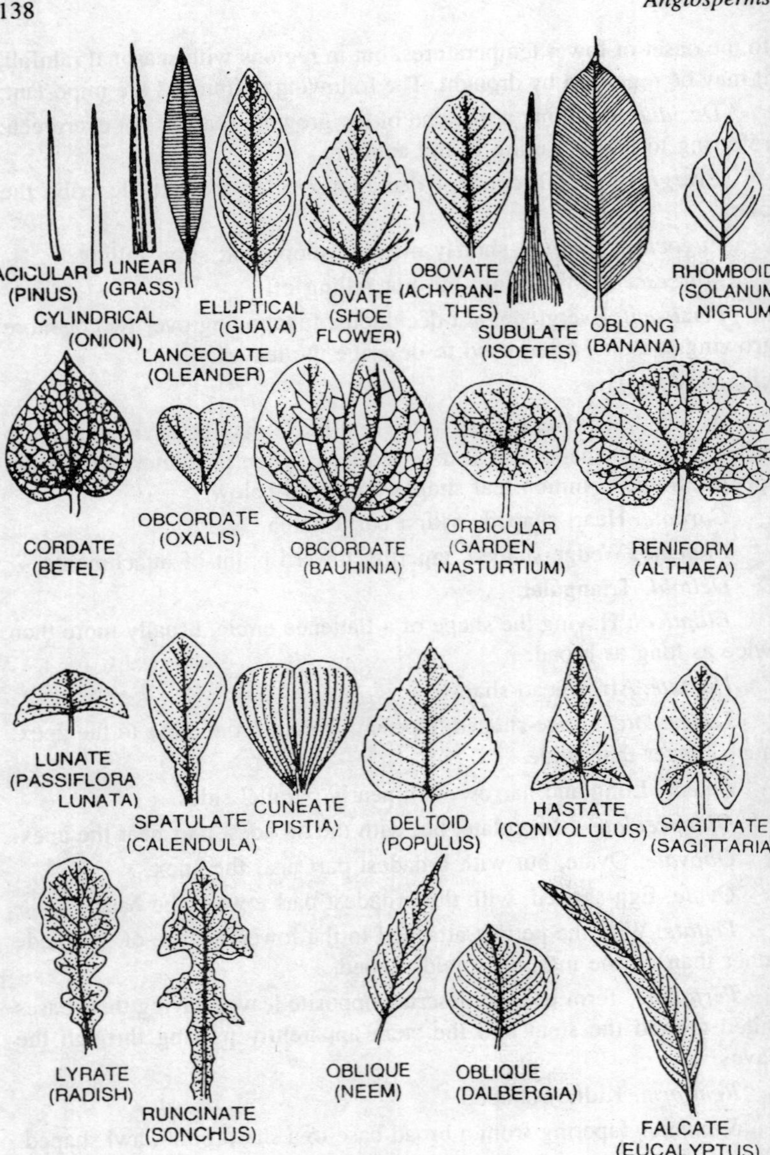

Fig. 5.16. Types of Lamina outline.

Acute. Ending in a point that is less than a right angle, but one th is not acuminate; distinct and sharp, but not drawn out.

Aristate. With a bristle at the tip.

Cuspidate. Tipped with a sharp and rigid point.

Emarginate. Shallowly notched and indented at the apex.

Mucronate. Abruptly tipped with a small point, project from the midrib.

Obtuse. With a blunt or rounded tip.

Rounded. With a broad arch at the apex.

Truncate. Cut squarely across at the apex.

Base of leaf blades

The part of the leaf blade where the petiole is attached is the *base*. Note the following definitions.

Auriculate. With earlike appendages at the base.

Cordate. Heart-shaped, with a notch at the base.

Cuneate. Wedge-shaped, gradually narrowed toward point of attachment.

Oblique. Asymmetrical, having one side of the blade lower on the petiole than the other; inequilateral.

Rounded. With a broad arch at the base.

Sagittate. Term describing basal lobes drawn into points on either side of the petiole, like the base of an arrowhead.

Truncate. Cut squarely across at the base.

Margins of leaf blades

The edge of a leaf blade is referred to as the margin. Margins fall into these categories:

Crenate. With low rounded or blunt teeth.

Dentate. Having sharp, marginal teeth pointing outward.

Dissected. Cut into more or less fine divisions.

Divided. Cut into distinct section, extending to the midrib or base.

Doubly serrate. With small serrations on larger serrations.

Entire. Smooth; devoid of any indentations, lobes, or teeth.

Lobed. Divided into parts separated by rounded sinuses extending one-third to one-half the distance between the margin and the midrib.

Parted. Cut or dissected almost to the midrib.

Repand. Slightly wavy.

Revolute. With the margin rolled inward toward the underside of the leaf.

Serrate. Having marginal marginal teeth pointing toward the apex.

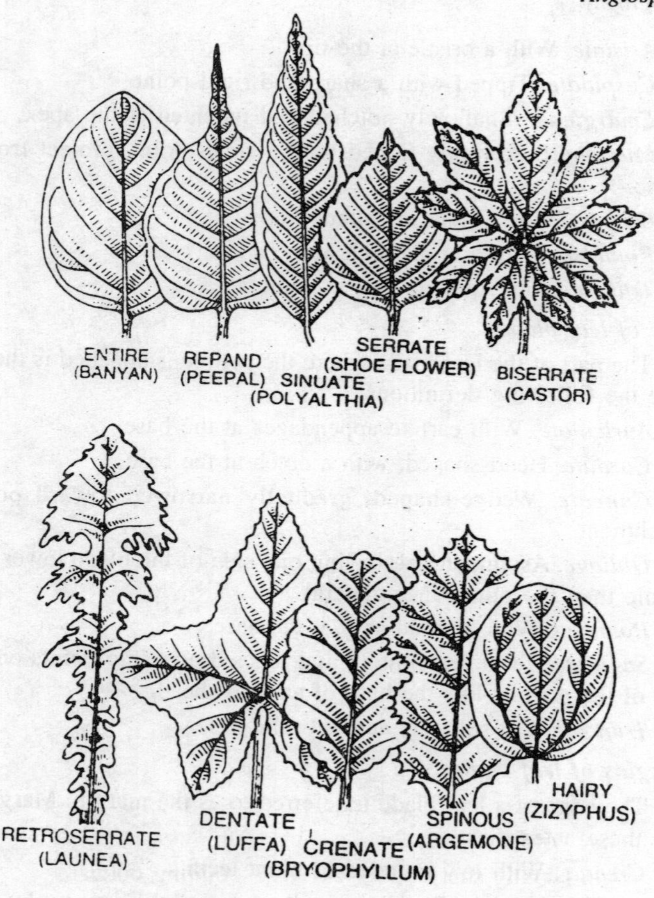

Fig. 5.17. Types of leaf margin.

Sinuate. Having a deeply wavy margin.

Undulate. Having a slightly wavy margin.

Modification of leaves

Many plants produce leaves that are quite different from the customary foliage leaves. Some of these are not easily recognizable as leaves. Among the modification are the following:

Bract. A greatly reduced or highly modified leaf often found in the inflorescence or subtending a flower; e.g., lemma and palea of grasses, or the brightly coloured bracts of poinsettias.

Bud scale. A small scale surrounding the bud.

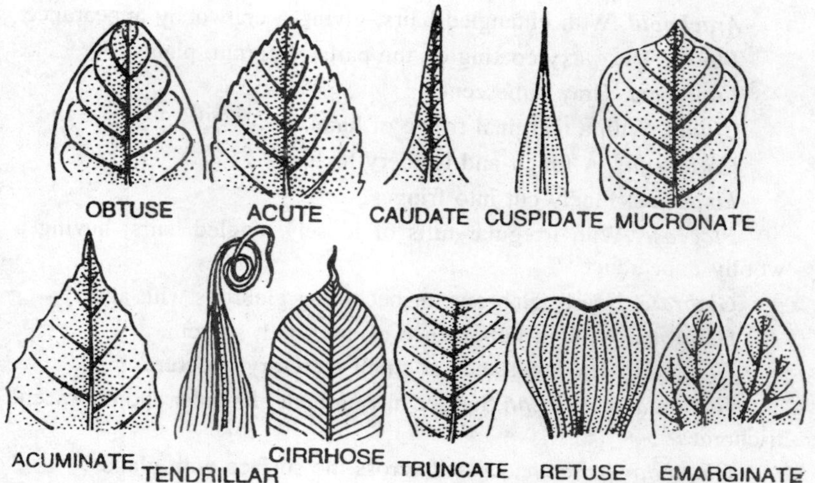

Fig. 5.18. Types of leaf apex.

Bulb scales. Fleshy, succulent storage leaves; e.g., the bulb of an onion.

Chaff. A bract at the base of the achene in the family Compositae; i.e., a pale.

Glume. A bract, usually found in pairs at the base of the grass spikelet.

Involucral bract. Subtending the head of the family Compositae; i.e., a phyllary.

Lemma. The outer of the two bracts subtending the grass floret.

Pale. See *Chaff*.

Palea. The inner bract subtending the grass floret. See *Lemma*.

Phyllary. A bract subtending the head in the Compositae family, sometimes called an involucral bract.

Pitcher. A tubular, insectivorous leaf able to hold water; e.g., *Sarracenia*.

Spathe. An enlarged bract enclosing an inflorescence; e.g., *Calla* and *Anthurium*.

Surfaces, Surface covering, and Texture

The surface features of angiosperms provide many valuable taxonomic character. The terminology tends to be somewhat subjective and in practice is difficult to apply:

Arachnoid. With entangled hairs, giving a cobwebby appearance.

Bloom. The waxy coating on the parts of certain plants.

Canescent. Gray pubescent.

Ciliate. With a marginal fringe of hairs.

Coriaceous. A tough and leathery texture.

Fimbriate. Finely cut into fringes.

Floccose. With irregular tufts of loosely tangled hairs; having a woolly appearance.

Glabrate. Nearly glabrous, or becoming glabrous with age.

Glabrous. Without pubescence of any kind; smooth.

Glandular. Having glands or small secretory structure.

Glandular-pubescent. Surface having both glandular and pubescent trichomes.

Glandular-punctate. Dotted across the surface with glands.

Glaucous. Having a waxy appearance due to a bloom or powdery coating of wax, i.e., a waxy bloom.

Hirsute. With long, shaggy hairs, often stiff or bristly to the touch.

Hispid. With stiff, rough hairs.

Lanate. Woolly, with long, intertwined, coiled hairs.

Pilose. With scattered, long, slender, soft hairs.

Puberulent. Somewhat or minutely pubescent.

Pubescent. Covered with short, soft hairs.

Rugose. Wrinkled.

Scabrous. Rough to the touch.

Scarious. Thin and membranous, usually dry.

Scurfy. Covered with scales.

Sericeous. With soft, silky hairs, usually all pointing in one direction.

Stellate. With star-shaped hairs.

Stinging hairs. Hairs which cause a stinging or burning sensation when they come into contact with the skin.

Strigose. Stiff hairs often appressed (i.e., pressed next to the stem) and pointing in one direction.

Tomentose. With densely matted soft hairs; woolly in appearance.

Velutinous. Velvety.

Villous. Covered with long, fine, soft hairs.

Reproductive Morphology

Much of the classification of flowering plants is based on reproductive structures, and a knowledge of the terminology of flowers, fruits, and seeds is essential to identify plants and to understand classifications. A *flower* is a highly modified shoot with specialized appendages. Flowers may arise in the axil of a leaf or, more often in the axil of reduced leaf called a *bract*. Ripened ovaries in the flower develop into the *fruit*. Fruit may have other floral structures associated with them and normally contain *seeds*, which are ripened ovules. The seed germinates and produces a new plant.

Parts of a Generalized Flower

The major components of the unmodified flower are the perianth, the *androecium*, and the *gynoecium*. These structures are attached to the *floral axis* or receptacle. A *hypanthium* may be present and is sometimes wholly or partly receptacular in nature; at other times it is composed wholly or partly of perianth and androecial tissue. The *perianth*, or *floral envelope*, is subdivided into the *calyx* (composed of the sepals) and the corolla (made up of the petals). A group of *stamens*, where the pollen is produced, is called the *androecium*. The *gynoecium* consists of one or more carpels, when two or more, they may be either separate (simple pistils) or united (compound pistils). Some terms associated with the generalized flower are defined below:

Androecium. The male part of the flower; a collective name for the stamens of a flower and the parts derived from stamens. The androecium occupies a position inward from the perianth.

Anther. That part of the stamen where the pollen is produced.

Calyx. The collective name for the sepals. The calyx occupies the outermost position in the flower. Together with the corolla, the calyx makes up the perianth. The calyx is usually green but is sometimes coloured; e.g., *Fuchsia*. Sometimes the calyx intergrades into the corolla; e.g., water lily.

Carpel. A simple pistil; one part of a compound pistil. The carpel is the innermost part of the flower, bearing one or more ovules. It is thought to have evolved from an ovule-bearing leaf, i.e., megasporophyll.

Corolla. The inner part of the perianth, composed of the petals. The corolla is usually larger than the calyx and is often brightly coloured. The petals may be fused or separate.

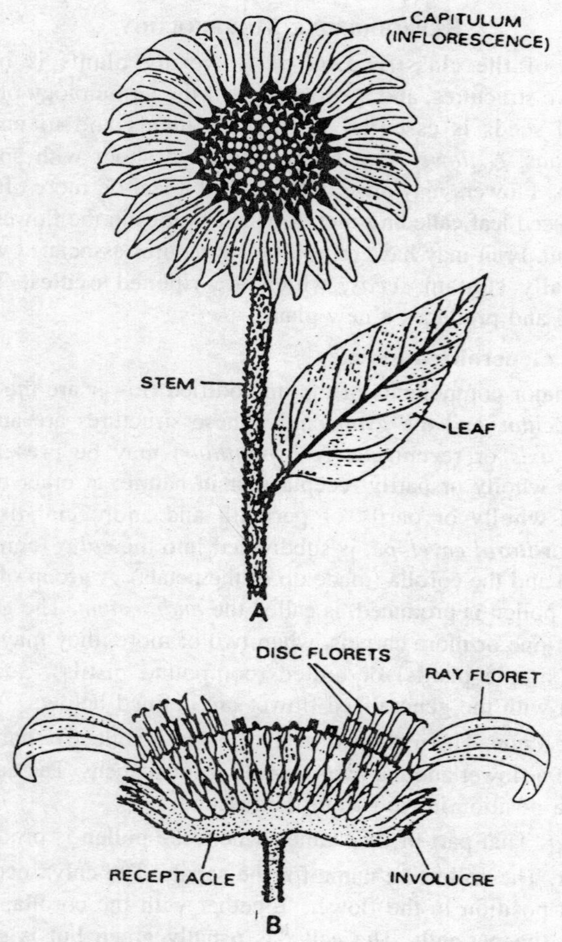

Fig. 5.19. Capitulum of Sunflower. A—entire capitulum; B—vertical section of capitulum.

Filament. The stalklike part of the stamen the supports the anthers and which is attached to the receptacle. It may be fused to the petal(s) for part of its length.

Fruit. Mature ovary or ovaries containing seed.

Gynoecium. The famele part of the flower; the collective term for the carpels or pistils. It is the innermost part of a flower and is composed of one or more carpels, which may or may not be united into a compound pistil.

Hypanthium. A floral cup or tube formed either from the receptacle or from fusion of the bases of the sepals, petals, and stamens or both. The hypanthium appears to bear sepals, petals, and stamens or both. The hypanthium appears to bear sepals, petals, and stamens on the rim of its cup or tube.

Ovary. The usually enlarged, basal portion of the pistil where the ovules are borne. The ovary, sometimes with associated floral structures, develops into the fruit.

Ovule. The structure in the ovary which develops into the seed.

Pedicel. The stalk of each individual flower.

Peduncle. The stalk bearing the entire inflorescence or a solitary flower.

Perianth. The colective term for the floral envelope, i.e., calyx and corolla.

Petals. The individual parts of the corolla occupying a position between the sepals and stamens.

Pistil. A structure composed of one or more carpels and usually havng a stigma, style, and ovary. The carpels in a flowar are collectively temed the *gynoecioum.* A *simple pistil* is composed of a single carpel; a *compound pistil* is composed of two or more united carpels.Rachis. The central axis of an inflorescences, e.g., grasses.

Receptacle. The portion of the stem which bears flowers parts. It consists of several short nodes and internodes.

Seed. A matured ovule consisting of a seed coat (integument), an enclosed nucellus, embryo, and the remains of the megagametophyte.

Sepals. The individual components of the calyx; the outermost whorl of the flower. The sepals are usually green, but may be petalled and coloured.

Stamens. The pollen-producing part of the flower, located just inside the corolla. Stamens usually have an anther and a filament. The stamens in a flower are collectively known as the *androecium.*

Staminode. A sterile stamen.

Stigma. The portion of the style that is receptive to germination of pollen.

Style. The elongated stalk connecting the stigma and ovary.

Inflorescences

The term *inflorescence* refers to the arrangement of the flowers on the plant. Some inflorescences are simple and readily distinguishable, but other are complicated aggregations that are difficult to characterize.

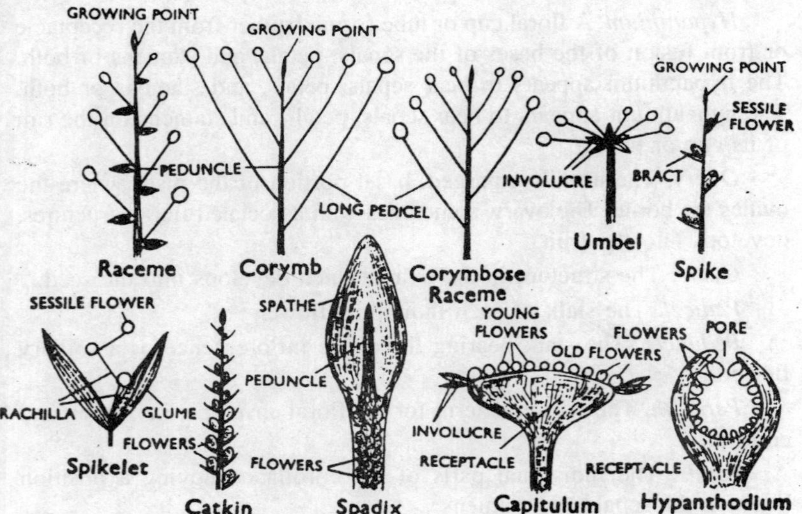

Fig. 5.20. Diagrammatic representation of various types of simple racemose inflorescence.

Several terms are associated with inflorescences and are essential to their understanding. *Determinate* inflorescences are those where the flowering sequence begins with the terminal flower at the tip of the stem or at the center of the flower cluster. In contrast, *indeterminate* inflorescences have a flowering sequence which starts at or near the base of the inflorescence, or the outside of the cluster, and proceeds upward or toward the center. The following are some common types of inflorescences:

Ament. A deciduous, erect or lax, spikelike inflorescence, with scaly bracts and unisexual, apetalous flowers.

Capitulum. See *Head.*

Catkin. A soft spike or raceme of small unisexual flowers, the inflorescence usually falling as a unit. See Ament.

Compound. Composed of two or more simple inflorescences aggregated together, e.g., panicle.

Corymb. A broad inflorescence in which the lower pedicels are successively elongate, giving the inflorescence a flat-topped appearance; indeterminate.

Cyme. A broad, more or less flat-topped inflorescence with the main axis terminating in a single flower that opens before the lateral flowers; determinate.

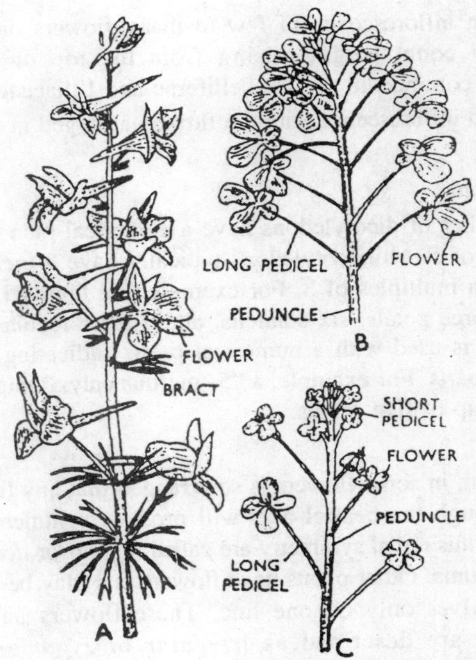

Fig. 5.21. A—typical raceme of Larkspur; B—corymb of young Candytuft; C—corymbose raceme of Mustard.

Head. A dense cluster of stalkless flowers; a capitulum.

Helicoid. Curved or spiral inflorescence.

Panicle. A compound inflorescence in which the main axis is branched one or more times and may support spikes, recemes, or corymbs.

Raceme. An inflorescence with a single axis and the flowers arranged along the main axis on pedicels; indeterminate.

Scorpioid cyme. A cyme that appears to coil, like a scorpion's tail.

Solitary. With a single flower.

Spadix. A thick or fleshy spikelike inflorescence with very small flowers which are massed together and usually enclosed in a spathe, e.g., Araceae.

Spike. An inflorescence with a single axis and flowers without pedicels.

Thyrse. A compound, compact panicle with an indeterminate main axis and laterally determinate axes.

Umbel. An inflorescence of few to many flowers on pedicels of approximately equal length arising from the top of a peduncle; indeterminate; common in the Umbelliferae and Liliaceae.

Verticil. An inflorescence with the flowers arranged in whorls at the nodes.

Numerical Plan

Most families of dicotyledons have a numerical plan or 4 to 5, or multiples of 4 or 5. Monocotyledons typically have 3-merous flowers or with parts in multiples of 3. For example, the family Liliaceae has three sepals, three petals, six stamens, and a three-locular ovary. The ending-*merous* is used with a numerical prefix indicating the number of each of the parts. For example, a "5- merous calyx" implies that the calyx is made up of five sepals.

Symmetry

The perianth in some flowers is so arranged that any line bisecting the flower through the central axis will produce symmetrical halves. Flowers having this radial symmetry are called *regular* or *actinomorphic*, e.g., rose or petunia. Other plants have flowers that may be divided into symmetrical halves only on one line. These flowers have bilateral symmetry and are described as *irregular* or *zygomorphic*; e.g., snapdragon.

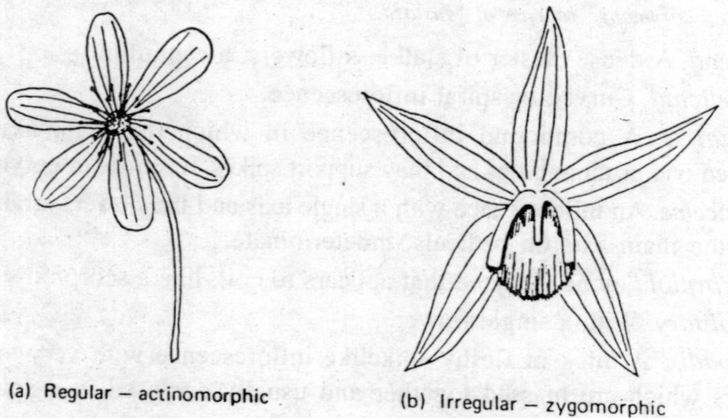

Fig. 5.22. The symmetry of flowers: (a) regular; (b) irregular.

Insertion of floral Structures

The attachment and arrangement of floral parts is called *insertion.* The term *inserted* implies that one part is attached to or growing out of another part - e.g., stamens inserted on the petals. The parts of a flower may be fused together but eventually are attached to the receptacle. The petals normally alternate with the sepals in point of attachment, and the stamens alternate with the petals. The *insertion of the ovary* relative to the position of the other floral parts is a widely used taxonomic character. If the ovary is attached to the receptacle above the attachment of the other floral parts, it is *superior*. The ovary is called *inferior* when it lies below the attachment of the perianth and androecium and is embedded in receptecle tissue.

Flowers may be classified into one of three groups: *hypogynous*, *perigynous*, or *epigynous*, depending upon the ovary, perianth and androecium position. In a *hypogynous* flower, the perianth and androecium are inserted around the base of the gynoecium. In some hypogynous flowers, the stamens may be adnate to the base of the petals, but the bases of the perianth are not fused together to form a floral cup. In addition, no floral structures are fused to the ovary. Hypogynous flowers have superior ovaries.

The *perigynous* flower has the perianth and stamens united by their bases to form a cuplike hypanthium, or else has a hypanthium developed from receptacle tissue. The hypanthium is free from the pistil,

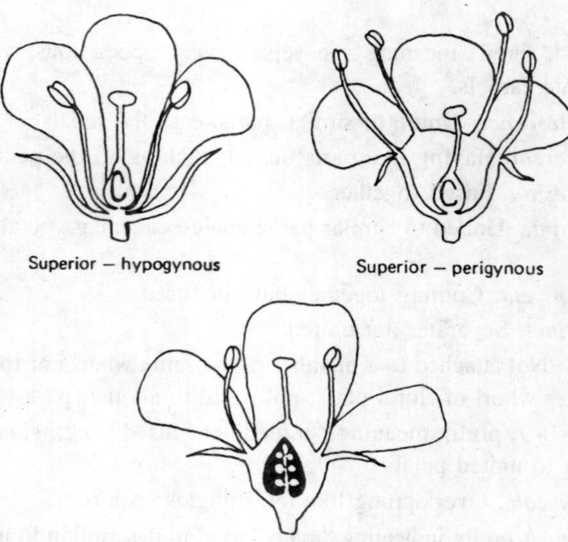

Fig. 5.23. Ovary position and insertion of flower parts. Superior and inferior describe the ovary; hypogynous, perigynous, and epigynous describe the flower.

although it surrounds the ovary. The upper part of the sepals, petals, and stamens are inserted upon the rim of the hypanthium and appear to grow from it. The ovary is superior in a perigynous flower.

An *epigynous* flower has a hypanthium fused to the ovary, and therefore the sepals, petals, and stamens appear to arise from the top of the ovary. The ovary is obviously inferior in an epigynous flower, since it occurs below the point of attachment of the perianths and androecium. In some instances, the ovary is sunken in the receptacle. In some groups, the hypanthium extends only partially up the ovary, and the ovary is termed *half-inferior*. The hypanthium may extend beyond the top of the ovary, forming a cup or tube around the style. Many intergradations exist among the idealized flower types and ovary positions. To determine whether the flower is hypogynous, perigynous, or epigynous, view the flower after making a longitudinal cut.

Union of floral parts

Manuals use number of terms to describe the connection (fusion or lack of fusion) among floral parts. Some of the more frequently encountered terminology include the following:

Adhesion. Union of unlike parts.

Adnate. United to a part of a different kind; e. g; stamens united to petals.

Apo- A prefix meaning free, separate; e.g; apocarpous, which refers to separate carpels.

Coalescence. Union of similar parts; e.g., the sepals.

Coherent. Having like parts united, such as all the petals united.

Cohesion. Fused together.

Connate. United to similar parts; coalescence, e.g., petals united to petals.

Connivent. Coming together but not fused.

Distinct. Separate; not united.

Free. Not attached to a member of the same whorl nor to a member of another whorl of floral parts; not fused to another part.

Gamo- A prefix meaning "unitied" or "fused", e.g., *gamopetalous*, referring to united petals.

Imbricate. Overlapping like the shingles on a roof.

Poly - A prefix indicating "many" used in description to imply many separate petals or sepals; e.g., *polypetalous* and *polysepalous*.

Syn- A prefix meaning "united" or "fused", e.g., *syncarpous*, "having fused carpels." The prefix become *sym-* before the latters b, m, and p, as in *sympetalous*, "having united petals."

Valvate. With margins of adjacent structures touching only at their edges.

Presence or absence of floral parts, and sexuality

The following are the essential terms in describing the sexuality and the presence or absence of floral parts:

Apetalous. Without petals, or with only a single perianth whorl.

Bisexual. Having both stamens and pistils in the same flower.

Complete. Having calyx, corolla, stamens, and pistils.

Dioecious. Having staminate and pistillate flowers on different plants; e.g., American holly or willow.

Imperfect. Refers to a flower having either stamens or pistils but not both; i.e., one sex but not both.

Incomplete. Used to describe a flower that lacks one of the four whorls of a typical flower, but usually with reference to the absence of petals or sepals or both.

Monoecious. Having stamens and pistils in different flowers on the same plant; e.g., corn.

Naked. Lacking a perianths or enclosing bracts.

Obsolate. Rudimentary; or not evident.

Perfect. Having both stamens and pistils in the same flower.

Pistillate. Term applied to plants or flowers with pistils, but no functional stamens.

Polygamous. Term applied to plants or flowers with stamens, but no functional pistils.

Unisexual. Having only stamens or pistils, but not both, i.e., imperfect.

Perianth

A calyx or a corolla of many separate sepals or petals is classified as *polysepalous* or *polypetalous*. If the sepals or petals are united, they are described as either *synsepalous* or *gamosepalous* and either *sympetalous* or *gamopetalous*. The corolla and sometimes the calyx may be formed into a basal cylindrical structure called the *tube*. The inside opening of the tube is referred to as the *throat*. The spreading part of a sympetalous corolla is the lobe or *limb*. The calyx often has prominent veins called nerves. The narrowed, petiolelike base of some petals or

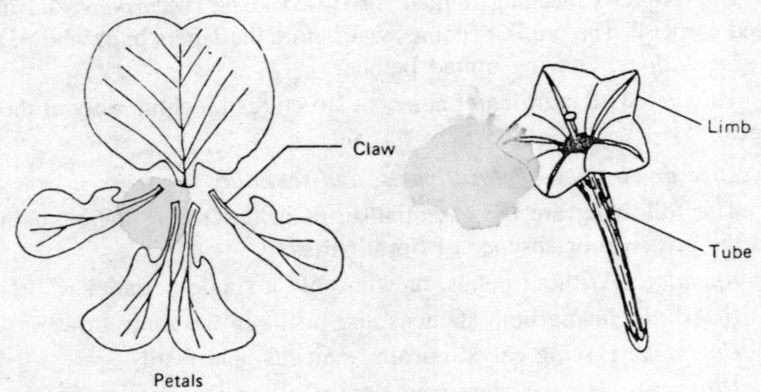

Fig. 5.24. Corolla. Claw, limb, and tube.

sepals is the *claw*. The spur is a hollow protrusion of the corolla or calyx; e.g., the corolla spur or *Linaria*. Several terms are used to refer to the shape of sympetalous corollas.

Bilabiate. Two-lipped, with two unequal divisions.

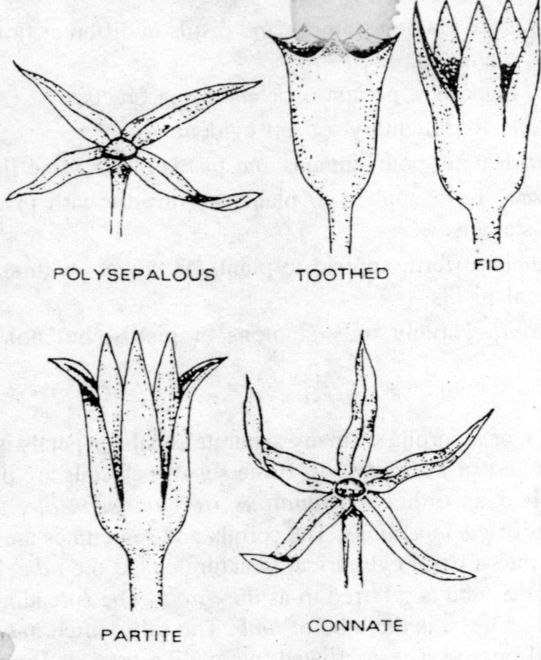

Fig. 5.25. Types of fusion in Calyx.

Campanulate. Bell-shaped.

Funnelform. Having the shape of an elongate funnel.

Rotate. Radiately spreading in on plane, with a short tube.

Salveform. A term used to describe a corolla with a cylindrical elongate tube and spreading rotate limbs that are abruptly flared.

Tubular. Cylindrical and hollow; having an elongated tube and short limb.

Urceolate. A term used to describe a corolla with united petals forming a tube that is expended below the middle and narrow at the top, urn-shaped.

Androecium

The collection of stamens in the flower is termed the *androecium.* Stamens are composed of the pollen-producing part, the *anther*, and the *filament.* Some taxonomic characters that are derived from the androecium include; number, fusion, and insertion. Terms used to describe the androecium include:

Basifixed. A term to describe an anther attached by its base to the filament.

Centrifugal. Developing first at the center and then gradually toward the outside, used to describe the sequence of stamen maturation in a single flower with many stamens.

Centripetal. Developing first at the outside and then gradually toward the center, used to describe the sequence of stamen matruation in a single flower with many stamens.

Connective. Term describing the tissue between the two locules of an anther.

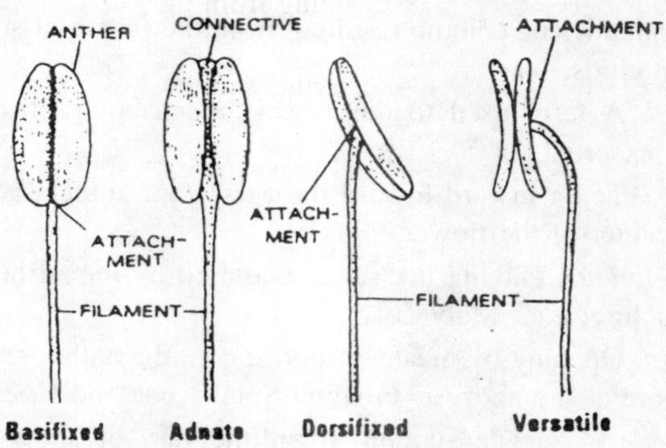

Fig. 5.26. Forms of fixation of anthers.

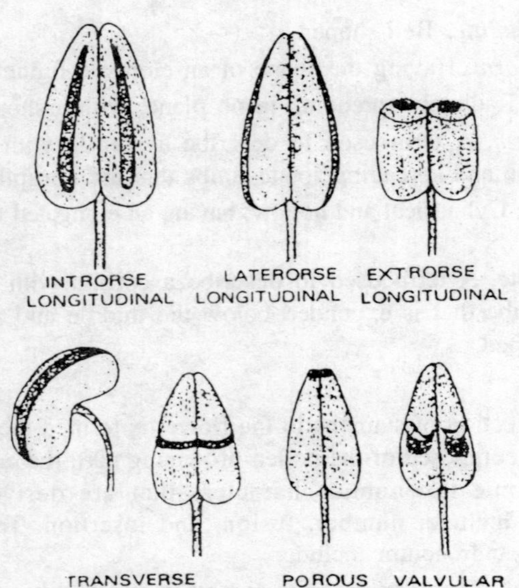

Fig. 5.27. Dehiscence of anthers.

Diadelphous. The term applied to stamens which form two groups by the union of their filaments.

Didynamous. A term used to describe flowers with two long and two short stamens; common in the Labiatae.

Epipetalous. Having stamens inserted on the petals.

Exserted. Having stamens extending beyond the corolla.

Extrorse. Facing outward from the axis; e.g., anther sac opening away from the center or the flower.

Gynostemium. The column resulting from the fusion of stamens and pistil; e.g., orchids.

Included. A term used to describe stamens not protruding from corolla; not exserted.

Introrse. Facing inward toward the axis; e.g., anther sacs opening toward the center of the flower.

Monadelphous. Having the stamens united by their filaments into a single structure; e.g., Malvaceae.

Poricidal. Opening by means of pores; e.g., the anther sacs opening through a pore at the apex, as in some Solanaceae and Ericaceae.

Staminode. A sterile stamen; a rudimentary or showy structure arising from the stamen whorl; e.g., *penstemon*.

Synatherous. Term applied to stamens fused together by their anther; e.g., the Compositae family.

Tertradynamous. Having four long and two short stamens, as in most members or the family Cruciferae.

Versatile. Attached at the middle; describing an anther attached at its midpoint to the apex of the filament; e.g., lily.

Gynoecium

The collection of carpels in the flower is the *gynoecium*. The *carpel* is generally considered to have been derived from an enrolled leaflike structure bearing ovules. A gynoecium of one or more separate or individual carpels is described as *apocarpus*, and the individual carpel is a *simple pistil*. If the gynoecium is composed of two or more united carpels, it is *syncarpous* and is a *compound pistil*. A typical gynoecium is composed of pollen-receptive area at the apex called the *stigma*, a stalklike *style*, and an enlarged basal portion bearing ovule called the *ovary*. A simple or *unicarpellate* ovary has one chamber within it where the ovules are borne. The chamber is called either a "*cell*" or more properly, the *locule*. A *multicarpallate* ovary may have one or more locules. If an ovary has two or more locules, the interior wall separating the locules is the septum (plural, *septa*).

The stalk that attaches the ovule to the ovary wall is called the *funiculus*. The placenta is the point of attachment of an ovule wall. It is

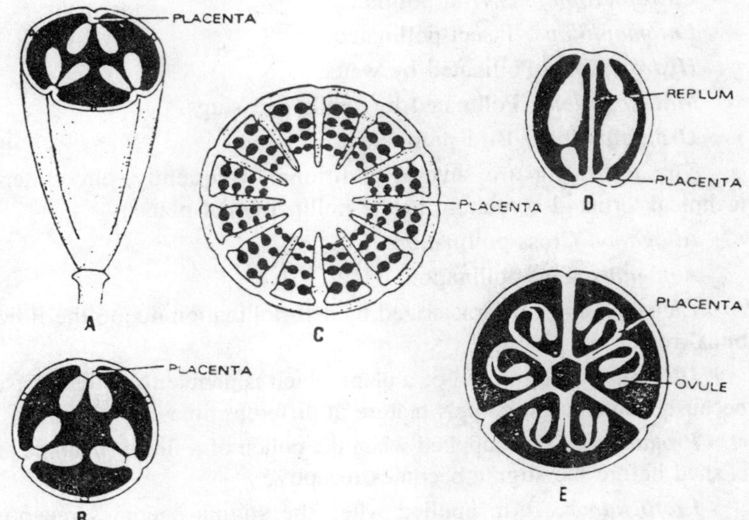

Fig. 5.28. Parietal placentation. A-B—normal parietal placentation. C—parietal placentation of Poppy. D—parietal placentation in Brassica. E—modified parietal placentation of Cucurbita.

sometimes difficult to determine the number of carpels in a gynoecium, but in many instances each group of placentae represents location of a separate carpel. The arrangement of the *placentae*, or *placentation*, described the way in which the ovules are arranged in the ovary. The five main kinds of placentation are marginal, axile, parietal, free-central, and basal. In simple (unicarpellate) pistils, the plecentation is marginal. The four other basic types of placentation are found in compound pistils. When there are separate locules for each carpel, with placentae at the center, the placentation is *axile*. If the septa between the carpels are absent, resulting in a one-chambered ovary and several ovules usually has parietal plecentation. A syncarpellate ovary in which the septa are absent and the placentae are on a central stalk arising from the base of the ovary is said to be *free central*. If the central stalk is absent and the placentation is directly on the floor of the locule, the placentation is *basal*.

Pollination

The movement of pollen from the anther to the stigma is known as pollination. Some of the terms frequently related to the agents of pollination of flowers include;

Anemophilous. Wind-pollinated.

Cantharophilous. Beetle-pollinated.

Chiropterophilous. Bat-pollinate.

Entomophilous. Insect-pollinated.

Hydrophilous. Pollinated by water.

Malacophilous. Pollinated by snails and slugs.

Ornithophilous. Bird-pollinated.

The following are several additional, frequently, encountered technical terms also relating to the pollination of plants:

Allogamy. Cross-pollination.

Autogamy. Self-pollination.

Cleistogamous. Characterized by self-pollination due to the flower buds' remaining closed.

Dichøgamy. Term used of a plant which is unable to pollinate itself because stamens and carpels mature at different times.

Protandrous. Term applied when the pollen of a flower matures and is shed before the stigma becomes receptive

Protogynous. Term applied when the stigma becomes receptive before the stamens mature in the same flower.

Pollen, Fertilization, and Embryo Development

Microsporogeneis (production of microspores by meiosis) occurs in the anther, eventually giving rise to pollen or male gametophytes. Pollen is shed either in the *binucleate* (with two nuclei) or less often in the *trinucleate* (with three nuclei) stage. The pollen grain of an angiosperm consists of two cells: the *tube cell* and the *generative cell.* When a pollen grain lands on a receptive stigma, the tube cell grows rapidly and forms tube; then by digesting its way through the stigma and style and into the locule, the pollen grain enters the ovule through the micropyle. The generative cell produces two male gametes or sperms that follow the tube nucleus into the tube. Pollen grains are variously shaped. *Monosulcate* pollen grains have a single, long furrow that develops on one side of the grain and are *characteristic* of gymnosperms, monocotyledons, and certain families of dicotyledons. *Tricolpate* pollen grains have three meridionally placed furrows, and are characteristic of most of the dicotyledons.

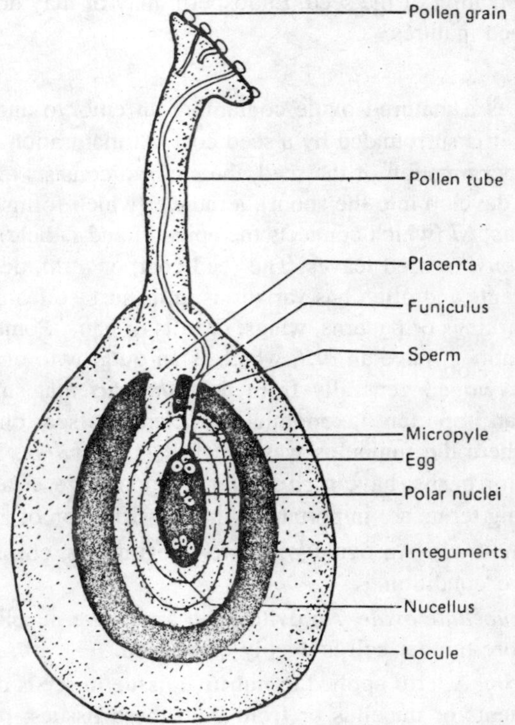

Fig. 5.29. Pollen germination and fertilization in angiosperms.

The outer wall of an angiosperm ovule is composed of one or two integuments. The *micropyle* is a minute opening through the integuments. The integuments later from the seed coat. The ovule contains a megasporangium called the *nucellus*. Megasporogenesis (production of magaspores by meiosis) occurs inside the nucellus, eventually giving rise to the *embryo sac* or female gametophyte. The embryo sac, which is embedded in the nucellus, has a group of eight nuclei. The three most important nuclei in the embryo sac are the *egg nucleus*, or female gamete, and the two *polar nuclei*. When the pollen tube enters the embryo sac, the sperms are discharged into the embryo sac. One sperm fuses with the egg nucleus to form a *zygote*. The zygote rapidly develops into the *embryo* within the ovule. The second sperm unites with the polar nuclei to form a primary endosperm nucleus that develops into a tissue called *endosperms*. The endosperm becomes a nutritive source for the embryo during embryo development and often during germination of the seed. Endosperm may or may not be present when the seed matures.

Seed

A *seed* is a matured ovule containing an embryo and sometimes endosperm; it is surrounded by a seed coat. At maturation of the ovule and before germination of the seed, the embryo consists of an *epicotyl* (which will develop into the shoot), a radicle (which forms the primary root), a *hypocotyl* (which connects the epicotyl and radicle), and one or two *cotyledons*, or seed leaves. The seed coat, or *testa*, develops from the integuments and often has variations in its surface features. Among these are markings or patterns, wings, or tufts of hairs. Some seeds, e.g., litchi and nutmeg, have an *aril*, which is an outgrowth or covering of the seed developed generally from the *funiculus* that supports it, is sometimes an important taxonomic character. The scar on the seed at the point where the funiculus was attached is the *hilum*. Some seeds, such as castor beans, have a *caruncle*, or appendage, near the hilum. The following terms are important in the classification of angiosperms:

Bitegmic ovule. An ovule with two integuments, considered to be the primitive condition.

Crassinucellate ovule. An ovule with a massive nucellus in which the megaspore mother cell is deeply embedded.

Perisperm. A term applied to nutritive tissue in seeds derived from the integuments or nucellus or from both these tissues; prominent in the families; prominent in the families Caryophyllaceae, Polygonaceae, Chenopodiaceae, and phytolaccaceae.

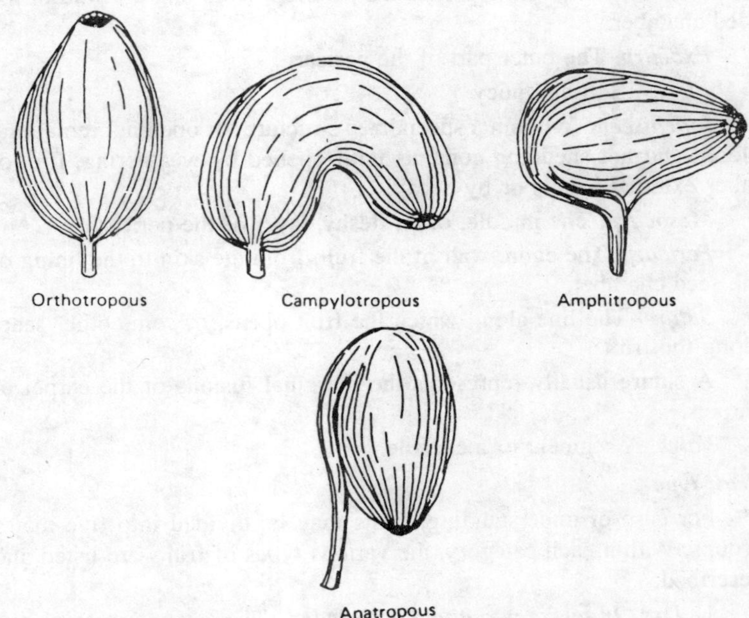

Fig. 5.30. Types of ovules.

Tenuinucellate ovule. An ovule in which there is a small amount of nucellus with the magespore mother cell located subepidermally.

Unitegmic ovule. An ovule with a single integument; considered advanced relative to the bitegmic ovule.

Fruits

A *fruit* is a matured ovary, including its contents. It may have other floral parts fused to it, including the receptacle, involucre, calyx, hypanthium, and so forth. Fruits may be diagnostic at the family level or used to delimit taxa within families. There numerous kinds of fruits, some easy to classify, other more difficult.

Definitions. The following terms should be noted:

Accessory structures. Parts of fruits derived from nonovarian tissue.

Beak. Persistent style of style or style-base; the slender elongated ending of a fruit; e.g., *Brassica*.

Circumscissile. Dehiscent horizontally at or above the middle, the top part falling away as a lid.

Dehiscent. Opening and shedding contents in a regular and often distinctive manner.

Endocarp. The inner part of the pericarp, including the wall of the seed chamber.

Exocarp. The outer part of the pericarp.

Fleshy. Soft and juicy.

Indehiscent. Without a specialized structure for opening; remaining closed and not shedding contents until opened by weathering, fire, or other external agents or by decay.

Mesocarp. The middle, often fleshy, layer of the pericarp.

Pericarp. The entire wall of the fruit, from the skin to the lining of the seed chamber.

Suture. The line along which the fruit opens, or some other seam along the fruit.

A suture usually represents the marginal fusions of the carpel or carpels.

Valve. A segment of a capsule.

Fruit type

For ease of understanding, fruits may be divided into five major groups. Within each category, the various types of fruits are listed and described.

1. *Dry, Indehiscent, and One-seeded*. The group contains the following the types:

Achene. A dry, one-seeded fruit with a firm, close-fitting wall, but with the pericarp free from seed; examples include sunflower "seeds."

Caryopsis. A fruit in which the pericarp is fused with the seed; a grain; e.g., the fruit of the grasses.

Grain. See *Caryopsis*.

Nut. A fruit developed from a pistil with more than one carpel and having a hard woody coat; e.g., hickory, pecan, or oak.

Nutlet. A small nut.

Samara. An achene with a wing; e.g., ash.

Schizocarp. A single, compound ovary that splits simulating separate fruits derived from the ovaries of a simple pistil, e.g., the dry seedlike fruits derived from an inferior ovary, as in the Umbelliferae.

Utricle. A one-seeded fruit with a thin, bladdery, persistent, sometimes inflated wall; found in some amaranths.

Winged Schizocarp. Splitting samaras that are winged and dry, e.g., maple.

2. *Dry, Dehiscent, and Having Several to Many Seeds*. In this groups are the following types:

Capsule. A fruit originating from two or more carpels. There are several types of capsules; *loculicidal*, dehiscent lengthwise down the middle of the carpels (e.g., *Iris*); *poricidal*, dehiscent by pores at the top (e.g., poppy) *pyxis*, dehiscent in a circumscissile manner (e.g., *plantago*); *septicidal*, dehiscent lengthwise at the junction of the carpels (e.g., *Yucca*).

Follicle. A fruit developing from a single pistil and dehiscing along one margin; e.g., *Delphinium*.

Legume. A one-locular fruit dehiscent on two sutures; e.g., bens and peas.

Loment. A flat modified legume that is constricted between seeds and falling apart into one-seeded joints; e.g., *Desmodium*.

Silicle. A short, two-locular fruit, composed of two valves which separate from the central partition, e.g., *Capsella* and other short-fruited members of the Cruciferae.

Silique. An elongated, two-locular fruit with two parietal placentae, and usually with two valves that separate from the partition on dehiscence; e.g., *Brassica* and other long-fruited member of the Cruciferae.

3. *Fleshy, Derived from a Syncarpous Gynoecium of a single Flower.* This group contains two types:

Berry. A fruit with a soft and fleshy pericarp; e.g., tomato, grape.

Drupe. A fruit with a fleshy mesocarp and a stony endocarp which forms a hard covering around the seed; e.g., peach, plum.

Hesperidium. A berry with a tough, leathery rind; e.g., a citrus fruit.

Pepo. A one-locular fruit developed from an inferior ovary with parietal placentae, having a hard outer covering; e.g., a gourd.

4. *Derived From a Single or Compound Ovary and Nonovarian Tissue Such as the receptacle or Hypanthium; Accessory Fruit*. This group contains two types:

Hip. An aggregation of achenes surrounded by a cup-like receptacle and hypanthium; e.g., rose fruit.

Pome. A fruit developed from a compound inferior ovary with the seeds encased within a cartilaginous wall, as in an apple. The fleshy part of the pome is largely developed from the hypanthium.

5. *Resulting From the Union of Many Separate Carpels of a single Flower, or the Fusion of Several Fruits of Separate Flowers*. A fruit in this category can be aggregate or multiple:

Aggregate. A cluster of fruits clearly traceable to separate pistils of the same flower and inserted on a common receptacle; e.g., *Rubus* and *Ranunculus*.

Multiple. Term describing fruits on a common axis that are usually fused and derived from the ovaries of several flowers; e.g., mulberry. The *syconium* of the fig, which is a collection of achenes borne on the inside of a hollowed-out receptacle or peduncle is a multiple fruit. The fruit of the pineapple results from a fusion of rachis, bracts, and ovaries.

Descriptive Methods

Floral Formulas

The floral formula methods has been used for many years to describe graphically the floral morphology of the angiosperms. Although the formulas used by different authors may differ in a few details, they are all designed to compare graphically floral structure. Morphological

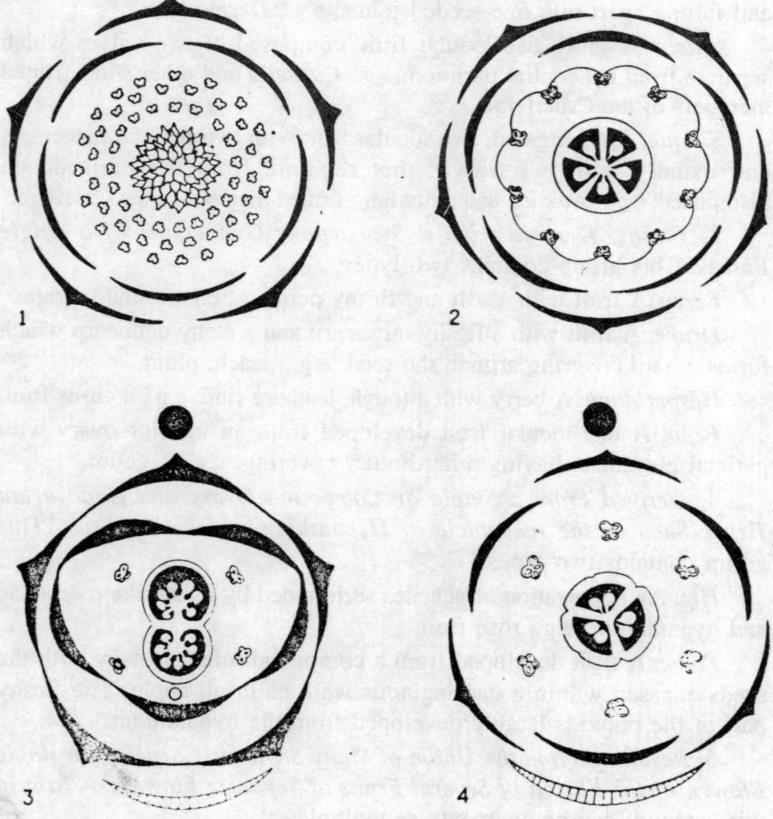

Fig. 5.31. Representative floral diagrams. 1, Ranunculus. 2, Geranium. 3, Penstemon (irregular). 4, Lilium.

features, such as the calyx (CA), corolla (CO), androecium (A), and gynoecium (G), are represented by letters. The number of parts of each whorl is shown by numerals. CA^3 and A^6 convey the massage that the plant has a calyx of three sepals and an androencium of six stamens. Symbols and other distinguishing marks, relatively few in number, are used to describe further the flowers; e.g., CA^5 a calyx of five fused sepals, or G, an inferior ovary, versus $\underline{G}$, a superior ovary. Most descriptive methods have advantages and disadvantages, and floral formulas are no exception. One problem is that there has been no attempt to standardize the symbols and systems.

Formulas, when broadly generalized, may not convey to the user all the variation present within a family. Families may contain genera which are somewhat atypical of the majority of the genera. A floral formula of $A^{3\text{-}6}$ conveys the fact that the group has some plant with three stamens and others with six stamens, but it does not convey the relative frequency of each. Actually, six stamens may be very common and three is very rare. That could be shown by $A^{6(3)}$. On the other hand if the formula system is made increasingly complicate by handling all special cases the purpose of the system is defeated. When allowances are made for its weakness, the floral formula may be a valuable tool.

Floral Diagrams

Floral diagrams (fig. 5.31) visually depict the essential features of a flower in cross section. The floral parts are represented by semi-diagrammatic symbols or ideographs. This illustrates both the number of whorls and floral parts and other features, such as fusion of structures and flower symmetry. Although ovary position must be shown in a second drawing using a longitudinal section, this drawback does not diminish the overall value of such diagrams.

6

BIOSYSTEMATICS

By observing the diversity, structural complexity, and adaptative natures of plants, systematists have described taxa developed classification, and unraveled the processes of evolution. Biosytematics attempts to discover the mechanisms and processes in flowering plants that: (1) direct their evolution, (2) influence their variation patterns, and (3) cause speciation. Classical taxonomists emphasize the evolutionary end products and differences between taxa, often using biosystematic information sources in developing their classification. *Organic evolution* is the series of transformations of the genetic materials of populations through time. This is primarily caused by natural selection in repose to changing environmental interaction. Evolution is not synonymous with speciation. Evolution is the gradual accumulation of small genetic changes through time which sometimes leads to speciation. *Speciation* describes the evolutionary divergence among similar population.

The term *biosystematics* is derived from biosystematy and was introduced by Camp and Gilly (1943). As originally conceived, biosystematics was an attempt to produce a system of classification to express explicit relationships. Since the 1940s the term has come to mean the application of experimental, genetic, cytologic, and population approaches to systematic problems, especially at the levels of species and infraspecific taxa.

SOURCES OF VARIATION

The complexity and diversity of plant life is apparent among species and between individuals of the same species. This variation forms the basis of both evolution and classification. An understanding of the evolution of plants may be discovered by careful analysis of this variation. Systematists are confronted with variability in population possessing three components: (1) developmental variation, (2) environmentally induced variation, and (3) genetic variation.

Developmental Variation

Adult plants are often strikingly different from the immature seedling. This developmental variation that is genetically detemined can cause confusion in the identification of flowering plants. For example, G. N. Jones (1968) found that specimens indentified as *Tilia neglecta* were nearly always stump-sprout leaves of *T. americana*. Likewise, Crawford (1973) noted that the type specimen of *Chenopodium palmeri* was an immature plant and that key characters used to separate that species from *C. arizonicum* were based merely on differences in maturity. The adult leaves of gum tree (*Eucalyptus*) are alternate and usually pendulous, whereas those on young seedling shoots of many species are opposite and horizontal to the erect axis. The seeding leaves of ash (*Fraxinus*) are simple, but the mature leaves are pinnately compound. Because the seedling stage is the most critical in a plant's life, the characters present during this period surely have survival value. The needlelike seedling leaves of red cedar (*Juniperus*) and arborvitae (*Thuja*) may carry on more photosynthesis than the small scalelike leaves on the mature branches. The biosystematist studies these developmental differences in the garden and greenhouse, observing them at all stages of the life cycle. Developmental variation has been primarily of interest in understanding morphology and evolution.

Environmental Variation

Plants are quite variable, and some specie may very by altering their pattern of growth in response to environmental differences. *Phenoplasticity* (change in appearance) is caused by such variables as light, water, nutrients, temperature, and soil. A classic example of a type of water-controlled variation known as *heterophylly* may be found among some species of the aquatic plant arrowhead (*Sagittaria*). These plants have arrow-shaped leaves when emergent and long strap-shaped leaves when submerged Similarly, the aquatic water butttercup (*Ranunculus aquatilis*) has dissected submerged leaves, but on the same stem will have lobed emergent leaves.

Other environmental influences may produce variation in the sizes of many annuals. Chickweed (*Stellaria media*), for example, will reach only 3 centimeters in height in poor, dry soil but will exceed 20 centimeters in rich, moist soil. Systematists must detect environmentally induced variation. This is usually done by growing cloned (genetically

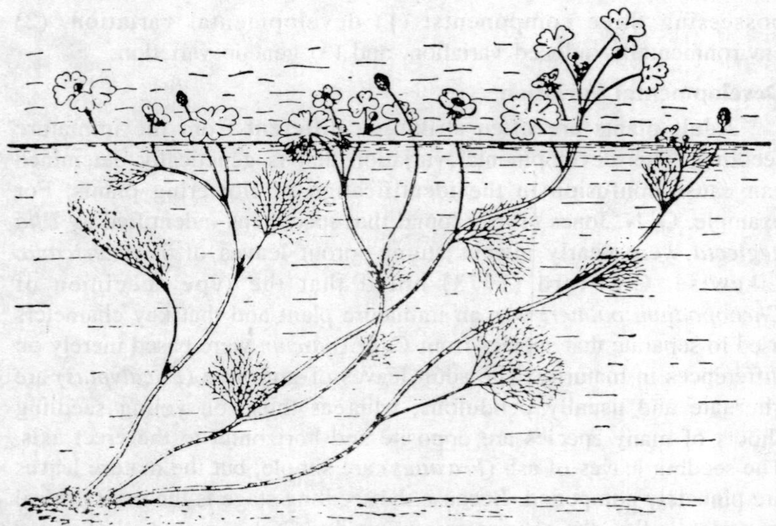

Fig. 6.1. Habit sketch of a water buttercup (Ranunculus aquatilis) with submerged and emergent leaves.

identical) plants under several different environmental conditions. When experimenting with wild plant material, the phenotype (the appearance of the plant) is usual rather plastic when grown in different environments. Characters that are easily modified are size of plant and individual organs, and number of branches, leaves, or flowers. Those characters that are less easily modified are lobing, serration, and pubescence of leaves, and the size, shape, and number of floral parts. Plasticity may have survival value and contribute to the success of an organism because it allows the plant of adjust to its environment.

Genetic Variation

The third type of variation in plants is genetic variation (genotypic variation). Genotypic variation, which is heritable, has two sources: (1) mutation and (2) gene flow and recombination. *Mutation* is a transmissible change in the hereditary material. It is the ultimate source of variation in species and replenishes the supply of genetic variability. When broadly used, the terms mutation includes both genic or point mutations (nucleotide substitutions) and chromosomal mutations. By way of contrast, *gene flow* describes the exchange of existing genes among the populations within a species or, through interspecific hybridization, between populations of two distinct but related species. *Recombination* brings together via meiosis and fertilization the novel genes or alleles carried by different individual and forms a new single genotype thus multiplying the possible number of genotypes in a population. Although mutation is the ultimate source of genetic variation, gene flow and recombination provide an immediate source of variability in sexually reproducing plants.

Gene mutation

A gene mutation is an alteration in the sequence of nucleotides that brings about a change in the action of the gene. In higher plants, genic mutations are known to affect many features of plants, for example, the hairless peach or nectarine. The role of genic mutations in evolutionary change depends on several factors: (1) the amount of their effect upon the organism, (2) the adaptive advantage or disadvantage of the mutant individuals, and (3) the role of the mutant gene in population-environment interaction.

The environment in which a mutation occurs will have a great effect on its incorporation into the local population. Unless the environment is changing, mutations will most likely lower the level of adaptation of the species. If a species is adjusted well to its environment, small mutation might possibly allow it to inhabit new or changing environments, but drastic changes are almost certain to make it function poorly. Analysis of hybrids between two distinct ecotypes of *potentilla glandulosa* demonstrated that their features were controlled by a series gene called *multiple genes*. With multiple genes the hereditary differences in quantitative characters are uaually determined by two or more independent genes having similar and cumulative effects. Each of these local populations of *Potentilla* had become specialized, or adapted for living under one particular set of conditions. Adeptedess in *Potentilla*, therefore, was achieved by the accumulation or many small changes through time rather than by large mutations.

Chromosomal mutations

These includes: (1) *Polyploidy*, or multiplication or chromosome sets; (2) *aneuploidy*, or loss or gain of gain of chromosomes from the set; and (3) gross structural changes in the chromosomes themselves. Structural changes may be any of the following : deficiencies or loss of segments from a chromosome; duplications of chromosome segments; translocations, or interchange of chromosome segments among nonhomologous chromosomes; or inversions of segments of chromosomes. Structural changes in chromosomes do not produce new genes but function in providing unique gene combination, new gene arrangements, and different linkage groups.

Gene flow and recombination

The movement and exchange of genes between breeding populations is described as *gene flow*. *Recombination* is the result of new combination of previously existing genes. Gene flows and recombination involve genes already existing in nature. These processes are the main, immediate sources of variability in nature and are brought about by cross-fertilization between individuals, populations, and species. Recombination produces new gene arrangements by cross-fertilization and by the interchange (crossing over) of segment of homologous chromosomes followed by independent assortment of the chromosomes at meiosis.

Many genotypic variations in a populations may be due to recombination of genic differences that have existed in the population system for many generations. Of major importance is that the potential for this variability is already present in the populations. Biochemical polymorphisms have clearly demonstrated that existence of great stores of variability in natural populations of various organisms. However, as Raven (1980) has pointed out, recent conceptual and practical advances have suggested that severe limitations are imposed on gene flow as it actually occurs in nature. Work over the last several decades has fully established the fact that local populations of species are highly adapted to local conditions and only rarely interbreed with other nearby populations. This suggests a limited short term evolutionary significance for gene flow, but the rare instances of interbreeding may have considerable evolutionary significance in the long run, particularly when the environment is changing.

Reproductive systems

The method or methods by which a plant reproduce itself are inseparably connected to the patterns of variability found in the species. Reproduction is an obvious necessity for the perpetuation of a plant through time, to provide the means for an increase in numbers, and for the colonization of new territories. The sexual process of cross-fertilization, involving meiosis and union of gametes, is recognized as a mechanism for promoting gene recombination. Recombination is, in turn, the chief source of hereditary variation in a sexually reproducing species and provided the raw material for response of the species to heterogeneous or changing environmental conditions.

In contrast, it is generally recognized that self-fertilization and vegetative or other nonsexual means of reproduction are advantageous in a more or less constant local environment in which the populations of a species have become well adapted (Grant, 1981). In natural populations of plants species, reproduction takes place by four common methods:

1. Cross-fertilization on *allogamy*, predominantly between neighbouring individuals that are presumably related.
2. Self-fertilization or *autogamy*, resulting in highly inbred offspring;
3. Vegetative propagation
4. Production of seeds whose embryos develop from maternal tissue without fertilization, a process called *agamospermy*.

The first two methods involve meiosis and fertilization and are thus sexual; in the last two, sexual reproduction is circumvented and reproduction is asexual. Of the two system of sexual reproduction, cross-fertilization promotes heterozygosity, resulting in a pattern of continuous variation with considerable diversity among individuals, similar to that in human population. Self-fertilization on the other hand promotes homozygosity, yielding more uniform populations comprised of many similar individuals. Vegetative reproduction may occur by fragmentation of individuals connected by stolons or rhizomes, by adventitious buds, by offshoots of corms or bulbs, or by the production of vegetative propagules arising within a flower or inflorescence. Widespread among perennials, vegetative propagation can produce clones of identical individuals that sometimes form large stands covering several acres, with all plants clonal in nature.

Agamospermy has appeared among numerous, often unrelated, families of angiosperms. The effect of agamospermy on the variation pattern is to produce series of groups of identical individuals that differ from other such groups by rather minor characteristics. These asexual reproductive systems, collectively called *apomixis*, are considered in greater detail in the final section of this chapter. It is generally agreed that the mode of reproduction of flowering plants influences the observable variation patterns of both local populations and the entire species. Among the flowering plants the spectrum of breeding systems ranges from obligate outcrossing at one extreme to extreme to almost complete autogamy at the other. Many species simultaneously employ two, three, or all of the four reproductive systems just described. It follows plants may be relatively uniform or highly variable and generalizations about species as a whole are difficult to make.

Natural Selection

In order for evolutionary change to occur, there must be a source of genetic variation that is mutation or gene flow and recombination and

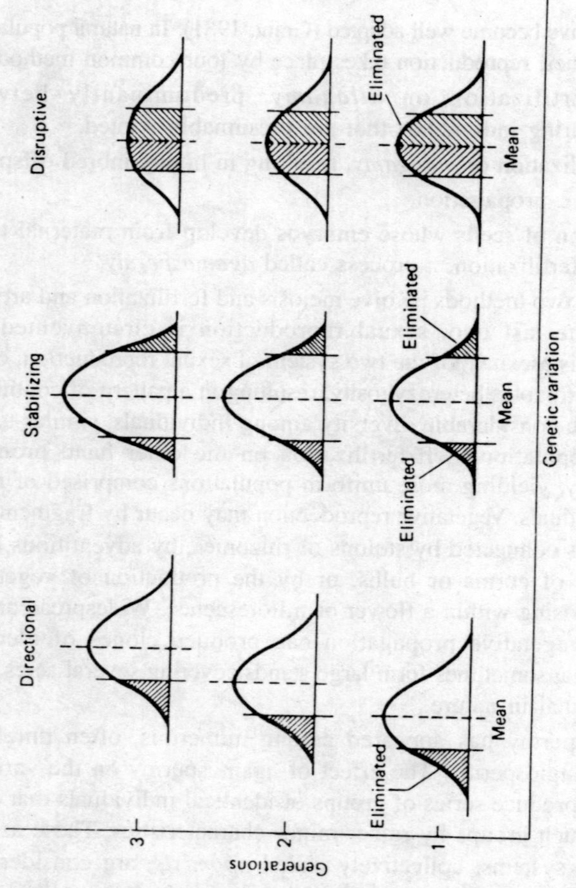

Fig. 6.2. The three kinds of selection: directional, stabilizing, and disruptive. Directional selection operates in a changing environment, stabilizing selection in a stable environment, and disruptive selection in a mosaic environment.

a driving force. The prime force is called *natural selection*, and it is the most important of the evolutionary forces or processes. The consequence of natural selection is adaptation, or adjustment to the environment. Natural selection is the mechanism by which populations become modified in response to the environment. Natural selection operates by differential reproduction of certain genotypes. The favoured genotpes are those that have higher reproductive success and consequently contribute a disproportionate share of individuals to the succeding generation. The adaption of plant that permit them to live successfully in given environments generally involve a combination and coordination of characters.

Natural selection establishes and preserves favourable gene combinations ranging from flower colour to changes involving the habit of the plant. Evolutionists have described three kinds of selection: (i) stabilizing, (ii) directional, and (iii) disruptive. These are determined by the organism's environment. Stabilizing selection favours "normal" individuals-that is, it eliminates variants, or deviations from the norm-and operates in a more or less constant environment. Directional selection operates in a unidirectional changing environment, that is changes from a moist to dry climate following a modification of the weather pattern. Divergent selection pressure in a heterogeneous environment produce disruptive selection. This breaks up a uniform population. An example is the formation of ecotypes adapted to their own local environmental conditions.

Random Events

Theoretically, in the absence of natural selection, gene frequencies will remain relatively constant in natural population. However, gene frequencies may fluctuate because of the random sampling of genes transmitted in each generation. Thus effect is greatest in small, isolated populations. It may be verified by tossing coins; that is, a few tosses may give a disproportionate number of heads or tails, but in a large sample the number of heads and tails will be almost equal. Random change in gene frequencies resulting from sampling error between parents and their offspring is known as *genetic drift*. In small populations, genetic drift may play a role by permitting novel genotypes to become more frequent than selection would ordinarily permit.

The overall importance of genetic drift in the evolution of flowering plants has long been debated, but it undoubtedly occurs in nature. It is difficult, if not impossible, to find good examples of drift in natural populations because of the interaction of drift with natural selections. Some of the best examples are with human blood group, but even here recent evidence indicates selection may be involved. Minor recial variation can be found in a number of plant species that have a population system consisting of a series of small, colonial populations in environmentally diverse areas (Grant, 1981). Certain species of *Cupressus* in California, for example, exit in a series of small, isolated stands. Certain of these have plants that can be recognized as unique to the particular site.

The *founder principle* is a random event related to the chance distribution of individual seeds or propagules that colonizes new areas. For example, a colonizing seed that arrives on an island or some other isolated area obviously cannot bring a complete sample of the genetic diversity of that species to the new area. Simply by chance, that seed could carry some genetic variant and establish a new population distinctly different from the original populations. The gene frequencies of a popuolation system may also be altered by a *genetic revolution* associated with a sudden reduction in the size of a population. By chance, the few survivors may possess system of variability different from those found in the original population. There has been an unresolved controversy regarding the adaptive nature of much biochemical or molecular variation. One school of population genetics holds that gene frequency changes in biochemical traits have occurred solely through random processes.

Classical evolution theory has demonstrated beyond any doubt that the basic mechanism for adaptive evolution is natural selection acting on variations produced by genetic changes. Our understanding has been greatly enriched by data from molecular biology, which has revealed many new and unexpected details about the genetic material of living organisms. The most surprising new information from molecular studies is that the great majority of nucleotide changes and hence of nucleotide variability in populations are selectively neutral or nearly so. Those who favour random processes, or the *neutrality theory* of protein evolution, generally agree that the evolution of most morphological and ecological traits is governed by natural selection, but they suggest that the evolution of most proteins, and the genes coding for them, is for the most part due to chance.

The neutrality theory of protein evolution assumes that a large proportion of all mutations is deleterious and that genes resulting from these harmful mutations are kept at low frequency by natural selection. On the other hand, a certain percentage of the naturally occurring mutations is adaptively equal, and the mutations do not affect the fitness of their carriers. Thus, the latter are not subjected to natural selection. According to the neutrality theory, evolution at the molecular level consists of a gradual replacement of one amino acid sequence with another sequence that is functionally equivalent to the first. The neutrality theory also predicts that molecular evolution takes place over long time periods at relatively constant rates (similar to radioactive decay). Therefore, it follows that the degree of protein differentiation among species is indicative of their phylogenetic relatedness. For example, two species with similar is indicative of their phylogenetic relatedness. For example, two species with similar cytochrome *c* amino acid sequences are more closely related to each other than they are to a third species with a very different cytochrome *c* amino acid sequence.

When such data are related to the paleontological record, this so-called molecular clock can be used to measure the time of occurrence of events in a phylogeny and greatly increases knowledge of the evolutionary record. Selectionists adhere to the theory that mutant substitutions in nucleotides must be adaptive and are caused by natural selection. They suggest that protein polymorphisms, as revealed by comparative studies of protein and DNA sequences, are adaptive and suggest further that the variation is maintained in species as some of balancing selection acting as a buffer against the environment.

Catastrophic selection, whereby an entire population is suddenly eliminated by an environment extreme such as drought, except for one or more exceptionally adapted individuals, may provide the conditions necessary for the establishment of a distinct population. Evidence for catastrophic selection has been reported by Lewis (1962) based on observation that ecologically marginal populations of *Clarkia* (Onagraceae) have become extinct during dry periods but is in one populations a rare genotype, marked by flower colour, survived. By elimination of the parental population, catastrophic selection isolates the few survivors into small distinct eco-geographic areas. Raven (1964) suggested that since marginal populations are very likely to be growing on some soil type unusual for the species as a whole, catastrophic selection, caused by drought in arid regions, may be a potent force in the creation of new edaphically restricted endemics regions such as the southwestern United States.

Variablity in Populations and Racial Differentiation

Many years of experimental work have clearly demonstrated that local populations of species of flowering plants are highly variable. This variation is in morphological features and in cryptic differences such as allozymes, chromosomes, natural plants products, and so on. Such local populations are often called *ecotypes*. For present purpose, an ecotype can be defined as a local pupulation or a series of populations adapted to particular ecological conditions. Both *clinal variation* (gradual changes in particular characters) and mosaics of distinctive ecotypes may be recognized.

The type of pattern depends on the geographical and ecological distributions of the species concerned. Forest trees, prairie grasses, and other plants found over large areas where environmental conditions change gradually tend to have clinal variation. However, plants occupying territories broken into mosaics of sharply distinct ecological conditions tend to have *mosaic variation*. Mosaic patterns are often related to soil and moisture conditions. *Experimental cultivation* is a major technique fro exploring phenotypic plasticity of species and for detecting differences between ecotypes. In the early middle 1800s A. P. de Candolle advocated cultivation of critical forms side by side. In middle 1800s the French botanist A. Jordan used comparative culture to study many species native to France. Finding much ecotypic variation, Jordan spilt many good Linnaean species. For example, he divided *Erophila verna* into 200 entities which are now refered to as "Jordanons." This was the ultimate in splitting. In the 1920s G. Turesson pioneered the use of experimental cultivation to show the genetic differentiations of populations.

Turesson coined the word *ecotype* for genetically adapted ecological adapted ecological races and proposed the terms *genecology* for studies dealing with the ecology of ecotypes. Turesson investigated servers common species of plants from southern Sweden that grow in diverse habitats. When samples of these species were grown in a uniform garden, Turesson found they differed in a number of features, indicating genetic differentiation. He was able to demonstrate climate and edaphic (soil) ecotypes in a number of herbaceous species. Shortly afterward at Edinburgh, J. W. Gregor, working with *Plantago*, was able to demonstrate an *ecocline*. This shows genetically based gradual changes in variation of a species, correlated with an observable gradient in environment conditions. The classic work in the United States was by a team of scientists: Jens Cleusen, a cytologist; David Keck, a taxonomist; and William Hiesey, an ecologist. In the 1930s they established experimental plots at three locations in California which differed greatly in climate and altitude.

One experimental plot was at Stanford University at an elevation of 10 meters; another at Mather on the western side of the Sierra Nevadas, at 1400 meters; and the third at Timberline at an elevation of 3000 meters, near the summit of the Sierra Nevada. For their studies, they selected perennial plants that grew along the transect from sea level to alpine conditions. To facilitate their work, they selected plants that grow in clumps. These could be divided to produce genetically identical clones that were transplanted into each of the three gardens. Thus, genetically identical plants were compared in the differing environments of each location. Their carefully documented work clearly demonstrated that local populations are distinctly adapted to their own environments.

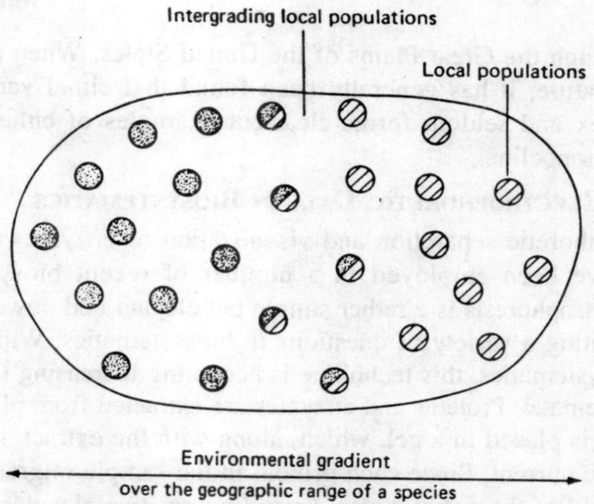

Fig. 6.3. A diagrammatic drawing of the range (large circle) of a species and its local populations showing geographical differentiation in morphological and ecological features and intergrading local populations. A species is generally not continuous throughout its range about occurs in local populations in suitable habits.

The experimental work of the California transplant experiments paved the way for much recent genecological work and had great influence on plant taxonomists. Most species are subdivided by their distributions and breeding patterns into small local populations. This provides an ideal situation in which natural selection may operate on the genetic variation without interference from gene flow from nearby populations. These local populations may differentiate physiologically, morphologically, ecologically and geographically and eventually form genetically different ecotypes. Such ecotypes or races may be the basis of a taxonomic subspecies or variety. Pure races in the sense of homogeneous population systems do not exist in nature. Instead, geographical and ecological gradients are found in the variation pattern.

Races normally intergrade if they continuously inhabit territory that provides intermediate ecological zones. If races are to be described and given names, they are usually delimited by sharp discontinuities in their variation pattern. Where the habitats of ecotypes are not distinct but form an environmental continuum ecologically adapted species exhibit a continuous range of variation termed a *cline* Two types of clines have been recognized: (1) an *ecocline*, formed along an ecological gradient, such as from a seacoast beach to the adjacent inland area; and (2) a *topocline* situated along a geographic gradient such as a north/south transect through the Great Plains of the United States. When carefully studied in nature, it has generally been found that clinal variation is very complex and seldom forms clear-cut examples of either a pure ecocline or topocline.

Electrophoretic Data in Biosystematics

Electrophoretic separetion and visualization of enzymes and seed proteins have been employed in a number of recent biosystematic studies. Electrophoresis is a rather simple but elegant and powerful tool for investigating a variety of questions in biosystematics. Widely used in animal systematics, this technique is becoming increasing important in plant systematic. Proteins and enzymes are extracted from plant parts. This extract is placed in a gel, which, along with the extract, is subject to an electric current. Since each protein in the sample migrates in the electrical field of the gel at a rate depending on its molecular size and charge, it is possible to make its position visible by using stains that are specific for certain proteins or enzymes.

It is assumed that seed proteins and enzymes from different plants that migrate to the same place in electrophoretic gels, and which after staining yield bands similar in appearance, represent homologous proteins. Several types of biosystematic questions have been addressed by using seed protein profiles. One of the more common problems considered is determining the identity of the progenitors of taxa of hybrid origin. This approach is based on the assumption that certain proteins found in each of the supposed progenitors will also appear in the offspring. In a similar manner, seed protein profiles have also been used for testing hypotheses of phylogenetic relationships in a genus.

At the interspecific level, seed protein profiles have been used to document differences among plants that have been traditionally treated as morphologically based species complexes and as a measure of genetic variation within species. Several factors may limit the utility of electrophoretic data of seed proteins in biosystematics. In most instances, the genetic bases of the seed proteins compared on the gels are not known. For this reason caution must be exercised in any interpretation of seed protein profiles. Technical problems and interpretation of the location and intensity of the bands may also cause difficulty in using this technique. Enzyme data are significant because they allow the direct comparison of the products of individual genes.

It has been well documented that different alleles at a single locus code for different molecular forms of an enzyme of an enzyme of different subunits in the more complex enzymes. These genetic variats can easily be made visible as coloured bands in electrophoretic gels. These bands represent localized areas of an enzyme catalyzing a particular reaction. In many instances the genetic basis of the enzyme banding patterns may be determined by crossing plants with known enzyme banding patterns and examining the progeny by electrophoretic techniques. From electrophoretic analysis of enzymes, allelic frequency data are obtained, and these data are employed to measure genetic similarities among species within a genus (congeneric populations), between population within a species (conspecific populations), or among subspecific taxa (Subspecies or varieties). Such comparisons have indicated that conspecific populations are very similar in their enzymatic allelic frequencies. For the most part, allelic frequencies of different subspecific entities are comparable to the values obtained for conspecific population of the same taxon.

Cross-pollination plants typically exhibit greater interpopulational variation of genes coding for enzymes than di self-pollinating species. Congeneric plant species exhibit lower similarities of enzymatic allelic frequencies than do conspecific or subspecific plants. However, this is not surprising since congeneric taxa are morphologically distinguishable, that is, divergent at genes determining morphological features and normally genetically isolated for a period of time. Allelic frequency data are useful for examining populations within a species or genus or perhaps for comparing closely related genera. On the other hand, at higher taxonomic levels such as families, it is usually not possible to compare allelic frequency data for technical reasons related to homology of the enzymes.

In biosystematics, one potential value of studies using allelic frequency data is to examine the cause or causes of problem species complexes in certain genera, that is, groups of species that are morphologically confusing. They may also be useful in investigating instances of hybrid speciation if the supposed parental taxa are extant and available for comparison. Enzyme and seed protein electrophoresis has value for approaching a number of problems related to phylogenetic relationships in closely related species. In many modern studies in plant biosystematics, electrophoretic evidence is taken into account as part of a broad based reevalution of evolutionary or systematic relationships in plant groups. Undoubtedly, its use will continue to increase in the future.

Isolation and the Origin of Species

In his now classic *variation* and *Evolution* in plants, stebbins (1950) noted that the most important principle derived from evolutionary studies was that evolution and the origin of species are not synonymous as implied by the title of Darwin's classic book. The present definition of species stress the importance of genetic and morphological continuity within species and the discontinuity among closely related species. Yet systematists realize that an array of genetic variation may exist within the species. Infraspecific taxa are based on incomplete morphological separation as reinforced by ecological and geographical isolation. These two factors are the keys to recognizing genetically based on distinctions. Therefore, taxa are maintained by limited gene exchange and adaptations to local environmental conditions.

Reproductive Isolation

Isolating mechanisms are phenomena that prevent interbreeding of closely related taxa. They are based on the following factors: (1) spatial distance, (2) environmental factors, and (3) reproductive biology. Isolating mechanisms are usually placed in one of two categories: *prezygotic* mechanisms, which prevent fertilization, or *postzygotic* mechanisms, which function after fertilization. A classification of reproductive isolating mechanisms is given in Table 6.1. In general, most instances of isolation in nature involve combinations of one or more mechanisms rather than a single mechanism.

Table 6.1. A classification of reproductive isolating mechanisms.

Prezygotic mechanisms
Geographical. spatial distance prevents interbreeding because the taxa do not occur in the same area.
Ecological. The plants live in the same region but grown in distinctly different habitats.
Seasonal. The plants occur in the same region but flower at different times.
Ethological. The plants are isolated from one another by pollinator behaviour patterns.
Mechanical. Pollination is prevented by structural differences in flower parts.
Postzygotic mechanisms
Hybrid Inviability. The hybrid zygote dies after fertilization or inviability occurs at some developmental stage before flowering.
Hybrid Sterility. First-generation hybrids are produced which flower, but the stamens about or the gemetes are infertile.
Hybrid Breakdown. First-generation hybrids are produced are normal and produce fertile pollen, but second- or later-generation hybrids are weak and have reduced fertility.

Geograophical isolation exists between two species whose respective geographical ranges are separated by gaps greater than the normal radius for dispersal of their pollen and seed. *Platanus occidentalis* and *P. orientalis* occur naturally in eastern North America and in a the Mediterranean region. They are morphologically distinct and effectively isolated by spatial distance. Yet, when brought together in the garden, they form fully fertile hybrids. When the ranges of two taxa do not overlap, as in the previous example, the species are said to be *allopatric*. When the ranges of two species coincide, the species are described as being *sympatric*.

Ecological isolation results from differentiation in habitat requirements. Although plants might live in the same region, interbreeding is prevented because they grow in different habitats. Two species of *Varnonia* occur in middle Georgia: *Vernonia noveboracensis* in sunny, moist roadsides and meadows; and *V. glauca* in shady, welldrained, matura hardwood forests. Where the two species are sympatric, hybrids are rarely found, although the cross may be easily made in the greenhouse.

Genetically controlled differences in reproductive features and habits may also block gene exchange. Sympatric species may be *seasonally isolated* and flower at different times. The ecological isolation between *V. noveboracensis and V. glauca* is reinforced by differences in flowering time. One species blooms in midsummer and the other on late summer.

Ethological isolation, or incompatible behaviour patterns, is a very important isolating mechanism in animals. When interpreted to include plant reproduction, it involves the flower-visiting behaviour or some insects and birds. Certain hummingbirds, bees, and hawkmoths as well as other insects and birds are species-specific. They have sensory perceptions enabling them to distinguish different flowers on the basis of odor colour, from, and so on. For example, hummingbirds are attracted to red flowers and hawkmoths to white flowers. Certain pollinators are flowers-constant and have instincts that cause them to feed preferentially on one species during a succession of feeding visits.

The advantage of flower-constant behaviour to the insect is a matter of efficiency and the floral mechanism learned, an insect can obtain more nectar and pollen in a shorter time by continuing to visit the same species. The structural difference in flower may prevent cross-pollination. This *mechanical isolation* occurs in families with complex floral mechanisms, such as the Asclepiadaceae and Orchidaceae. In both of these families, pollen is released in pollen sacs of pollinia of various sized and shapes. In the Asclepiadaceae, the proper pollinia must be inserted by the pollinator into slits in the gynoecium.

Likewise, the location of the pollinium on the insect, whether on the head or thorax, may be very important for successful insertion of the pollinium in the Orchidaceae. Interspecific pollination does not necessarily lead to the formation of successful and fertile hybrids. Many developmental steps occur between pollination and production of mature fertile hybrids. Developmental block may occur in various stages in the production of sexually mature plants: for instance, the pollen tubes fail to reach the ovules, the sperm fails to unite with egg, or in *hybrid inviability* the hybrid zygote dies, the endosperm disintegrates causing death of the embryo, the fruit fails to set, the number of seeds is reduced, seed fail to germinate, seedlings die, anthers about, or gametes are sterile (Grant, 1963). In *hybrid sterility*, first-generation hybrids are produced which flower, but they have aborted stamens, or more typically, infertile pollen.

Hybrid sterility may be in part developmental, involving the development of reproductive structures. It may also result from abnormal segregation of chromosomes at meiosis. Doubling the chromosome number may partly overcome problems with meosis, but it has no effect on developmental sterility. Segregational sterility is caused by translocations, inversions, and other chromosomal rearrangements, as well as by cryptic structural differences in the chromosome. Hybrids between two species may exhibit any degree of reduction in fertility. Some hybrids will be fully fertile, others semifertile, and still others completely sterile. The final barrier to gene exchange between species *hybrid breakdown*, occurs in second-or later-generation porogeny.

In some interspecific crosses in *Vernonia*, the first-generation hybrids are completely fertile, but the second-generation progeny exhibit continuous variation from fully fertile and vigorous to completely inviable. The causes of hybrid breakdown are not fully understood but are probably related to chromosome structure, gene content, and not fully understood but are probably related to chromosome structure, gene content, and also to cryptic structural differences in the chromosomes. These causes developmental and segregational difficulties which interfere with the physiology and the formation of gametes. Isolating mechanisms develop as a byproduct of evolutionary divergence. As parts of population system differentiate ecologically, morphologically, and biochemically, barriers to gene exchange begin to develop. For example, the presence of a particular pollinator in large numbers in one part of the range of a species might, through natural selection, channelize the adaptation of floral structures toward those best-suited for that insect.

The gradual accumulation of structural differences in the chromosomes in subpopulations will eventually lead to the origin or chromosomal barriers. The origin of isolating mechanisms is no different from any type of evolutionary change.

Speciation

Speciation is the evolutionary divergence and differentiation of a formerly homogeneous population system into the two or more separate species. Speciation requires genetic variations, which are universal in population systems, plus isolation. In the late 1860s the German Zoologist Wagner (1889) and the Austrian botanist Kerner (1896-1897) formulated the idea that geographical and ecological radiation of populations is paralleled by morphological and physiological differentiation. Kerner also suggested that new races originate in areas marginal to the main population. The American zoologist Mayr and other evolutionists have stressed the need for spatial distance as a prerequisite for speciation. This concept is designated the *geographical theory of speciation*. Basic to this idea is the principal that before populations can differentiate, there must be any divergent pressure of natural selection. These conditions generally must be met for any divergence, whether it is the formation of isolation mechanisms or that of races or species. Furthermore, the divergent element must be isolated enough so that the initial genetic divergence will not be overwhelmed by gene exchange with neighbouring population; for instance, on lead mine tailings there is selection for lead tolerance in grasses within 1 square meter, but speciation does not occur because of the proximity of nontolerant plants of the same species.

There are basically two modes of evolutionary change: phyletic evolution and divergent evolution. A change from one state to another state through time in which a species might evolve into something different from its ancestor is called *phyletic evolution*. In contrast, a single population system that differentiates into two evolutionary lines is called *divergent evolution*. As populations become genetically differentiated, isolating mechanisms also begin to develop. Partial reproductive isolation is commonly present between local population of a species because of spatial distance, season of flowering, pollination agents, ecological requirements, and so on. Differentiation of ecotypes in response to their environment reinforces the development of isolating mechanisms.

As an ecotype becomes specialized for its environment isolating mechanisms may build up reducing gene exchange. As divergence

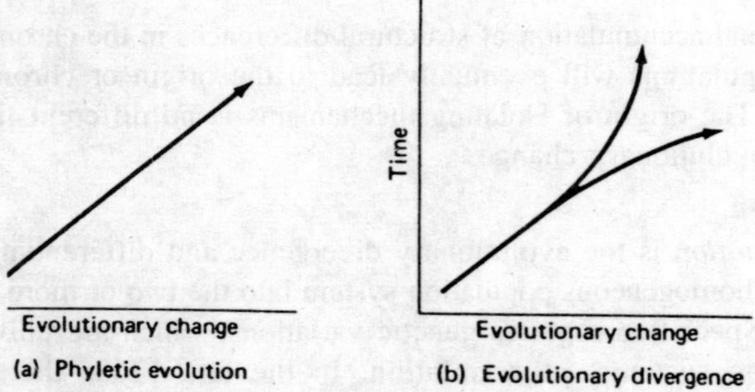

Fig. 6.4. The two modes of evolutionary change: (a) phyletic evolution and (b) evolutionary divergence. In phyletic evolution, the species moves in time from one state to another state, but in evolutionary divergence one species diverges through time into two species.

increases, the elements of the populations become less and less able to exchange genes. Divergence leads to isolation, and isolation promotes speciation; that is, one event reinforces the other. There has been unresolved debate about whether divergence without geographic isolation, or *sympatric speciation*, can occur. Sympatric speciation involves the differentiation of closely related population that coinhabit a geographical area- that is, are sympatric. The answer to this is not yet clear. Theoretically, the possibility if sympatric speciation seems more likely if the plant are self-fertilizing, because this would tend to restrict gene flow.

Although many evolutionists have minimized sympatric differentiation, it might happen under strongly divergent ecological differentiation, reduction of outbreeding, or structural changes in the chromosome. In one example, there was rapid (40 to 60 years) differentiation of populations of *Anthoxanthum odoratum*, an outbreeding grass, on small 20 by 35 meter plots in response to pH. Variation in chromosomes not correlated with ecogeography has been reported in *Lasthenia* in California. Not all evolutionists accept the classical geographic theory of speciation. Fro example, Raven (1980), in a recent discussion of speciation in plants, argues that few examples are available in which plants fit the classical model by the separation of formerly continuous population. Raven's assertions are based on many reports of severe restrictions on gene flow and the high level of adaptation of local populations. It has been observed in some plants, such as *Clarkia*, that morphological and physiological novelties constantly arise in a mosaic environment, providing raw material for adaptation of local populations and for evolutionary, novelties.

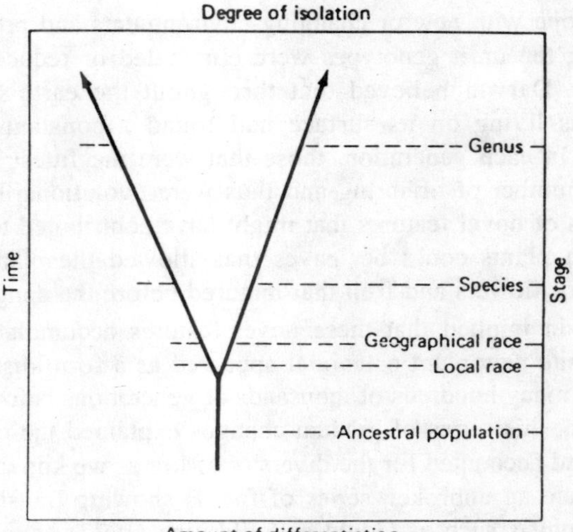

Fig. 6.5. States of divergence. As the divergence increases, the amount of isolation increases.

Evolutionists who support this view contend that reproductive isolation is not necessary for survival of these unique features. Furthermore, it appears from the available evidence that chromosomal rearrangements (such as translocations) are not necessary for the survival of evolutionary novelties. Apparently, local populations acquire a set of distinctive features through a complex variety of evolutionary mechanisms and then the adaptive novelties spread over certain areas and replace less adaptive genotypes. The spread of novel features is undoubtedly influenced by mode of reproduction, hybridization, or polyploidy, and by the environmental circumstances that confront the population. Thus, it is impossible to fit rigid concept of the geographic theory of speciation to the differentiation of populations of plants. For these reasons, Raven (1980) suggests that the mechanisms of speciation in plants may be highly compatible with the theory of punctuated equilibria.

Darwinian gradualism versus Punctuated equilibria

More than a century ago, Charles Darwin believed he had the answer to what has brought about the astounding diversity of plant and animal life. His answer was an eloquently simple one: Darwin suggested that if the fossil record were examined, it would show that today's plants and animals had emerged as survivors during a period of environmental change. The predecessors of the present-day flora and fauna were less able to cope with new or changing environments and produced fewer offspring; the unfit genotypes were eliminated or reduced by natural selection. Darwin believed that throughout the earth's history, the organisms living on its surface had found a constant struggle for survival. In each generation, those that were the fittest produced the greatest number of offspring and thus were evolutionarily successful. Examples of novel features that might have contributed to the success of certain plants could be leaves that allowed the plants to endure drought, or flowers and fruit that matured before the danger of frost.

Darwin implied that these novel features accumulated gradually until a quite new plant or animal appeared as a form distinct from its ancestors many hundreds of thousands of generations before. According to Darwin, this series of gradual changes explained the origin of new species and accounted for the diversity of life as we know it. Unable to demonstrate an unbroken series of fossils showing transitional forms, paleontologists such as Gould and Eldridge (1977) have attacked the concept of Darwinian gradualism and have attracted much attention in the popular press. They suggest that gradualism should be replaced by what they call *punctuated equilibria*. That is to say, the history of organic evolution is not one of gradual unfolding, but rather a story of homeostatic equilibria that are disturbed only rarely (relative to the length of geologic time) by rapid and episodic events of speciation brought about by catastrophic worldwide events such as the rapidly expanding effects of a disrupted ecological food chain caused by droughts, flooding, temperature changes, meteorite impact or increased ultraviolet radiation.

The survivors of such catastrophes might have novel features that in the face of relaxed competition enabled new species to emerge and survive. Where recent climatic changes have occurred, where mountain uplifts have taken place, where new oceanic islands have arisen, or where altered environments have occurred, rapid speciation of plants has been observed and documented. The association of rapid speciation of plants with such appears compatible with the idea of punctuated equilibria.

Hybridization

The role of hybridization in evolution has been a controversial topic in systematics. Linnaeus suggested a hybrid origin for several plants species. In the late 1800s and early 1900s, it was believed haybridization could be the starting point for new species. Until the early 1950s many floristic botanists were reluctant to recognize hybrids in nature. Today plant systematists regard interspecific hybrids to be common in many genera. Sunflowers (*Helianthus*) and oaks (*Quercus*) have species that are difficult to identify, because there has been interspecific hybridization blurring species lines. Natural hybridization is more frequent in perennial plants than in annals. The perennial nature ensures that occasional hybridization between species will not interfere with their existence, since the genotypes are preserved by the perennial habit.

Therefore, there has not been strong selection against hybridization in perennials. Interspecific hybrids are more common in plants than in animals. There several reasons for this: (i) ethological isolation is more effective with animals; (ii) zoologists may have overlooked some hybridization because animals are not as easily collected as plants; and (iii) because of the complex structure and sequence of development of animals in relation to plants, hybridization is more likely to cause problems in successful gene interaction and development in animals than it is in plants. First-generation hybrids generally have characteristics that are intermediate between the parental species. Such hybrids represent crosses between normally well-adapted parents.

First-generation hybrids are also frequently intermediate in their habitat requirements. They are usually at a competitive disadvantage unless the environment is changing or unless intermediate habitats are available. Two sympatric species may be ecologically isolated, with one species growing in a wet site and the other in a dry site. They may hybridize when intermediate habitats are located between adjacent populations of the two, such as along a wet to dry transect. An example of this occurs between *Vernonia fasciculata* and *V. baldwinii*; another between *Lysimachia quadrifolia* and *L. terrestris*. Following natural hybridization, first-generation hybrids typically backcross to one of the parental species. There are several reasons for this. More plants with the parental genotype are present, increasing the changes of crosses with a member of the parental group.

Fertility of the hybrids may be lower, so that they produce less-viable pollen than do the parents. Therefore, the most common effect of interspecific hybridization is successive backcrossing of hybrids with the parents. This results in the reversion of hybrid offspring toward the prental types. Backcross hybrid are usually more successful because they tend to approach the adaptive peaks of their parents. Backcrossing may result in the movement of genes from one species to another via the hybrids and backcrosses. This is referred to as *introgressive hybridization*. Introgressive hybridization may lead to changes in the variation pattern because of gene flow. The overall effect of hybridization depends largely on the environment where it takes place. In stable environments, it is likely to have no important effects. In unstable, new, or changing environments, hybrid segregates may be better-adapted than are the parental taxa.

Because hybrids are often associated with disturbed habitats, Edgar Anderson (1949) coined the phrase *hybridizaed habitat*. Because of hybridization, plant systematists must often deal with the phylogeny of reticulate networks rather than the dichotomously branched or phyletic tree that is the typical model commonly presented for presented for various groups of animals. This in itself, can make botanical classification more difficult than zoological classification. Additionally, the existence of numerous hybrids between many species of flowering plants can cause practical problems in identifying plants because the hybrids are not referable to any on species. Sometimes only an occasional hybrid between two species is found, but in many other instances hybridization between two, three or more species has become so prevalent that the majority of the individuals in a population appear to be of hybrid origin. Working especially with animal groups, naturalists and population biologists formulated some years ago what is called the *biological species* concept. That view suggested that individuals that can interbreed and form fertile hybrids constitute a single species, while those that cannot interbreed or that form sterile hybrids represent separate species.

In zoology, the use of sterility of progeny as the basis for the biological species concept can be useful in delimiting species. The sterility of the mule, the offspring of a horse ´ donkey cross, is a well-known example. With a long history of extensive practical knowledge about hybridization in plants, botanists have generally viewed with misgivings the efforts to apply the biological species concept to plants. Typical so-called biological species are found among flowering plants, but they represent only one type of population organization in angiosperms. Departure from typical, well-developed biological species is common among plants and should not be considered abnormal. In many genera, morphologically distinct species freely interbreed both in nature and tin the garden. For this reason there is little adherence to a strict biological species concept as a definitionn of a species among higher plants. In recent years, botanists carrying out evolutionary studies have developed an increased appreciation of the role of hybridization in the adaptation and evolution of plant population.

Polyploidy

The multiplcation of the chromosome set or genome, called *polyploidy*, is one of the most widespread and distinctive cytogenetic processes affecting the evolution of flowering plants (Stebbins, 1971a). Polyploidy refers to a numerical series based on multiple sets of chromosome. In a diploid species, the basic haploid set of chromosomes, the genome, is $x = n$, but in the polyploid species, n is a multiple of x. For example, *Chrysanthemum* (Compositae) has closely relateed species that differ in chromosomes, number to form a *polyploid series*, with the basic set of chromosomes of $x = 9$ in the diploid ($2n = 2x = 18$) species, but among the polyploid species there are tetraploids ($2x = 4x = 36$) hexaploids ($2n = 6x = 54$), and decaploids ($2n = 10x = 90$). *Aneuploidy* refers to instances where the chromosome number is not an exact multiple of the basic number. *Clarkia* (Onagraceae) has species with an aneuploid series of $n = 5$, $n = 7$, $n = 8$, and $n = 9$ chromosomes. Such variation in chromosome number among plant taxa has provided an important source of information for biosystematics. For a more complete review of the subject, the reader is referred to Grant (1981) and Stebbins (1971b). Around 40 percent of the species of flowering plants and a higher percentage of the ferns are polyploids.

The morphological and physiological effects of polyploidy may include an increase in cell size, a reduction in number of cell division, changes in shape and texture of plants parts, lowered fertility, and physiological imbalance. Traditionally, polyploids derived from on especies are known as *autopolyploids* and those from two parental species as *allopolyploids* or *amphiploids*. However, the sitution in nature is much more complex than this, and natural polyploids often have various combinations of chromosomes. Unreduced gametes (2n), resulting from meiosis failure, may combine to yield a 4´ polyploid (Harlan and de Wet, 1975). Polyploidy may also arise from mitosis failure diuring vagetative growth that later could develop a polyploid flower. Like hybridization, polyploidy is often related to the growth habit. Annuals tend to have the lowest frequency of polyploidy, and herbaceous perennials have higher frequency than woody plants. Interspecific hybridization is often involved in groups having polyploidy, with crossing taking place between parental species and hybrid progeny at various ploidy levels, blurring distinctions between taxa.

Apomixis

The replacement of sexual by asexual reproduction is referred to as *apomixis*. Many groups of plants, especially those wholly or partially sterile because of complex patterns of polyploidy or aneuploidy, may survive primarily because of their ability to reproduce asexually by apomixis. Apomixis is often associated with polyploid, but it does appear in a few diploids such as Citrus.

Many apomictic plants are adated to microhabitats, as was clearly demonstrated by Solbrig and Simpson (1974) with dandelion. Some groups of facultative apomicts are capable of both sexual and asexual reproduction, for example, *Poa* and *potentilla*. Some have geographical variation in sexual and asexual population, for example, *Taraxacum* and *Bouteloua*. With apomixis, irregular meiosis and sexual sterility are no barrier to seed production, and triploids, pentaploids, and other odd-chromosome-number apomictic clones may be found. The close association between agamospermy (embryo production without fertilization) and hybridization has been recognized for many years.

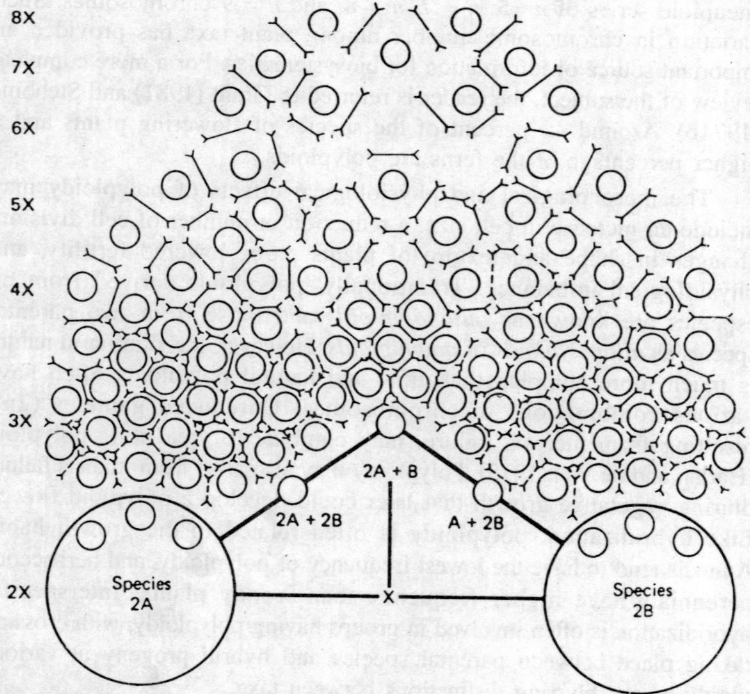

Fig. 6.6. The structure of an agamospermic complex with both hybridization and polyploidy (2X, 3X, and so on). The small circles represent morphologically distinct, isolation, apomitic populations that provide a continuum of variation between the two parental species.

Agamospermy is most commonly found among complex hybrid derivation or species hybrids. Typical agamospermous groups contain large number of poorly demarcated populations often termed "microspecies."

The numerous, slightly distinct, microspecies are so interrelated morphological as to make taxonomic classification virtually impossible. The variation pattern is similar to that observed in complex hybrid swarms and show a continuum of variation between the two parental species. Since the population structure is based on hybridity, agamospermy permits continued reproduction of the successful genotypes of the presumably well-adapted "microspecies." Classic examples of such situations occur in two taxonomically difficult genera of the rose family *Rubus* and *Crataegus*. In both these genera, it is difficult to organize satisfactorily the variation pattern into the taxonomic categories provided by the International Code of Botanical Nomenclature. Vegetative propagation by buds, bulbs, rootstocks, and so on, is widespread among many perennials. Vegetative propagation in a hybridizing plant groups permits multiplication of a well-adapted genotype that might disappear in the course of sexual reproduction. It has been found that many woody and herbaceous perennials can expand through clonal growth. Hybrid, vegetatively reproducing clones are known in *Quercus*, *Iris*, *Populus* and many other genera. It is generally agreed that clonal, vegetatively reproduced complexes generally exhibit a simpler variation pattern than do agamospermous "microspecies."

Counting Chromosomes by the Squash Technique

Freshly collected flower buds are used for studies of chromosomes at meiosis. Many beginners tend to collect bud material that is too old. For most plants, an easy way to tell when bud material is past the appropriate stage is to examine the colour of the anthers. If the anther sacs have already turned yellow, the pollen grains have been formed and the material is too old. When collecting bud material, collect an entire inflorescence, including buds just forming to those with anthers already turning yellow The buds are dropped into a killing solution. Small, wide-mouth, screw-top vials are best for collecting buds in the field. *Voucher specimens must be made of the material collected.* Root tips for studying mitosis are usually collected from newly germinated seeds or from rapidly growing roots of pot-bound plants. A modified *Carnoy's solution* is used by many botanists as a killing and fixing solution:

For meiosis (4 : 3 : 1)

4 parts chloroform

3 parts absolute alcohol

1 part glacial acetic acid

Bud Squashing

The bud material in the 4 : 3 : 1 solution be poured into a petri dish. The material need not be washed in water or other solutions. With tweezers or needles, select a promising bud and place it on a glass slide. If the buds are small , select several of approximately the same size. Add 1 or 2 drops of *acetocarmine* to stain the chromosomes. To make acetocermine, boil 2 to 4 grams of carmine dye in 100 cubic centimeters (cc) of 45 percent acetic acid for several minutes, and then cool and filter. The resulting saturated solution is diluted with 45 percent acetic acid to obtain the required strength. For best results, use a 1 percent acetocermine solution for meiotic material. Squash the buds with the smooth end of a glass rod.

Observe the material under the lowpower objective of a compound microscope, and make a rapid search. It is quite easy to see both tetrads and diads and even chromosomes in pollen mother cells (PMCs) under low power without a coverslip. If diads are observed, then some PMCs will be found to have metaphase divisions and perhaps a few cells at late prophase. If, in superficial observation, the material is found to be either too young or too old, then the macerated material is wiped off immediately and a new bud is selected and placed in the acetocermine, and the whole process in repeated.

It is quite possible to look at 20 to 30 unpromising buds in a period of 15 or 20 minutes. If the bud material is adequate, the proper meiotic stages should be found soon. If metaphase figures or diads are observed, remove the slide and macerate the tissue with a little more care. Using a pair of flattened needles, remove any remaining large floral tissues. Place a No. 1 coverslip over the squashed material. A blotter is placed over the top of the cover-slip (to absorb the extruded stain following pressure), and with the thumb, pressure is applied so that the cells are flattened. The degree of pressure that must be applied to the cover-slip varies and must be learned by experience.

Interpretation

Interpretation is the greatest difficulty in counting chromosomes from meiotic material. Beginners may feel frustrated when typing to make an accurate count from meoitic cells. The bivalents (paired chromosomes) are sometimes torn apart, and it is often difficult to determine the exact number present in any given cell. Another complication is that some species may be apomictic and show asynapsis with the chromosomes existing as univalents. If one chromosome is superimposed on another chromosome, it may appear as a single unit.

Chromosome that are superimposed will be about twice as dark as those in which the arms are separated and flattened in single plane. Fine-adjustment manipulation can reveal the true nature of such chromosomes. Some species will form chains or rings of four and six chromosomes at meiosis. These formation are best detected at diakinesis when their ring structure becomes apparent. At metaphase it is much more difficult to detect rings of four, since the extreme shortening will often make them appear as a single large bivalent. Chromosome fragments can usually be recognized by their very small size and their irregular pattern in the cell. The best stage at which to count chromosomes in PMCs of nearly all species is that of diaknesis. While the chromosomes at diakinesis do not take such a dark stain as those at metaphase, they are widely separated and do not tend to stick, and their bivalent nature can be easily ascertained. However, in some species, particularly those with small chromosomes, metaphase is the best stage for chromosome counts.

Preparing Squashes of Plant Root Tips

Seeds or rootstocks of plant material collected in the wild should be potted and placed in a lathhouse or greenhouse until the plants become established in pots. Actively growing plants should be watered heavily in the late afternoon on the day before collecting the root tips. If the next day is bright and sunny, remove actively growing root around noon of the day after the heavy watering. The actively growing root tips should be white, and usually are translucent just back of the root cap. The time when the root tips are ready for removal will depend on the time of active mitosis and varies with plant material, the time of year and probably on other factors as well. The roots may be placed directly in fresh (3 parts 95 percent ethyl alcohol: 1 part 45 percent acetic acid) fixative and fixed overnight, or placed in a 0.1 percent to 0.5 percent (usually 0.2 percent) colchicine solution for a three-to five-hour pretreatment to shorten the chromosomes.

After the pretreatment, the root tips are then fixed overnight in the 3 : 1 fixative. One of the difficulties encountered in preparing root tip squashes is securing the satisfactory separation of cells, or of single layers of cells. Excessive numbers of cells or cells stacked on one another will often obscure the details of division. Various chemicals or enzymes have been used for maceration to secures separation of cells.

One relatively simple technique is the use of hydrochloric acid (HCl) to soften the root tips. In this technique, the root tip is removed from the fixative, placed in concentrated HCl for three minutes and then washed several times in distilled water. Correct concentration, times, and procedures with the HC1 will very according to the nature of the plant material. Accordingly, dilute solutions of HC1, for comparatively brief periods, are advisable for some material. Following the softening treatment, remove about 1 mm of the growing point of the root tip and place it on a slide in a drop of 1 percent aceto-orcein (prepared from Gurr's Orcein following the same method for preparing acetocarmine). Use a dissecting needle to tear up the growing point of the root tip thoroughly. This will take some time to insure that no sizable bits of tissue remain. If thin pieces of root epidermis resist tearing, remove them from the drop of acetoorcein stain and discard. Apply a cover slip; with sometimes additional 1 percent aceto- orcein stain may be needed under the cover slip. With the cover slip in place, heat the slide by passing it back and forth through the gentle flame of an alcohol lamp.

It is important not to allow the acetic-orcein to boil, since this will ruin the cells; watch carefully for small bubbles, and stop heating at once; The heating of the cells (1) intensifies the staining of the chromosomes and (2) helps to clear the cytoplasm, giving maximum contrast of the chromosomes relative to the other cellular contituents. Following the heating process a blotter is placed over the top of the cover slip (to absorb the extruded stain following pressure), and thumb pressure is applied to the cover slip so that the cells are flattened. The degree of pressure needed to keep from breaking the slide and ruining the preparation varies and must be learned by experience. After squasing, additional stain may be needed, but add sparingly or the cover slip will float off the preparation. Examine the slide under the microscope and look fro single cells, or for layers of tissue that are mostly one-cell thick. If chromosome divisions can be seen, the slide should be sealed with paraffin or clear nail-polish. The division figures may be examined at once, or after an hour or so. A root tip squash is often clearer after several hours. Observation of the chromosomes in root tips is usually make from metaphase figures. Information can be obtained on both chromosome number and morphology by using the root tip squash method. An advantage of the root tip squash technique is that vagetative plant material can be used, while it is often difficult to obtain flower buds in the greenhouse, especially with tropical meterial or with perennials.

EXPERIMENTAL HYBRIDIZATION

Experimental crosses are used to help define and position taxa. Such experiments require time and greenhouse and garden space. Plants must be maintained daily. Six months to several years may be required to obtain flowering progeny. Some plants require environmental conditions that are difficult, if not impossible, to maintain in a greenhouse or garden. Other species flower for a brief period each year. The various species in a genus may be seasonally isolated, requiring timing on crosses. Synchronization of crossing may become critical in hybridization experiments. The problems are mentioned to provide information for planning, not to discourage one from performing crossing experiments. If collection of living plant material are not available, travel to the area where the plants grow may be required. Before the collecting trip, it must be determined where populations may be located and when they flower and produce seed.

Growth habit-that is, annual, biennial, perennial, or woody- will influence the methods used to gather living material. This information may often be obtained from herbarium specimens. Annuals are best started with seed. Rootstocks of perennials and cutting from woody plants are useful in propagation. Horticulturists who have worked with native plant material may give pointers for handing the plants. Care must be taken with living material to ensure survival during movement from the field into the greenhose and garden. Seeds should never be dried with heat.

It is usually best to cut a large plant back to ground level. This keeps the root and shoot system in balance, since most roots are lost in digging. Rootstocks can be washed, tagged, and placed in heavy plastic bags. Never allow bags of living plant material plant material to be directly exposed to the sun or to stay in a hot automobile. Voucher specimens should be collected form all population sampled. The plant material can be identified with a collection number. Seeds or plants obtained from other collectors may be given an accession number and recorded in a bound data book. One such system used is "85-33," for the thirty-third accession of 1985. The germination of seeds or establishment of rootstocks is most critical, and competent advice about these processes should be sought from horticulturists. Once the plants are established and flowering, the first step is to determine whether the taxa are self-incompatible, self-fertile, or apomictic. Individual plants should be isolated in a screened greenhouse or enclosed in insectproof cages. Bagging of individual flowers or flower clusters to prevent pollination often causes problems with plant disease because of high temperature and relative humidity. If seeds are not set in selfing tests, it is likely that the plants are self-incompatible, which can make crossing rather easy. If isolated plants set seed, they are either self-fertile or apomictc. This result should be followed with a progeny test. If the progeny are homogenous apomixis can be verified by detailed embryo sec studies.

Another approach might be to emasculate the flowers and to check for seed set. Self-fertile flowers pose difficulties in crossing experiments because they must be emasculated. The degree of difficulty depends on the size and morphology of the flower. Experimental crosses should be carefully planned, and F_1S, F_2S, and backcrosses should be made, if time permits. The experiments should be designed with controls available for comparison with the experimental individuals. When fruit from the crosses is mature, estimates can be made of seed set percentages in the hybrids versus the controls. Seed set, however, does not in itself indicate if the seed will produce a viable hybrid. Seeds from the hybrids should be planted and controls established to determine their viability.

The squash technique may be used to estimate sterility by checking chromosome pairing at meiosis in pollen mother cells. Another estimation of fertility can be obtained by staining pollen grains. For this, the pollen grains are stained fro 24 hours in a drop of 1 percent aniline blue in lactophenol on a microscope slide with a coverslip. Two hundred grains are counted and scored as either stained or not stained. The stained grains are used to estimate the frequency of fertile grains.

Local Population and Character Analysis

To study variation in plants, populations must be sampled and their characters analyzed. Considerations must be given to the objectives of the study, the proposed techniques, methods for obtaining the samples, the parts of the plants to be sampled, and the characters to be analyzed. Collecting a local population sample involves pressing and drying a number of whole specimens or the critical parts of individual plants. Although nor usually feasible, the best sample of a local population would be all the plants in the colony. Since this is not practical, a sample is taken that gives an estimation of the variation in the colony.

Local population samples consisting of 25 to 75 plants will yield an estimate of the variation within and between population and taxa. The number of plants sampled depends upon the variability and size of the populations. With little other than normal variation, 25 individuals will be adequate; but if two species are hybridizing, 50 to 75 individuals are required. Care should be taken not exterminate the colony. A sampling procedure should be used to ensure that the population sample is a random one representative of the local plants. This precaution is often difficult to implement in the field because the terrain or vegetation may restrict the mechnics of sampling. Collect only the parts of the plants essential to the study, along with voucher specimens. This saves press and storage space and eases the drying problem.

A voucher specimen is collected from each population. It is given a collection number, and data are recorded in the field book. Each local population sample receives a collection number. There are several basic approaches used to analyze local population samples. First, an intensive study may be made of one or two characters, especially those that show geographical or ecological variation, using a large number of population samples. Second, an analysis of the interrelationships of 8 to 20 characters on fewer samples may be performed. The samples may be compared by means of diagrams, statistics, or numerical methods. Each approach has its own advantages and disadvantages. With few characters and large number samples they may be analyzed with ease, but great care is required in selecting characters.

The use of many characters in the second approach helps with the critical problem of character selection. Interrelationships of characters can be shown, and it is sometimes possible to distinguish populations and taxa using a combination of characters. Visual presentations of data are helpful when communicating with others. The include polygons, bar diagrams, scatter diagrams, ideographs, line drawings, graphs, dice grams, and so on. Benson (1962) and Davis and Heywood (1963) provide excellent examples. Frequently, population data are visually presented in relation to a background map or physiographic profile in order to convey not only the features of the populations but also their geographical relationships.

7

PREPARATION OF HERBARIUM

Collection of plant specimens are essential for taxonomic research. They circumscribe species and document their variability, they are the prime sources for floristic studies, and they are vouchers for experimental investigation. Plant materials must be carefully selected, prepared, and preserved, since herbarium specimens become a permanent record for later investigators to examine. The taxonomist in the field observes plant locations and records habitat information. The variation within a single population or among populations of the same specimens in different environments may be measured. Observations may be made of patterns produced by hybridization or by changes in soil, moisture, slope, light, and so on. Features may be evident in living plants that are not easily observed in dried materials, for example, flower colour or fragrance. Species response to environmental disturbances such as fire, grazing, and clearing, may be noted.

There is no substitute for firsthand knowledge of plants as living entities though field observations. Student collections are fundamental to study and training in plant systematics. As a beginner's collection grows, methods change from those of a collector interested in natural history to those of a researcher interested in the taxonomic problems of a particular groups. Ideally, the best specimen for identification and research is an intact and complete plant. Attempts to identify a specimen form a single flower or leaf usually fail. Such specimens have little or no scientific value.

It is possible to collect entire plants of small annual species and some herbaceous perennials. However, one would not attempt to press an entire tree. Underground parts of herbaceous perennials such as rhizomes, root, and bulb should be collected. Representative leaves and reproductive structures are essential. The flowers, fruits, and seed of flowering plants are especially important, since most keys for identification use reproductive characters. With large herbs, shrubs, and trees, the different kinds of foliage will be helpful. Individuals should be selected that are representative of all phases of the natural populations. Insect-damaged plant material or monstrosities should be avoided. When pressed and dried, the specimen should yield the maximum amount of information concerning the living plant species and be representative of the population.

Pressing and Drying Plant Specimens

The materialise needed to press plant consist of five items: The press, straps or ropes, blotter, corrugate ventilators, and torn newspapers. The *press* may be purchased from a biological supply house or may be inexpensively constructed from 3/8-inch plywood cut into two 12 by 18 inch pieces to be used for either end of the press. *Straps* or *ropes* may be constructed from webbing with claw buckles or from window sash cord. The straps or ropes should be at least 5 feet long and are used to tighten the press. *Blotters* are heavy blotting papers that will absorb moisture from the plant specimen. They are 12 by 18 inches and may be purchased from paper sheet of corrugated 12 by 18 inch cardboard. They provide space for air passage through the press to remove water vapor.

The corrugations must run the short distance rather than the long distance of the cardboard. This is because the press is usually dried with the 18-inch side down over the heat source, and the 12-inch corrugations act as vents. The *torn newspaper* receives and contains the specimen while in the press and until the time it is mounted. A double- page sheet of newspaper is torn or cut in half and then folded over. The newspaper should be just slightly smaller than the press. Useful tools include a heavy -duty trowel for digging roots, pruning shears for clipping tough plants, and pole pruners for collecting specimens from tall trees.

Plant specimens should be pressed as soon as possible after they are collected. The best preserved specimens are obtained by using a field press. This is arranged to hold 100 sheets of torn newspaper, allowing the pressing of the 100 specimens. At the end of the day or by the next morning the specimens must be transferred (still in their own sheets or newspaper) to a drying press. By this time, the specimens will have relaxed and may be slightly rearranged to improve the quality of the specimens.

Table 7.1. Checklist of Field equipment and Supplies.

I. For both beginners and advanced students

1. Bound field notebook (for collection data)
2. Field press (complete)
3. Drying press (complete)
4. Digging tool (such as trowel, geological pick, bricklayer's hammer, dandelion digger)
5. Pruning shears (avoid cheap shears; Wiss or Seymour Smith give years of service)
6. Heavy-duty plastic bags (50-lb. fertilizer bags are excellent)
7. Newspaper (for wrapping bundles of speciemens and for pressing)
8. Plastic milk jug with water (for moistening specimens)
9. 10X hand lens
10. Pocket knife
11. Soft lead pencils

II. Additional items that may be needed fro advanced students and professionals

1. String taps
2. Plstic pot labels (for making waterproof labels fro living material)
3. Collecting vials and jars
4. FAA and Carnoy's killing and fixing solutions
5. Seed envelopes
6. Plasitc bags for living material
7. Highway, topographic maps; conty road maps; aerial photograps
8. Camera and film
9. Portable plant drier
10. Insect repellant
11. Pole pruner (for trees and large shrubs)
12. Semipermanent aluminum tag labels (for tagging plants in the field)
13. Plastic flagging (to relocate areas)
14. Altimeter
15. Compass
16. Entrenching tool (war surplus)
17. Folding pruning saw
18. Paper sacks

It is also possible to obtain suitable specimens by wrapping and rolling the freshly collected plants in newspaper while in the field. Be sure that the specimens are totally covered by the wrapping, or damage will result. The bundle should be tagged or labeled and tied with string. Water is poured through the open ends of the bundle and the excess drained away. Plastic milk jugs are excellent containers for taking water to the field. The ends of the bundle should be closed and the bundle placed upright in a large, heavy plastic bag. A second bag is placed over the end of the first bag to keep the plants moist. Specimens collected by this method will keep up to 20 hours. Specimens not pressed in the field are usually pressed directly in the drying press. Taxonomists formerly used a metal collecting can called a *Vasculum* to collect plants, but these have largely been replaced by heavy-duty plastic bags such as a 50-pound fertilizer bag. Plants pressed in the field will yield specimens of higher quality than those wrapped, bagged, and pressed later.

Table. 7.2. Organization of the Field press and the Drying press

Field press	*Drying press*
Plywood press	*Plywood press*
Corrugate	Corrugate
Blotter	Blotter
10 sheets of torn newspaper (each sheet will be used for one specimen)	Newspater (with specimen)
	Blotter
	Corrugate
Blotter	Blotter
Corrugate	Newspaper (with specimen)
Blotter	Blotter
10 sheets of torn newspaper	Corrougate (continue the sequence untill the press reaches the maximum of 3 feet in height.)
Blotter	
Corrugate (Continue the sequence until 10 groups of 10 newspapers are present.)	*Plywood press*
Plywood press.	

The disadvantage of the field press is that fewer specimens may be collected in a day's time. Around 100 specimens may be collected in a day using the field press, but over 300 could be collected using the bagging technique. The *placing arrangement* of the specimens in the torn newspaper is a matter of great importance and require careful attention to details. The final appearance of the specimen depends on a how it is pressed and dried. Each specimen should be arranged to look more or less natural and show the essential botanical details. Unnecessary overlapping of leaves and other plant parts must be avoided, since this slows drying and lowers specimen quality. Whenever possible, at least one leaf (or parts of a compound leaf) should be arranged with the lower side uppermost. This will allow observation of the lower leaf surface even when the specimen in mounted. All soil and trash should be removed from underground parts of plants before pressing. Maximum efficiency of the plant press is obtained when the surface of the newspaper is loosely covered with plant materials, but the plant specimen should never be overcrowded. Only one collection should be pressed in each newspaper to avoid the confusion of data. It is possible and desirable with small plants to have more than one plant of the same species per sheet if they are not overcrowded. Slender plants may be folded in the shape of a V or W. Large plants require cutting into segments.

Occasionally, two sheets for a single specimen may be required, but this should be avoided because of the problems of keeping the parts together in the herbarium. Bulky organs such as fruit may be reduced in thickness by slicing. But cross and longitudinal sections of fruits are useful. Pads of folded newspaper or foam rubber may be placed on the leaves to keep them flat and to avoid crushing the bulky stems, fruits, or other materials. Cones may be tagged with the collection number on a stringed label and set aside. Cacti and succulents should be split and the inner parts removed before pressing. Some collectors salt the cut surface to aid drying.

In humid tropical regions specimens are slow-drying, and decomposition of specimens may be a problem. This may be avoided by brushing or soaking the specimens with formaldehyde or alcohol to control decomposition. This treatment may, however, pose problems it fragments of the specimen are later removed for analysis of chemical constituents. Tall grasses, sedges, and reushes are difficult to press and will not stay bent during the pressing operation. A fold may be maintained in such material by slipping the bent end of the plant specimen throught a 1-inch slit in a card. The card may later be removed and resued. Because of pressing and size consideration, specimens should never extend byond the press.

Filamentous or filmy aquatic plants may be arranged by floating them in a pan of water over $8^1/_2$ by 11-inch sheet of bond paper. Slowly lift the paper out of the water with the specimen on it. The specimen will assume a fairly natural shape and will make an attractive specimens. The sheet of paper with the plant resting on it is then placed in the newspaper and the specimen is pressed and dried. The sheet of paper with the specimen adhering to it is often mounted on the herbarium paper. Paper with mucilage or with delicate flowers that stick to newspapers may be pressed between waxed or tissue paper. Some plants with large flowers should have a few flowers spilt open pressed. Actually, it is helpful to collect a few extra flowers and fruits and fill in vacant spaces to maximize the efficiency of the press and to increase the valve of the specimen.

Once the plant specimens are arranged, the press must be tightly closed with ropes or straps to prevent wrinkling of the specimens. Place the press on the floor or ground and kneel on one end while tightening the ropes on that end. Repeat the process several times until the press is tight. The press is now ready for the drier. For the best results, the specimens should be dried as rapidly as possible. When traveling in arid regions, the press may be placed on the luggage rack of a vehicle with the corrugation oriented to funnel the dry air through the press. In humid regions, a drier must be used. A wooden box 18 inches wide and 3 feet long made of 1 by 10-inch boards with five 60 -watt light bulds as a heat source will usually dry one press in 12 hours. Some small openings should be present in the bottom of the box to allow air to enter. The air will be heated and rise by convection through the corrugates. The press becomes loose when the plants are dry.

Field Notebook

A permanently bound field notebook with horizontal rulings is an indispensable item for the collector. Full data on each collected plant should be recorded in the field at the time the collection is made. The data should be so complete that the label may be prepared directly from the field notebook. Data recorded at the time fo collection should include locality, elevation, habitat, infromation about the plants, date, and *collection number*. It the specimens were collectied in the same location on a given day, the locality data may be recorded only once. Additional data for each species can be added down the page.

The collection number is a numerical series with number series starting with number 1 and continuing throughout the collector's lifetime. Duplicate collections of one species that are taken on the same data at the same place are given the same collection number. Otherwise, the number are not duplicate. Taxonomists routinely collect duplicate specimens in sets of 5 or 10 or 20 fro exchange to other herbaria. At the time the collection is pressed, the collection number should be recorded on the long margin of the newspaper opposite the fold. Later, when the specimen is identified, its complete scientific name (genus, specific epithet, author) is written in the long margin of the newspaper and added to the field notes. Additional notes about the specimen are sometimes written on the margin.

The specimen will stay in this same newspaper until it is mounted, so the collection number serves as an identification number associating the specimen with data in the field book. If not pressed in the field, the specimens must be tagged or labeled so that errors of associating specimens with incorrect collection localities with not be made. Pocketsize tape recorders are sometimes useful in the field. The tapes are then used to edit the field notes that day. Never delay recording data in the field notebook or editing the field notes.

Collecting, Conservation, and Law

Increasing human population and changing land-use patterns in North America during the past century have reduce or destroyed some habitats. This has had serious impact on the size of the population system of species that have specific habitat requirements. Those critical habitats that remain are often under constant threat from development, forestry practices, agriculture, actions of governmental agencies, and pressures from the increasing populations. Many plant species that were once common are now rare. It is essential that collectors be thoughtful conservationists and use good judgment when collecting, and not collect rare or uncommon plants. The wise collector will not dig the only plant of a species at a locality.

Some species have become extinct at particular sites because of thoughtless collecting. Some conservationists advocate a ban on all collection, which would severely limit specimen collection. There are many laws that protect native plant species. The federal Endangered Species Act of 1973 protects plant determined to be threatened or endangered, and federal and state agencies have obligation under this act. Collectors should be aware of laws that apply to the areas where they are gathering specimens. Collecting is generally prohibited in city, county, state, and national parks. States may have lows giving further protection to specific endangered plants. Obtain the necessary permission, including collecting permits in advance, it at all possible. For some foreign countries, this may require six months to a year and probably collecting in national forests or on lands managed by the Bureau of Land Management. Whenever possible, it is advisable to obtain permission in writing before collecting on private lands.

Identification

Professional usually identify their specimens after they have been pressed and dried. Beginners should attempt to identify their collections by using fresh material. If it is necessary to make dissections, herbarium material may be softened by boiling or by using a wetting agent. One such wetting agent is prepared by the following formula: aerosol O T (dicotyl sodium sulfosuccinate), 1 percent; distilled water, 74 percent; methanol, 25 percent. Once identified, the names are typed on the labels and the specimens are mounted.

HERBARIA

A collection of pressed and dried plants arranged in some order and available for reference or study is known as a *herbarium* (plural, *herbaria*). Many large research and educational institutions serving as basic resources fro systematic botany had their beginning as gardens which included herbaria. A herbarium may contain a few hundred specimens collected locally or millions of gradually accumulated specimens which document the flora of one of more continents. The overall goal of herbarium management is to collect and preserve plant specimens with adequate label notes and to collect the literature of taxonomy in the herbarium library. *Herbarium New*, a valuable monthly newsletter published by the Missouri Botanical Garden, communicates current events in the herbarium community. Herbaria began early in the sixteenth century in Italy as collections of dried plants sewn on paper. Luca Ghini probably initiated the first collection and may have passed the concept long to his students and associates. Its value was appreciated by those interested in plants, and the technique quickly spread over Europe.

By the mid-1500s the Englishmen John Falconer Turner are reported to have had collections. Up to the early part of the nineteenth century, plants were sewn or pasted on plain pages and bound into volumes. Linnaeus apparently popularized the current practice of mounting specimens on single sheets of paper and storing them horizontally. From such simple beginning, herbaria have developed into facilities housing millions of specimens, usually in steel cases. A list of the herbaria of the world and their standard abbreviations can be found in Holmgren, keuken, and Schofield (1981). The large herbaria tend to be national institutions supported by governmental funds. University herbaria generally are smaller because of limited resources.

Functions of Herbaria

Herbaria are permanent respositories of plants specimens and are sources of information about plants and vegetation. Practically all taxonomic research involves the use of preserved specimens that yield vast amount of data when properly prepared. For practical purposes, only in a herbarium can all the related species of a genus or family be gathered together for study. The classification of the world's flora is primarily based on herbarium material and the literature associated with in (which, in turn, was derived from herbarium materials). A recent outgrowth of the environmental movement in North America has been the use of herbarium specimens to compile list of endangered species. Many states and provinces now have published lists of their endangered plant species. These lists help in education both professional and the public on the species not to the disturbed.

Table 7.3. Some of the major herbaria of the world and the number of their specimen holdings

Outside of North America	
Museum of Natural history, paris	6.5
Royal Botanic Gardens, kew	over 5 million
Komarov Botanical Institute, Leningrad	over 5 million
Conservatiory and Botanical Garden, Geneva	5 million
british Musem of Natural History, London	4 million
Universitiy of Lyon, Lyon	3.8 million
Natural History Museum, Vienna	3.5 million
University of Uppsala	2.2 million
Botanical Garden and Botanical Museum, Berlin (Largely destroyed in World War II.)	
National Botanical Garden of Belguim, Brussels	over 2 million
Royal Botanic Garden, Edinburgh	1.7 million
National Herbarium of Voctoria, Melbourne	1 million
Botanical Survery of India, Calcutta	1.3 million
North America	
Combined herbaria, Harvard University, Cambridge	4.5 million
New york Botanical Garden, Bronx	4.3 million
U.S National Herbarium (Smithsonian), Washington, D.C	4.1 million
Field Museum of Natural History, Chicago	2.4 million
Missouri Botanical Garden, saint Louis	2.9 million
Academy of Natural Sciences, Philadelphia	2 million
University of California, Berkeley	1.5 million
Universty of Michigan, Ann Arbor	1.4 million
University of Texas, Austin	900 thousand
Vascular Plant Herbarium, Ottawa	675 thousand
University of Montreal, Montreal	630 thousand
University of North Carolina, Chapel Hill	561 thousand

More importantly, these lists also serve private conservation groups such as *The Nature Conservancy* and governmental agencies, guiding them in decisions on which critical habitats are most in need of conservation. Herbaria supplement the limited individual field studies by using the results of travel and collecting by many botanists. Monographers of genera and families study not only the specimens of their own herbaria but also those in other herbaria. This is accomplished through loans of specimens or by visits to other institutions. No longer are taxonomists content to study a multitude of specimens has becomes of each species. The importance of studying a multitude of specimens has become evident as the complexity of species has bocome better-understood.

Ideally, specimens studied from several herbaria will include most of the range of the geographic and ecological variations of the species and will reveal the trends (constancy or instability) of characters. Many distinct populations are represented since herbaria usually contain a series of specimens collected in different places at different times by many individuals. Herbaria are the repository of "original documents"- that is, specimens upon which all our knowledge of the taxonomy, evolution, distribution, and so on of the flora rests. All manuals, monographs, and wild flower books eventually stem from herbarium resources. Question about the nature of species, classification, identification, and nomenclature ultimately lead to the herbarium. The activities of herbaria are categorized as follows:

1. Providing sample of the flora of an area. For example, it is possible for an ecologist to go to the herbarium and put together a series of specimens that represent the major vegatational components of a region. By studying these specimens, much time may be saved while in the field. Also, a taxonomist writing the flora of a region would go through the herbarium to ascertain which species are represented in the flora.
2. Pointing out the existence of classification problem. A preliminary examination of herbarium specimens may indicate that a species contains plants that do not combine the characters normally listed in manuals, therefore suggesting the need for additional studies.
3. Providing a standard reference collection for verifying the identification of newly collected plants. This is a major function of many small herbaria.
4. Preserving type specimens and serving as a repository of chromosome, chemosystematic, and experimental vouchers. Examinations of type

material allows a researcher to determine which plant was described by the original author and to find the exact specimen associated with the name. Modern taxonomic literature is documented by reference to individual plant specimens according to herbarium, collector, and collection number. If a mistake is make in the identification of a plant used for chromosome counts, the mistake may later be corrected if a voucher was prepared and deposited in a herbarium. It is now impossible to publish chromosome counts of chemosytematic data without reference to a deposited voucher.

5. Documenting the presence of a species at particular locations and providing data on its geographic range. It is often possible to go back to the exact spot where a plant was originally collected collected and to find again the plant material.
6. Serving as a reference collection for plant taxonomy and other botany courses.
7. Providing plant material and data for analysis. Data are available in the form of vegetative and reoproductive morphology; pollen samples; leaf samples for chemical analysis; anatomical samples; data for distribution maps; and ecological, economic, and ethnobotanical data from the labels.
8. Training graduate and undergraduate students in herbarium practices.

Labels

Herbarium labels are an important and essential part of permanent plant specimens. The label provides pertinent collection data that cannot be determined by examination of the specimen itself. Ideally, a herbarium label should be a miniature essay on the plant and its habitat. It is essential that labels be legible, neat, and permanent. The paper should be of high rag content, preferably 100 percent rag. Unless done by offset printing, labels should be typewritten. Xeroxed labels are acceptable only if reproduced on high-rag-content paper. Labels should be approximately $2^3/_4$ by $4^1/_4$ inches, which is adequate to accommodate the necessary data. Larger labels take up space needed for mounting the specimen. They should never be folded. Although often used by governmental agencies, printed labels that designate the data to be provided are cluttered and are seldom satisfactory.

Each herbarium label should contain the following information:

1. *Collection number*. The literature of plant systematics identifies and refers to the specimen by the collector's name and collection number.

2. *Habitat.* Describe the kind of place where the plant was growing-vegetation type, moisture, parent material, soil, elevation (if critical), direction of slope, side of mountain range (because of differences in moisture due to rain shadow), physiographic region, and so forth.
3. *Name of collector.* The full name of the collector should be used. If several people are on the field trip, place their names on the label.
4. *Heading.* State or province; country or parish; country (if necessary); name of the institution (or person) with which the specimen originated.
5. *Scientific name.* Genus specific epithet, author or author. When the complete scientific name (including the author) is known, the reference to the original publication may be found in *Index Kewensis.*
6. *Date of collection.*
7. *Locality.* Be specific with localities. Refer to some town, latitude and longitude, or a permanent geographic feature. Indicate the place of collection in terms of distance in particular direction from the nearest permanent and readily located point of reference. Some labels have a map with a spot or arrow indicating the location. Ideally, someone else should be able to read your label and by following your directions go back to the same locality.

The following additional information may be placed on labels when appropriate: (i) associated plants, (ii) flower colour, (iii) pollinators, (iv) bark, (v) abundance, (vi) height or diameter at breast height (DBH), (vii) life form, (viii) economic uses, (ix) folklore, and so on. Labels are glued by one edge to the lower right corner of the specimen sheets.

In addition to the collection label, other labels may be found on herbarium sheets. *Annotation labels* are added when an expert examines the sheet and checks its identificaton. They are usually 2 by 11 centimeters in size. The annotation label will carry the full name of the person who examined the specimen along with the date examined. Other labels may indicate that pollen has been removed from the specimen or that it is a cytological or phytochemical voucher.

Mounting

Mounting is the process by which a specimen is attached to a sheet of mounting paper and label affixed at the lower right corner. Most herbaria in North America use a standard-size herbarium or mounting sheet or $11^1/_2$ by $16^1/_2$ inches. Various paper companies and biological supply house can supply suitable mounting sheets. The quality of the paper used will very according to the needs of the herbarium. For research herbaria, paper approaching 100 percent rag content is used for durability. Teaching collections use less expensive paper with lower rag content. The paper should be relatively stiff to prevent damage to the specimens when the sheets are handled. A number of method are used for attaching specimens to the mounting sheets.

A good quality pate or glue such as Swifts Z-5032, Elmer's Glue-All, Nicobon B, or Wilhold 128 is adequate. "Yes brand paste is used by some to affix labels to sheets. Gule is applied to the back of the specimens, and then the specimens are pressed onto the mounting sheets. Bulky parts may be sewn or fastened down with strips of linen tape. Archer's (1950) plastic formula is sometimes used for strapping bulk materials. Resyn 35-6262 is useful for specimens with leaves of hard texture since in retains elasticity. Weight of scrap iron, heavy washers, large nails, and so on are used to hold down the specimen until the glue has set. Valuable loose parts such as seeds, fruits, or flowers, are placed in fragment folders glued to the mounting paper or placed in a small folded packet, which is glued to the sheet. After the specimen has been mounted the name of the herbarium is stamped or embossed on the mounting sheet, and accession number is added.

Filing

Plant specimens are filed in herbaria by some preselected arrangement. In small teaching herbaria, the filing system may be entirely alphabetical by family, genus, and species. In research herbaria, the specimens are arranged by family or genus according to one of the well-known systems of classification. When filed by families or genera, the folders are numbered, and an index is displayed in the herbarium. Many herbaria use family and genus numbers from Dalla Torre and Harms (1900-1907), in which each family and genus has a place in a numbered sequence based on the Engler system of classification. Specimens are field in manila genus folders $16^{5}/_{8}$ by 12 inches in size when folded Names of the included plant material are placed on the lower edge of the folder. Folders are often segregated geographically, and various colour schemes are employed to indicate political and physiographic areas.

Such segregation prevents unnecessary handling of specimens and facilitates locating the specimens. Species within a genus are filed alphabetically or by sections to show relatedness. Special arrangements are required for filing bulky material such as palm leaves and flowers or gymnosperm cones. These are often tagged are placed in boxes designed to be stored in the pigeonholes of herbarium cases. Standard herbarium cases are insect- and dust-proof filing cases made of welded steel construction with two tiers of pigeonholes, each 19 inches deep, 13 inches deep, 13 inches wide, and 8 inches high. Dummy sheets and folders may be used to refer the user to other folders when familiar names are changed in the course of taxonomic revision or to specimens filed elsewhere in the herbarium. Most herbaria will have a folder at the end of each genus for specimens not determined to species.

Insect Control

Precautions must be taken to safeguard herbarium collections damage by insects. All the compounds used as repellants or fumigants are hazardous to some degree. Mandatory heating, freezing, or fumigation upon entry of the specimens, containment and occasional use of fumigants in the collection, and a herbarium climate inhospitable to insects must be the cornerstones of any workable and safe (to humans) policy. Particularly troublesome insect pests of dried material are the cigarette beetle, the drugstore beetle, and the black carpet beetle, often collectively called *dermestid beetles.*

The most destructive pest is the cigarette beetle, which can complete its life cycle in 45 to 50 days with three to six generations per year. Any of these pests may ravage specimens during periods of unattended storage. Recommendations for control and precautionary measures very, but they may be summarized and grouped under one of the three following categories: (1) heating or freezing, (2) repellents, and (3) fumigants. Income specimens may be treated by *heating* in a specially constructed cabinet at 60°C for a period of six hours. This temperature will effectively kill eggs, larvae pupae, and adults of the dermestids. If the herbarium is kept clean, at below 21°C, and with a relative humidity of 30 to 40 percent and if all incoming specimens are heated, chances of infestation may be greatly reduced.

Recently, some herbaria have microwave ovens to treat incoming specimens, but it is not always effective and microwave treatment may also cause damage to the specimens (Hill, 1983). Ultracold freezers set at -60°C have proven successful for some herbaria in controlling dermestids on plant specimens.

Repellents are substance that keep insets away from the herbarium specimens because of their offensive odor or taste. Paradichlorobezene (PDB) and nephthalene have been widely used as repellents in herbaria. A 2- to 3 - ounce cloth bag refilled once a year is sufficient for a standard herbarium case. Contact with either repellent should be kept to a minimum. For people working in the herbarium eight hours a day, five days a week, air exposure should not exceed 75 parts per million (ppm) for PDB and 10 PPM for nephthalene. PDB is currently under review by the Environmental Protection Agency (EPA) for carcinogenic effects.

Fumigants are chemicals used in the form of a gas and applied in an enclosure to kill insects. Fumigation of herbaria has been a necessary and common procedure for many years. Recent information indicates that many of he chemicals routinely used for this purpose may be extremely dangerous to humans, and in many instances, the traditional methods of application of these chemicals are also hazardous. The Federal Insecticide, Fungicide, and Rodenticide Act prohibits the sale of fumigants except for precisely those uses for which they are labeld. Herbaria are required to comply with the Occupational Safety and Health Act. Dowfume-75 has been cleared by the Environmental Protection Agency as fumigant for control of dermestid beetles and other pests in herbaria. Fumigation of the herbarium with Dowfume-75 should be conducted only by a trained fumigator wearing a full-face gas mask.

Large herbaria use hospital-type sterilizers for fumigator chambers. All incoming specimens are fumigated with either ethylene dioxide, methyl bromide, or ethylene dibromide at a temperature of 38ºC for three to four hours. All such equipment mist be carefully designed and operated. Under controlled conditions, vapona resin strips (Raid Strips) are suitable for use as fumigant in herbarium cases. One-third of a resin strip should be applied in each case for seven to ten days twice a year. The cases must not be opened during fumigation. Work with the resin strips should only be assigned for short periods, perhaps around 30 to 60 minutes for cutting and placing an 15 minutes for removing each day. The Work location should be well ventilated and excessive exposure to vapona avoided.

Disposable plastic gloves should be used while handling the material. Workers must be trained to wash carefully after handling vapona or any toxic chemicals. If exposures are to be excessive, a full-face gas mask should be worm. Some herbaria have reported vapona resistance in dermestid beetles. Three parts of ethylene dichloride (1,2 dichloroethane) mixed with one part of carbon tetrachloride (CCI_4) was once used for the fumigation of herbarium cases. The mixture was applied at the rate of 6 ounces in a beaker placed on the top shelf of each case. The case remained sealed for four of five days. Ethylene dichloride is explosive without CCI_4 and has the property of causing injury to the human liver and kidney from either excessive single or repeated exposures. Carbon tetrachloride is extremely toxic to humans, causing liver damage. These two chemicals should be considered extremely dangerous to the persons conducting the fumigation.

Type Specimens

Type specimens (or *types*, as they are usually called) are specimens on which the name of a taxon is based. Type specimens are often housed separately from the general collection because of their inherent value. Every precaution should be taken to reduce unnecessary handling and to prevent damage or loss of type specimens. Some herbaria include type in the general collection, but they are filed separately in specially coloured and marked genus folders. Photographs of type specimens from some collections and herbaria are available on microfiche. In many instances, these photographs rather than the specimens themselves can be used to solve nomenclatural problems.

Loans

Revisionary studies require the examination of as much relevant material as possible. This may necessitate obtaining material on loan. Travel to other herbaria to study their holdings often is too costly or not practical; for this reason, specimens may be borrowed for serious taxonomic studies. This service by herbaria to research workers in taxonomy also benefits the lending institution since the individual requesting the loan has the responsibility of annotating the specimens and revising identifications. Herbaria lend to other herbarium for study by a specific researcher. Specimens arriving from lending institutions should be carefully checked for transit damage and the packing list verified. Many institutions number and code the sheets lightly with pencil in the lower left corner on borrowed specimens. Consecutive numbers and the herbarium abbreviation facilitate later return of the loan. Borrowers must be meticulously careful to adhere to all provision of the loan.

Specimens are studied only on the institutional premises and are housed in standard metal herbarium cases when not in use. Dissections are made with care and all fragments placed in fragment folders. Special permission may be required to remove pollen or other samples. The borrower must annotate the specimens before returning the loan. Annotations for revisionary studies are made not directly on the collection label or on the sheets, but on annotation labels glued above the herbarium label. Each annotation label should have the complete scientific name of the plant (genus, specific epithet, and author), full name of the person making the annotation, and the year in which the annotation was made. A convenient way to agree with an identification is by using an exclamation point (!), meaning "I have seen it and agree with the identification." Since specimens are irreplaceable, details of packing for return shipment are highly important. Ideally, the original boxes and packing material are saved and used for the return shipment.

Exchanges

Exchanges of duplicate specimens among herbaria and collectors are a means of augmenting a collection at minimal cost. Exchanges are generally made on a one-for-one basis. Specimens for exchange are normally unmounted and loose in the pressing paper and include adequate labels. Arrangements for an exchange should be made before shipping the specimens. Poor-quality specimens made for class purposes are not normally acceptable by most curators as exchange material.

Herbarium Ethics

Herbaria should provide visitors with policy statements including such information as: (i) hours, (ii) filing arrangements, (iii) reshelving, (iv) loan procedures, (v) annotation labels, (vi) use of library collections, (vii) microscopes, and so on. Visitors have an obligation to adhere to the policy statements and to act accordingly. Specimens are dry and fragile. They are mounted on paper that is flexible and should not be bent. The sheets should be kept flat. Folders should not be leafed through by turning sheets like pages in a book. Instead, they stack on a sheet of cardboard (e.g., a press ventilator). One should never attempt to force an extra-large stack of specimens into a herbarium case pigeonhole. Heavy objects, elbows, books coffee cups, and so forth must not be placed on the specimens. Specimens should be studied only with a long-armed microscope to avoid bending the sheet. Loose material clearly identifiable with a particular specimen should be placed in fragment folders.

Damaged specimens should be put aside for repair and called to the attention of curators. Flowers and fruits from herbarium specimens should be dissected sparingly. First soften by boiling or with a wetting agent; then examine the specimen; and then return the specimen to the fragment folder. Notes from the dissection may be recorded on fragment folder. Reshelve specimens only with permission and always with extreme care. Visits to other herbaria are often profitable in terms of examining critical specimens and collections, searching the undetermined folders, and exchanging ideas with other systematists. A not in advance will help the curator to prepare for your visit and to provide a place to work and microscope. Many herbaria are closed in the evenings or on weekends. The privilege of visiting after normal hours may be offered but should not be requested.

Challenges Facing Herbaria

Systematic collections of plants in herbaria provide a permanent record of the earth's flora and its diversity. They are vital for a wide range of basic activities and play irreplaceable roles in sustaining society. The collections provide information for identification services, conservation and land-use planning, education, national defense, medicine and public health, business and industry, and agriculture and forestry. Herbaria have an educational role in the training of undergraduates. At the graduate level, the student in systematics is an apprentice scientist engaged in original research.

Many herbaria maintain programs of education in natural history for the general public and attempt to create an understanding of and appreciation for living environment. Faced with both increasing costs and demands for services, herbarium managers have a constant challenge to maintain financial security. Thus, it is essential for herbaria to develop and use modern management techniques. Continued development of computer software for data-processing, data-base management, word-processing, bibliographic aids, and management aids will result in the development and increased use of coat-effective computer methodology in collection management.

8

PLANTS IDENTIFICATION

Identification is an integral part of all taxonomic work. Plants may be identified by comparing them with named plant specimens. This process, combined with the determination of the correct name applicable to that plant, is sometimes referred to as *specimen determination*. Aids to identification have been developed so that with some training and practice, one may readily identify most groups of plants in North America with no major problem. In order to make identification, it is necessary to possess: (1) knowledge of taxonomic methods, characters and terms; (2) knowledge of manuals and other resources, such as herbaria; and (3) experience in the identification of plants. Identification assumes that a classification scheme exists that has distinguished groups of plants and has applied names to them.

Once specimens are identified and scientific names applied, the information stored in the classification system becomes available. The names of plants permit retrieval of information, such as chromosome numbers, natural products, distribution map, and so on, from the system. Accurate identifications are a prerequisite to using classification as information-retrieval mechanisms, and are fundamental to such research fields as biogeography, biochemistry, ecology, genetic, and physiology-that is, the whole realm of science involving plants, including agricultural and biomedical disciplines.

Names are a vital part of our language and are ultimately important to everyone. They help us to communicate facts about the world around us. Most systematists spend at least some part of their time doing identification work for other or themselves. They use several methods to achieve correct identifications. Persons familiar with the flora of an area can usually recognize thousands of species and can apply a name at some rank to specimens. It may be necessary to aid and improve identification by comparing the specimen with previously identified plants in the herbarium. Unknown specimens are often identified by means of *keys*. A key is a device for easily identifying an unknown plant by a sequence of choices between two (or more) statements. If a herbarium is a available, direct comparison is used after an identification has been made by keying. If the identification cannot be checked locally, some collections may be sent to a person who has become an expert in the identification of the particular group of plant.

Observing Plant Identification

Examining a specimen carefully before beginning is a good habit to develop. The beginner should select a freshly collected plant with roots (on herbs), stems, leaves, flowers, fruits and seeds. A specimen with all these parts will facilitate identification. For practice, make sure the plant selected has large flowers on which the parts can be easily seen. It is frustrating for beginner to attempt to identify plants with small complex flowers. To observe the plant properly, you will need a good hand lens of 10X magnification, a pair of sharp-pointed forceps, a straight dissecting needle, and some single-edged razor blades. Steps in observing the plant include the following (refer to Chapter 10 for diagrams and definitions of morphological terms):

1. Observe the flower and name its parts.
2. Determine whether the sepals and petals are fused or separate.
3. Determine whether it is woody or herbaceous. If herbaceous, is it annual or perennial?
4. Count the number of pistils, styles, and stigmas of the gynoecium.
5. Count the numbers of sepals and petals.
6. Remove the perianth and stamens. Make a cross section of the ovary with the razor blade. Count the number of locules. Observe the number of ovules and the type of placentation.
7. Count the number of stamens. Observe there they are attached. Note any fusion of the filaments of anthers. Observe the disposition of the anthers.
8. Note the distribution and kinds of surface coverings.
9. Select another flower and make a longitudinal section of the entire flower through its center. Not the ovary position and any fusion of the perianth.
10. Note the leaf type, arrangement, and venation.

When these characteristics are determined in advance, the process of identification will be much easier and you will be confident of step-by-step decisions. After the plant has been carefully examined, the next step in identification is keying.

Identification With Keys

It is far more efficient to identify a specimen by the use of keys then to shuffle through a stack of preciously named herbarium specimens until a likely comparison is found. In North America, unknown plants can usually be identified by means of key. A key is fundamentally similar to an outline where the topics are arranged in order of descending importance. The use of a key provides the correct identity of a specimen by a process of elimination. Keys have long been used to identify plants, and most manuals contain them. Plants were the subject of description and illustration in classical and medieval writing, especially in those works accurate identification was a matter of practical importance. This eventually led to the development of bracketed diagrams in the seventeenth century which functioned both as a classification device or *conspectus* and as identification tool.

According to Voss (1952), the Latin *clavis*, or "key," was apparently not used in connection with such diagrams until Linnaeus used it in 1736 (and then with reference to a diagram in which he was classifying botanists, not plants). The use of modern keys for identification is usually credited to Lamarck (1778). Keys are constructed using contrasting characteristics to divide the possible names in the key into smaller and smaller groups. Each time a choice is made, one or more taxa are eliminated. Statements in the keys are based on the characters of the plants. For example, a key might separate taxa using the following choices: (i) herbaceous versus woody; if herbaceous, the woody plants are eliminated; (ii) the next choice, zygomorphic flowers versus actinomorphic; it zygomorphic, the plants with actinomorphic flowers are eliminated; and so forth. Each time a choice is made, the number of characters are contrasted, the number of possibilities is eventually reduced to one.

The use of a key is analogous to traveling a highway that forks repeatedly, each fork having roadside directions. If the traveler lacks information or if the road sings are faulty, the destination may not be reached except by trial and error. In most manuals, the first step in identification of an unknown plant involves the use of keys to determine the family. Next, a key to genera will provide the generic name. After the genus is determined, the keying process is repeated within the genus for a determination of the species. Descriptions are normally available in manuals for families, genera, and species. By consulting descriptions, the likelihood of errors due to mistakes in observations or "keying" will be reduced. A few manuals use keys with many characters in each choice and do not use descriptions. In this case, additional references with illustrations or descriptions are useful.

Suggestion for the Use of Keys

1. Select keys appropriate for the plant material and for the geographic area where the plant was obtained.
2. If the specimen doesn't seem to fit the key and all choices are unlikely, you probably made a mistake. Retrace your steps.
3. Verify your results by comparing the specimen with an illustration of with a herbarium specimen.
4. Obtain as much information as possible about the characters of the unknown plant before starting to key. Attempts to use the key are likely to fail if a specimen consists of a single leaf of just a flower.
5. Confirm your choices by reading description.
6. Be sure you understand all the terms found in each choice. Use a glossary.
7. Read the introduction to the keys for abbreviations and other details.
8. If both choices seem possible, try going both ways.
9. Always read both choices, observing punctuation.

Types of keys

Keys used in floras are usually diagnostic, that is, identifying an unknown plant by stating only the conspicuous features by which the various taxa can be recognized. The *diagnostic characters* used in such a key should be conspicuous and clear cut. Diagnostic characters are sometimes referred to as *key characters*. Diagnostic keys are generally artificial in that the sequence of taxa within the key does not necessarily reflect their phylogeny. Most keys used today are *dichotomous*. A dichotomous key presents two contrasting choices at each step. Each pair of choices is called a *couplet*.

The key is designed so that one part of the couplet will be accepted and the other rejected. The first contrasting characters in each couplet are referred to as the primary key characters. These are usually the best contrasting characters. Characters following the lead are *secondary key characters*. A few keys may not be dichotomous and may provide three or four choices, but paris of choice are preferred. Several different formats for arranging the couplets are used for plant identification. The *indented* or *yoked key* is the one most widely used in manuals for the identification of vascular plants. In the indented key, each of the couples is indented a fixed distance form the left margin of the page.

In the second type, the bracket or *parallel* key, the two couplets are always next to each other in consecutive lines on the page. At the end of each line there is either a name or a number referring to a couplet later in the key. Both the indented and bracket keys have advantages and disadvantages. When an indented key is used, the lines become more indented for each couplet. This can be uneconomical, since additional pages are required. In contrast, the bracket keys make good use of page space.

Indented keys may have the second part of the couplet on page much later than the first part of the couplet. This may cause confusion when one tries to find the second part of the couplet. In bracket keys, both parts of the couplet are adjacent. The indented key groups similar elements so that their morphology can be grasped visually, often making identification easier. In the example of the indented key, *Centunculus* and *Samolus* are grouped under leaves alternate. *Lysimachia* and *Anagallis* are grouped under leaves opposite. In the example of the bracket key, this grouping is not immediately apparent. Modern dichotomous keys are conventionally published in either the indented or bracketed style. Although various other styles have been proposed, these traditional formats serve their purpose well and have gained general acceptance. The major considerations in key construction are ease, quickness, and accuracy. When preparing a key, several techniques should be followed:

1. Use readily observable features. Avoid using geographical location as a sole separation character.
2. Avoid the use overlapping ranges or vague generalities in the couplets, that is 4 to 8 millimetres versus 6 to 10 millimeters long, or large versus small, or southern versus northern.
3. Couplets of a key may be numbered or lettered, or may use some combinations of lettering and numbering, or may be left blank in the case of indented keys.
4. The key should be dichotomous.
5. It may be necessary to provide two sets of keys in some groups; flowering versus fruiting material, vegetative versus flowering, or staminate versus pistillate for dioecious plant.

6. The leads of consecutive couplets should not begin with the same word, since this may cause confusion when one looks away to observe the plant specimen.
7. The first word of each lead of the couplet should be identical. For example, if the first lead of a couplet begins with the word *stamens*, the second lead of the same couplet must begin with the word *stamens*.
8. The couplets should be written to make positive statements. An example to avoid is " leaves narrow versus leaves not narrow."
9. The two parts of the couplet should be made up of contradictory statements so that one part will apply and the other part will not.

How to Locate Keys and Manuals

In addition, lists of books are provided for identification of special groups such grasses, ferns, legumes, and others. For most references, the title alone provides the geographic area. If not, annotations are provided at the end of the reference. Floras are constantly being written or updated, and new books for plant identification can be found in libraries and bookstores. Inquiries can also be made of herbarium curators or botany professors who teach plant taxonomy for suggestions of books to aid in plant identifications. Two excellent books are available to aid in the identification of flowering plants down to the family level on a worldwide basis. They are *Thonner's analytical Key* to *the Families of flowering plants* and *The Identification of Flowering Plant Families*.

Unconventional Indentification Methods

Keys are the traditional method of identification in systematic botany. If keys are well written, if adequate specimens are in hand, and it the person doing the keying is careful, then the specimen can be successfully identified. Keys, however, have several major disadvantages. The use of certain characters in required even it the character is not evident in the unknown specimen. Entry is usually possible only a one point in the key. Thus, the standard dichotomous key with one entry point is known as a *singleaccess* or *sequential* key. Attempts to improve on the traditional identification process by using keys have resorted to either polyclaves or computer techniques. These methods provide *multientry* (or *multiaccess*) opportunities for the user; that is, the place of entry is not fixed by the format of the key.

Polyclave Identification

A *polyclave* is a multientry, order-free key implemented in one of several different formats. One form is a diagnostic key that uses cards placed on top of one another to eliminate taxa that disagree with the specimen to be identified. Another form is a computer-stored multientry key. Still another polyclave is a printed table or matrix giving the status of various taxa and characters useful for separating the taxa. Polyclaves have a tremendous advantage over dichotomous keys, since the user of the key rather than the author of the key selects the characters to be used. The user is free to select appropriate characters. For each unknown specimen. The route taken to a particular taxon may differ considerably from one specimen to another.

The logic of identification with a polyclave is the same as that in a key, but the user is free to choose any character, in any sequence, thus avoiding the rigid format of traditional keys. This freedom of choice is particularly valuable in attempts to identify fragmentary material. A number of *field key polyclaves* are in existence: (1) card with holes, commonly referred to as "peek-a-boo" window" cards; (2) edge-punched or "key sort" cards; (3) semitransparent overlays; and (4) standard computer cards.

The peek-a-boo or window card key to world angiosperm families uses a card for each characters. Round holes are punched beside a family that have that character. The plant to be identified is examined and its characters noted. With a good specimen, it is possible to find 20 to 25 characters. One selects those cards corresponding to the characters found on the plant. The cards are then put on top of one another in a stack. Families having all the observed characters will be indicated by the holes, or "window," which are easily seen when the cards are held in front of a light. The logic is simple. Each time a card is added families not perforated in this new card are excluded. With 20 to 25 cards, it is likely that the family may be determined or the search narrowed to several families. Until recently, field key polyclaves has to be manufactured specially, but now they may be punched by computer on standard computer cards. Two general kinds of *computer-based polyclaves* have been developed.

One kind, developed for qualitative taxon-character data, employs elimination. The other, developed for qualitative taxon-character frequency tables, employs likelihood ratios or other probabilistic techniques. In the latter case, the computer, on the basis of the characters supplied, can indicate the taxa that have been eliminated as well as the taxa that remain. The computer may also prompt the user by suggesting additional characters that might be observed, for instance, locule number. If locule number is unavailable, the user could input other characters available form the specimen.

Computer Identification

A computer functions by commands derived from a computer program. The computer program is a machine-independent procedure or *algorithm*, which is a set of rules selected from some group of commonly understood rules or actions. An algorithm is evaluated by carrying out the action of a computer program in the specified order. This produces some described result, no less and no more. The result is produced in the last step of the evaluation of the algorithm. A traditional taxonomic key is actually on example of an algorithm. The basic format of a taxonomic key includes directions for continuing the sequence of evolution on the basic of directed observations of directed observations of a specimen. The result, the identification of a specimen, is given at the end point of the path taken through the key.

Computer are machines intended only for the evaluation of data. The computer can be programmed to identify or to construct a key only if an initial algorithm has been successfully prepared by a taxonomist. The major research efforts in computerized identification may be grouped into four major approaches; (1) computer-stored dichotomous keys, (2) computer-constructed keys, (3) simultaneous character-set methods or matching coefficients, and (4) automated pattern recognition.

Computer-stored keys may be programmed to provide a dialog between the user and the computer. The computer asks a question, awaits an answer, then asks another question appropriate to that response. In general, computer-stored keys offer no advantage over printed keys and have the disadvantage of requiring access to a computer for each identification. Until computer are cheaper and easier to use than books, this approach will attract little interest. The second major approach, *Computer-constructed keys*, shows promise as a technique. Traditional dichotomous keys may be constructed by computer from machine readable files of taxonomic data. The challenging aspect of this topic lies more in the realm of data collecting and coding than in the computer algorithm itself. If these problems are overcome, computer aids to construction keys and diagnostic could be helpful.

Simultaneous character-set methods are those in which values or states or a predefined set of characters are recorded for the unknown plant and then submitted to a computer program. The computer program suggests one or more possible identifications or concludes that the specimen differs from those considered as possibilities. Because of the time and expense involved, it is likely that character-set methods will prove useful only in very difficult situations when traditional identification methods are not adequate, for example, hybrid or apomictic swarms, or with cultivated plants.

Automated pattern-recognition systems providing fully automated specimen identification may be possible for a least certain organisms. This technique combines automated character observation by optical scanning and pattern-recognition procedures. Such programs and techniques have been developed for analysis of chemical spectra and photomicrographs of chromosomes, recognition of abnormal cells in cytological samples, pattern recognition of remote-sensing imagery in vegetation and agricultural surveys, and so on. Its successful use in plant identification will most likely combine analysis of the chemical constituents with pattern recognition.

Computer programs are available to aid in the identification of plants and handle related aspects of plant taxonomy. These programs include routines for specimen identification, key construction, and description preparation. Also included are programs for comparing taxa, inverting data files, and producing—punched-card field keys. The routine application of computer-stored data matrices to specimen identification presents problems: (1) terminals must be located in herbaria; (2) a network of accessible taxonomic data matrix files must be prepared and be available. Until a critical mass of such data files are accessible via a network of computer terminals, there will be insufficient use to justify the cost. Identification aided by the computer is possible with present technology. In the future, libraries of computer programs and taxonomic data matrices may play a role in information processing and data retrieval in taxonomy. In the immediate future, identifications will be done with traditional keys and manuals.

9

RANUNCULACEAE

Generally erect herbs (rarely climbing shrubs), leaves exstipulate. Flowers hypogynous; floral parts on a well-developed thalamus (convex). Sepals often deciduous and petaloid. Petals sometimes wanting. Stamens free, indefinite. Carpels numerous and usually free. The family is spread over 35 genera and 1,500 species (Lawrence 1964). Rendle includes 30 genera and 1,900 species in the family. Willis (1966) mentioned 50 genera and 800 species whereas Porter recorded 40 genera and 1,500 species. Cronquist (1968) enumerated 2,000 species in this family. In India the family is represented by about 165 species.

Distribution

The family is mainly distributed in colder and temperate regions of the world especially at high altitudes in mountains. Plants are abundant in north temperate zones and not uncommon in south temperate regions of the globe. Several genera have also been recorded from tropical and sub-tropical parts of the world. A few members of the family are found in Arctic and Alpine regions too. In India members of the family are found in plains, submountaneous regions and at higher altitudes in Eastern and North-western Himalayas and Nilgiri hills. The plants flower during winter.

VEGETATIVE CHARACTERS

Habit

Generally the plants are annual or perennial herbs. They perennate either by means of rhizomes or by tuberous roots e.g., *Ranunculus*, *Aconitum Paeonia* etc. In many plants the primary root perishes and is replaced by adventitious roots. These roots become tuberous and contain reserve food material in them e.g., *Ranunculus*, *Aconitum*, *Paeonia* etc.

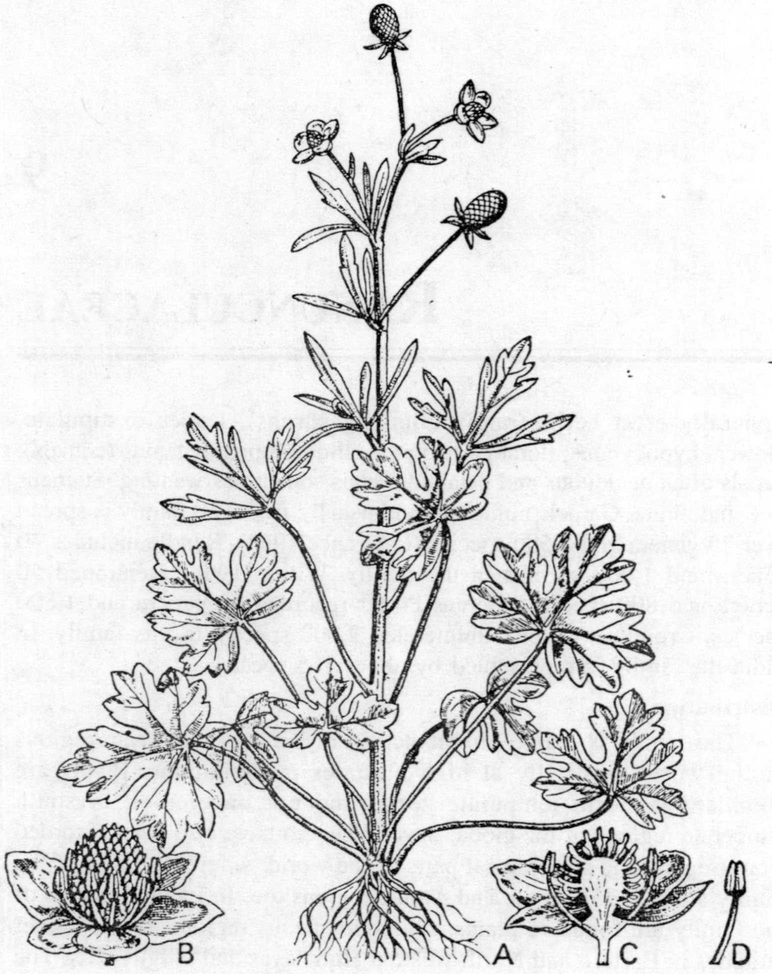

Fig. 9.1. Ranunculus sceleratus L. A—flowering plant; B—flower; C—flower longitudinal section; D—stamens.

A few species are annuals. Undershrubs, and shrubs are also found in this family. Usually the species of *Clematis* are climbing undershrubs. The species of *Naravelia* are also climbing shrubs.

Leaves

The leaves are alternate, exstipulate and more or less divided. In some cases the leaves may be opposite or even radical. In *Clematis*, the leaves are opposite and compound. They possess twining petioles. In *Clematis aphylla* the whole leaf becomes modified into a tendril. In the case of *Naravelia*, the leaves are compound and trifoliate, here the terminal leaflet becomes modified into a tendril. In *Anemone* the leaves are radical.

The leaf base usually broadens into a sheath which sometimes elongates into a pair of lateral stipules, e.g., in *Thalictrum.* The leaves of *Thalictrum folilosum* are pinnately decompound with auricled sheaths. The leaf lamina is sometimes entire. It is narrow in some species of *Ranunculus*, however, in *Ranunculus ficaria* and *Caltha* sp. it is cordate. Heterophylly is found in the aquatic species of *Ranunculus*. In such cases the submerged leaves are very much dissected whereas the floating ones are simply lobed. In *Delphinium* the leaves are palmately lobed and much dissected. The leaves of *Clematis* are climbing.

Inflorescence

Inflorescence is mostly of cymose type e.g., *Ranunculus*. In some species of Ranunculus, however, the flowers are panicled. In *Anemone vitifolia* the flowers are in decompound cymes. In some species of *Anemone* they are solitary terminal. In *Clematis cadmia* also they are solitary but axillary. In *C. nutans* and its several other species they are in panicles. In species of *Nigella*, first a terminal flower is produced but later on a raceme is developed as the buds below develop in ascending order. In *Nigella*, *Anemone* etc., there is an involucre of green leaves below the flower, which usually alternate with the sepals. In Delphinium, the flowers are arranged in racemes.

Flowers

The flowers are ebracteate rarely bracteate, hermaphrodite and mostly actinomorphic (regular), rarely they are zygomorphic (irregular) such as *Delphinium* and *Aconitum*, hypogynous and sometimes a disc is found below the gynoecium, e.g., *Paeonia*. The flowers are usually developed on an elongated receptacle. The floral parts being arranged wholly or partly spirally. Flowers are usually borne on an elongated receptacle. They are hermaphrodite and mostly actinomorphic, sometimes zygomorphic e.g., *Delphinium*, *Aconitum* etc., hypogynous and only rarely bracteate. Rarely a disk is present below the gynoecium e.g., *Paeonia*.

In most flowers of this family the perianth is not distinguished into a distinct calyx and corolla. It is mostly simple, petaloid and variously coloured. In species of *Ranunculus*, however, there is a distinct calyx of five sepals and corolla of five petals. The, petals are coloured and possess pocket-like nectaries at their base. Nectaries of various, kinds also occur between the perianth and the stamens and are suppose to be the modified petals. The number of the perianth leaves varies greatly in different genera and even in different species of the same genus. The number may vary from 4 to 20. The flowers of *Delphinium* are extremely specialized in having zygomorphic spurred perianth, but are primitive in possessing follicle fruits. These features make it very difficult to assess the overall degree of specialization.

Perianth

In most of the flowers of this family, the perianth is not distinguished into calyx and corolla. It is a simple, petaloid and variously coloured. However, in *Ranunculus* there are two distinct whorls of calyx and corolla, each consisting of five sepals and five petals respectively. The perianth is often associated with nectar secreting structures of various forms. In *Ranunculus*, the petals possess pocket like nectaries at their bases. The nectaries are supposed to be the modification of the petals. In certain genera the nectaries develop in the association of sepals, stamens or even carpels. The number of perianth leaves varies from 4 to 20.

Androecium

The *stamens* are usually many, polyandrous and spirally arranged or in some genera (*Helleborus*, *Nigella*, Aquilegia) they are arranged in definite rings. In members like *Helleborus* some of the stamens are converted into honey leaves. The anthers are adnate, dithecous, extorse and dehiscing longitudinally. The filaments are beautifully coloured in *Thalictrum*.

Gynoecium

The *gynoecium* is usuallly of numerous free carpels arranged spirally on a distinct thalamus but in *Aquilegia* there are five carpels, *Delphinium* has three to one carpel, and in *Actaea* and *Cimicifuga* the number of carpels is reduced to one. In some species of *Helleborus* the carpels are connate at the base only, whereas in *Nigella* the five carpels are completely united. The ovary is superior, unilocular and has one

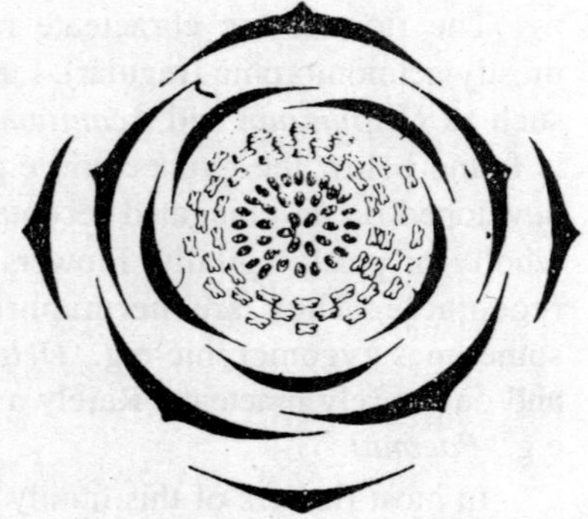

Fig. 9.2. Floral diagram of Ranunculus sceleratus.

to many anatropous ovules. The placentation may be basal (*Ranunculus*), apical (*Clematis*) or marginal (*Delphinium*). In *Nigella*, where gynoecium is syncarpous the ovary is pentalocular with axile placentation. The style and stigma are one.

Fruit

Aggregate, usually an etaerio of achenes (*Ranunculus*) or follicle (*Aconitum*) or capsular (*Nigella*) or berry (*Actaea*). The seeds are small, albuminous without aril. The endosperm is oily with an embryo.

Pollination

The flowers are usually protandrous but some are protogynous (*Thalictrum*, *Helleborus*) and those of *Anemone*, *Trollius* are homogamous. Pollination is entemophillous, the insects are attracted by coloured corolla and nectar concealed in the petals (*Ranunculus*) and staminodes (*Anemone* and *Clematis*). Flowers of *Thalictrum* and *Adonis* are wind-pollinated. In most of the genera self-pollination may also take place (*Anemone Trollius*). The most highly specialised flowers are those of *Aquilegia*, *Delphinium* and *Aconitum* that are suited for pollination by long-tongued bees (chiefly humble bees).

Economic Importance

The family is of little economic value. Many plants are grown in the gardens as ornamentals. Some are of medicinal importance. A list of few important plants is given below:

1. *Ranunculus aquatilis*; Verna—*Jal dhania*—It is used medicinally in the treatment of rheumatic pain and asthama.
2. *Delphinium brunonianum*; Eng.—Rocket Larkspur—This is native of South Europs. The plants are grown in the gardens as ornamentals. The seeds are insecticidal.
3. *Aconitum heterophyllum*; Verna.—*Atis*—It is found in the Himalayas. The roots of this plant are used medicinally as astringent, tonic and in cough and diarrhoea.
4. *Aconitum chasmanthum;* Verna.—*Banbalnag*—A herb. Its tuberous roots are the source of a drug known as *aconite*. This drug is used externally for rheumatism and internally to relieve pain and fever; usually found in Western Himalayas.
5. *Nigella sativa*; Eng.—Black cumin; Verna—*Kala jira, Kalonji*—The seeds are used as condiment and spice. The seeds mixed with sesamum oil are used externally in eruption of skin and for scorpion sting. The seeds are stimulant, carminative and diuretic. It is a herb. The plants are cultivated in the Punjab, Assam, Bihar and Bengal.

6. *Clematis paniculata*; A shrub, grown as an ornamental for its beautiful flowers.
7. *Thalictrum javanicum*; this plant is the alternate host of brown rust of wheat, caused by *Puccinia triticina*.
8. *Ranunculus ficaria*; Eng.—Pilewort—The plant is native of Europe. It makes a good medicine of piles. The patent medicine of piles, known as 'pileowrt ointment' is prepared from this plant.
9. *Anemone pulsatilla*—They are found in Europe and Russia. This is a herb, which yields a medicine known as *pulsatilla*. This medicine is given for the treatment of nervous and menstrual trouble of women.
10. *Clematis triloba*; Verna.—*Murhari*—It yields a medicine which is used for the treatment of leprosy and other blood diseases.
11. *Thalictrum foliolosum*; Verna.—*Mamira*—The roots are used as a tonic. The plants are found in the Himalaya.
12. *Nigella damascena*; Eng.—Love-in-a mist.—A herb, grown as an ornamental in the gardens. It is native of South Europe.
13. *Delphinium denundatum*; Verna—*Nirbisi*—The roots are used as a tonic.
14. *Delphinium caeruleum*; Verna.—*Dhakangu*—The root is used as an insecticide to kill maggots in the wounds of goats.
15. *Delphinium ajacis*; Eng.—Rocket Larkspur—This is native of South Europe. The plants are grown in the gardens as ornamentals. The seeds are insecticidal.

10

Rosaceae

Herbs, shrubs, scramblers or trees. Leaves stipulate: stipules sometimes adnate alternate, very rarely opposite, simple or compound. Flowers bisexual, usually actinomorphic, hypo-, epi- or perigynous, pentamerous. Sepals, Petals and Stamens perigynous or epigynous. Petals entire or at most shortly bilobed. Stamens indefinite, all fertile. Carpels one to many borne on expanded convex or hollow out thalamus in epigynous flowers, only the styles are free. Fruit aggregated achene, drupe or pome.

Distribution

The family consists of about 115 genera and 3200 species (Lawrence) of wide distribution in the temperate regions. In the Himalayas the following genera are common: *Potentilla*, *Geum*, *Agrimonia*, *Fragaria*, *Spiraea*, *Rubus*, *Pyrus*, *Prunus* etc.

Vegetative Characters

Habit

The plant may be annual or perennial herbs, or climbers of diverse habits showing variation in morphological character; may be scrambling bushes (*Rosa*) or trees and undershrubs (*Spiraea*, *Pyrus*, *Prunus*).

Root

Branched tap root taking part in the vegetative reproduction. The adventitious leafy buds are present on roots of Cherry (*Prunus* sp.).

Stem

Herbaceous or woody, spiny; sometimes some of the species are trailing by means of suckers (runner in *Fragaria* and sucker in *Rubus*). The branches of *Rubus fruticosus* touches the soil and develop into new plants.

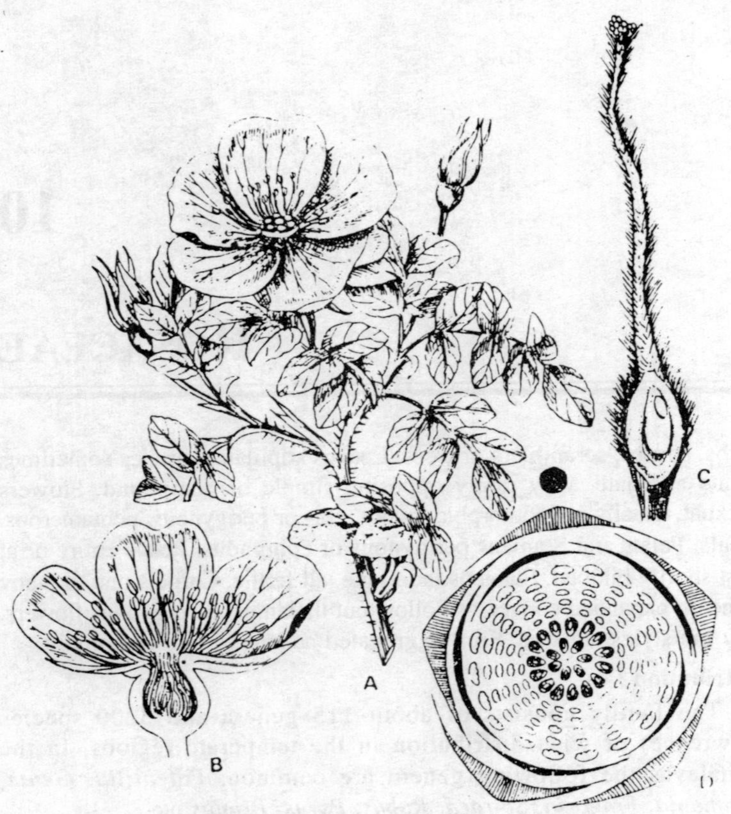

Fig. 10.1. Rosa species. A—Twig; B—L.S. of flower; C—Ripe carpel with ovary in L.S. and D—Floral diagram.

Leaves

Alternate, petiolate, stipulate, stipules adnate (*Rosa*) or caducous (*Pyrus*), simple (*Fragaria* species *Rubus* species, *Pyrus* and *Prunus* species) or compound (*Rosa*, *Fragaria indica* and *Potentilla supina*), sometimes simple and compound are found in some genus (*Fragaria*, *Rubus* and *Pyrus*), margin of the leaf or leaflet is entire or serrate, leaf base conspicuous, unicostate reticulate venation.

Flowers

Flowers are usually hermaphrodite. Rarely they may be unisexual e.g., *Spiraea aruncus*, where the male and the female flowers are borne on different plants (dioecious). They are mostly regular, but some; times

Fig. 10.2. Prunus persica Stokes. A—Flowering branch; B—Flower; C—Flower longitudinal section; D-E—Stamens; F—Ovary transverse section.

very irregular e.g., *Parinarium* and *Parastemon*. They may be hypo, e.g., *Prunus*, *peri* e.g., *Rosa*, or even epigynous e.g., *Pyrus*. The receptacle is usually more or less hollowed, resulting in various degrees of perigyny. In *Pyrus*, the carpels fuse with the receptacle, the latter forming a part of the fruit. The ovary thus becomes fully inferior in this case.

Calyx

Calyx consists of mostly five, but sometimes four sepals which are gamosepalous. In some species of *Rosa*, a few sepals may become foliaceous (leaf-like). The calyx tube may remain free or become adnate to the ovary wall. The aestivation may be *imbricate* or *valvate*. In a few genera e.g., *Fragaria* (Strawberry), *Potentilla*, *Alchemilla* and a few others an *epicalyx* of outer and smaller leaves, (described as *bracteoles* by Collet and Hooker) alternating with the sepals is present.

Corolla

Corolla consists of mostly five, but rarely four petals which are polypetalous and imbricate in bud condition. They are rosaceous and variously coloured. Sometimes the petals are entirely absent e.g., *poterium*, *Alchemilla*, *Pygeum parviflorum*, *P. gardneri* and a few other species of this genus. In some cases on the other hand e.g., species of *Rosa* the number of petals increase to indefinite. This is due to the fact that on account of the vegetative propagation by cutting etc., the outer one or more whorls of stamens get transformed into petals or petal-like structures.

Androecium

Usually the number of stamens is indefinite (15, 60), sometimes they are 5-10, the stamens are generally arranged in one to many whorls of five each; they are perigynous around the gynoecium and arise from the hypanthium, free; the anthers are small, dithecous (two—celled), introse and dorsifixed; the dehiscence takes place by means of longitudinal splits; the filaments are usually incurved in bud.

Gynoecium

The number of carpels is one to many; the gynoecium consists of either one compound carpel (syncarpous) or many simple carpels (apocarpous) arranged in cyclic or spiral way. The carpels are usually situated within the hypanthium or the hypanthium remains adnate to the compound ovary; the ovary is either superior or inferior or half superior half inferior (i.e., perigynous condition), when syncarpous, 2-5 locules are found, the placentation is axile, and the stigmatic lobes are as many as the number of carpels. The placentation is basal when one carpel is present (apocarpous); the ovules are one to many in each carpel. The style is free or connate, the stigma is simple lobed or copitate.

Fruit

Fruits of various types are common in the family, it may be a drupe (*Prunus prinsepia*) or an etaerio of drupe (*Rubus*), achene (*Potentilla*

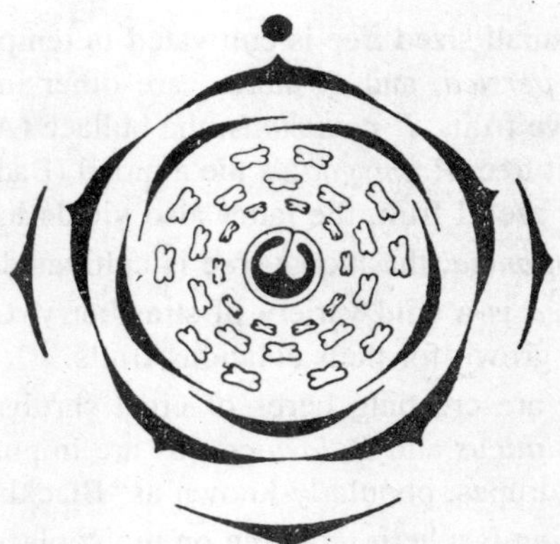

Fig. 10.3. Floral diagram of Prunus persica.

and *Genum*) or berry (*Eribotrya japonica*). In *Rosa* several achenes are enclosed within the calyx type forming a pseudobaccate fruit. The great variety of fruits in the family is due to various reasons depending on the persistence or non-persistence of receptacle, dryness or fleshyness of pericarp or receptacle, the number and form of the ripe carpel.

Pollination

Generally flowers are arranged in dense inflorescence and the petals are in attractive colour and so pollination is brought about by insects. Nectar also attracts the insects. However, the pollination through the ageney of wind may be found in *Poterium*.

Economic Importance

The economic importance of this family is too much on account of the presence of some important fruits and ornamental plants which may be described as follows:

1. *Pyrus.* It is another small sized tree of this family which yields very important and delicious fruits. *P. cummunis* is chinepear (Naspati) (syn : *P. malus*) while *P. sinesis* is nakh, (*Malus sylvestris*) is the apple, which besides being a delicious fruit, is very nourishing.
2. *Rosa*, many species of this plant are cultivated for the ornamental purposes. Everybody is familiar with the various varieties of roses, showing many colours like white, pink, red, purple, black or even green.
3. *Crataegus cranulata* is a large evergreen glabrous shrub or small tree growing on hill side.

4. *Prunus.* The small sized tree is cultivated in temperate climate for the fruits, *P. persica*, and *P. uddum* are other important species. which also give fruits. *P. communis*, the bullace (Alucha) is also an important fruit tree. *P. amygdalus* the almond (Badam) is cultivated for its highly useful fruit, the latter also yields important oil.
5. *Eryobotrya japonica*, the loquat tree is cultivated for its fruits.
6. *Fragaria indica* is a wild variety of strawberry. Other varieties of this plant are grown for their delicious fruits.
7. *Rubus.* These are creeping herbs or erect shrubs and are always prickly. *R. ellipticus* and *R. lasiocarpus* are important species and the fruits are drupes, popularly known as 'Blackberry'.
8. *Potentilla supina* is a herb, growing on moist places or near banks.

11

PAPAVERACEAE

This family contains 26 genera and 200 species distributed chiefly in north temperate regions. In India the family is represented by 5 genera and about 20 species which are mostly confined to the Himalayas.

Salient Features

Plants herbs, shrubs or rarely trees. Leaves alternate, lobed. Latex vessels present. Flower solitary or terminal, bisexual, actinomorphic. Sepals 2-3, caducous Petals 4. Stamens many, sometimes reduced to four. Carpels 2-many, parietal placentation, stigma distinct, rayed or lobed. Fruit capsule opening by valves or pores. Seeds albuminous.

Distribution

The plants are well represented in the subtropical and temperate regions of the Northern Hemisphere with centres of distribution in Western North America and eastern Asia. They also extend to the Arctic regions. Representatives of the family also occur in Europe, South Africa and Australia. Some species, however, grow as weeds throughout the world. In India, *Meconopsis* with about 26 species, grows wild in some regions of the Himalayas, *Papaver* with its two species is cultivated as an ornamental throughout India, *Argemone* occurs wild and *Eschsholtzia* is grown as a cold weather annual in the gardens.

Indian Representatives

Argemone, *Papaver*, *Meconopsis*, and *Eschsholtzia*.

General Habit

The plants are chiefly annual or perennial herbs, a few shrubs (*Dendromecon*) and rarely trees (*Bocconia*). The vegetative parts and fruits contain latex which is usually milky, white yellow or pinkish white.

Range of Vegetative Characters

Tap root extremely branched and usually a surface feeder. Stem erect branched, herbaceous or woody (*Dendromecon* and *Bocconia*), glabrous or sometimes hairy, hollow or solid, cylindrical. Leaves exstipulate, mostly alternate, in some cases upper leaves are subopposite, rarely whorled, petiolate or sessile (*Argemone*) simple, entire or lobed, cleft or cut margins with spiny teeth (*Argemone*), reticulate unicostate or multicostate.

VEGETATIVE CHARACTERS

Habit

They are mostly annual or perennial herbs having latex which may be white-yellow or pinkish coloured. Shrubs are rare (*Dendromedon* sp.) the trees are extremely rare (*Bacconia*). Some are climbing as *Corydalis* sp. and some sp. of *Fumaria* (*F. officinalis* and *F. carpreolata*) climb by means of their petiole.

Root

Branched tap root.

Stem

Erect, glabrous, branched, cylindrical and herbaceous.

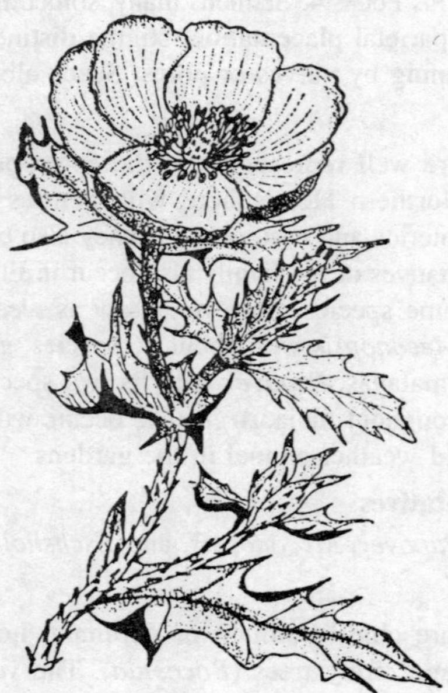

Fig. 11.1. A twig of Argemone.

Leaves

Mostly alternate, simple, upper once sometimes whorled, may be entire, lobed or cut margins with spiny teeth (*Argemone*) exstipulate.

Flowers

Flowers are mostly solitary. They may, however, be arranged in either racemose or cymose inflorescence with a cincinnal tendency. In species of *Macleaya* (China and Japan) the flowers are arranged in compound racemes (paniculate). Flowers are mostly very showy and attractive. They are bisexual, actinomorphic, complete and mostly hypogynous. In *Eschscholzia* (North America) they are perigynous.

Perianth is biseriate or triseriate.

Calyx

It consists of 2-3 sepals which are mostly free but sometimes fused e.g., *Eschscholzia.* They are usually caducous i.e., fall off in mature flowers. Sometimes the number of sepals is increased to four. They are hypogynous and concave with aestivation either twisted or imbricate.

Corolla

It consists of petals the number of, which is mostly double the number of sepals. Sometimes the number may be more (8-12). They are arranged in two whorls and are crumpled or rolled in bud. They are large, polypetalous and variously coloured, but very quickly fall off. In *Macleaya* and *Bocconia*, the corolla is absent and the flowers which are small, are arranged in compound racemes. The aestivation is imbricate.

Androecium

Androecium consists of indefinite stamens arranged in several alternating whorls. Stamens are all free and the filaments are usually slender with erect anthers which are 2-celled dehiscing longitudinally.

Gynoecium

Gynoecium consists of carpels the number of which varies from two to indefinite. It is syncarpous. Ovary is unilocular with parietal placen-

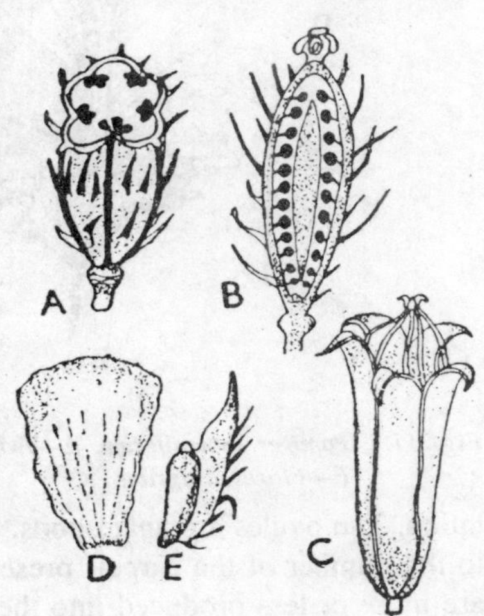

Fig. 11.2. Argemone. A—T.S. of ovary. B—L.S. of ovary. C—Dehisced fruit. D—Petal. E—Sep il.

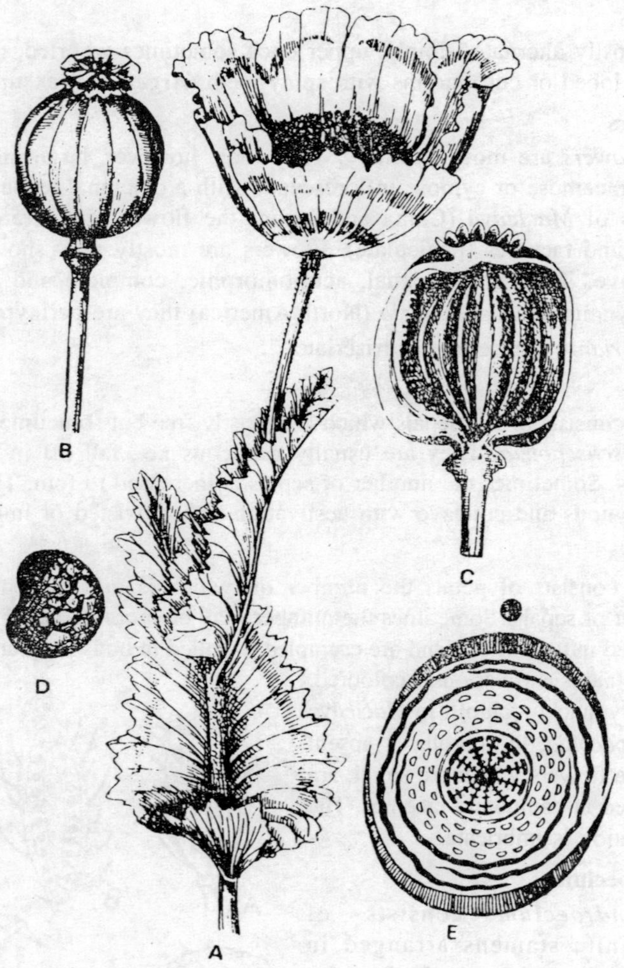

Fig. 11.3. Papaver somaniferum. A—Twig; B—Fruit; C—V.S. of fruit; D—Seed and E—Floral diagram.

tation. The ovules are anatropous. The number of placentae correspond to the number of the carpels present. Sometimes the parietal placentae are more or less produced into the ovary cavity and may in rare cases meet at the centre, thus forming a multilocular ovary. The placentation then becomes superficial e.g., *Papaver* (Poppy). Sometimes the ovary becomes spuriously bilocular by the formation of a false septum joining the placentae. The carpels rarely loosely unite, thus becoming free in fruit e.g., *Platystemon*. The stigmas lie opposite to or alternate with the placentae.

Fruit

A porous capsule (*Papaver*), prickly capsule opening by valves in *Argemone*, capsule with no spetum in *Glaucium* and indehiscent one seeded nut in *Fumaria*.

Pollination

Flowers are generally pollinated by insects as they are attractive and brightly coloured. Anthers of the flowers constitute food for the visiting insects. Stigmas of the flower are also beautiful to look at which also attract the insects. Occasionally self pollination may also take place in some flowers.

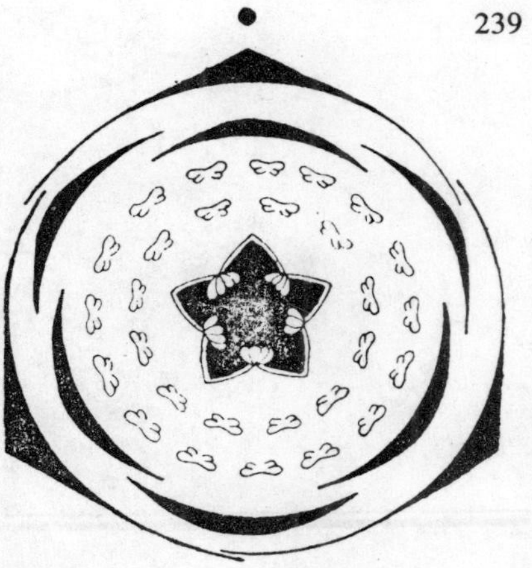

$\oplus$ ⚥ $K_3\ C_{3+3}\ A_\infty\ G_{\underline{(5)}}$

Fig. 11.4. Floral diagram of Argemone.

Economic Importance

The plants of this family are ornamental and of medicinal value. They are mentioned below:

1. *Papaver.* Its species are cultivated all over the world for the ornamental value. However, its species *P. somniferum* is economically very important as it yields the opium which is extracted from the unripe fruits. The incisions are made on the capsule which exudes latex. The latter hardens and constitutes the opium of the commerce. It is used in medicines and besides so many alkaloids it yields, morphine is most important. *P. rhoeas* is ornamental.
2. *Fumaria parviflora* is a common weed in the fields during winter.
3. *Eschscholtzia californica*, the Californian poppy is an annual cultivated during winter for its beautiful flowers.
4. *Dicentra spectabilis* is ornamental and its large drooping flowers of crimson colour look extremely beautiful.
5. *Meconopsis* is a prickly shrub with purple flowers and grows on hills.
6. *Corydalis* with spurred flowers is an erect or procumbent herb in the Himalayas.
7. *Aregmone mexicana* is a prickly herb containing yellow juice. It is a weed of the waste lands.

12

CRUCIFERAE

There is much diversity of opinion regarding the number of genera and species included in the family Cruciferae. Rendle considered 200 genera and 2000 species Shultz (1936) included 350 genera and 2500 species, Willis (1966) kept 375 genera and 3,200 species and Cronquist (1968) opined 3000 species in this family. In India the family is represented by 25 genera and 200 species.

Salient Features

Plants herbaceous with watery sap. Inflorescence typically raceme or corymbose raceme. Flowers bisexual. Sepals 4 (2 + 2). Corolla cruciform comprising 4 diagonally disposed petals which are clawed. Stamens tetradynamous. Gynoecium bicarpellary. Carpels transversely placed, syncarpous, bilocular due to the development of replum. Fruit siliqua or silicula with valves opening from base upwards. Seeds with large embryo.

Distribution

The plants are almost cosmopolitan, but the major centres of distribution are North temperate regions especially the Mediterranean regions. Some of the plants are alpine. In India, the members of this family chiefly occur in Western Himalayas and the plains of North West India. A few species, however occur in the Eastern Himalayas and plains of North India. In South India mostly cultivated species and a few weeds associated with them are found to occur.

Indian Representatives

Nasturtium, *Rorippa*, *Cornopus* (=*Senebiera*), *Cardamine*, *Farsetia*, *Cochlearia*, *Sisymbrium*, *Brassica*, *Eruca*, *Capsella*, *Lepidium*,

Raphanus, Iberis, Alyssum, Cheiranthus, Mathiolla, Malcolmia, Goldbachia, Labularia, Armoracia etc.

Vegetative Characters

Habit

The plants are mostly herbs with exstipulate leaves, may be annual as common weeds of cultivation e.g. *Brassica, Sinapis, Brassica nigra* and *Copsella bursa-pastoris* or biennials producing a rosette of radical leaves in first year and in the second year a flowering shoot which develops at the expense of tap root. Under cultivation the roots may swell as in Turnip, Radish etc. Examples of perennial plants are *Cherianthus* and *Cochlearia.* All the parts of the plant produce a pungent acid juice which is rich in sulphur content. There are speical means of vegetative reproduction in the family as by *bulbils* formed in the axil of upper leaves (*Dentaria*) or by *Coral roots* (scale bearing rootstock).

Root

Tap root swollen on account of stored food material. It may be conical, napiform or fusiform.

Stem

Herbaceous, erect, cylindrical, rarely woody, often reduced (Raphanus, Turnip) so much so that the leaves appear to be coming from the root. Hairy; hairs unicellular-stellate.

Leaves

Alternate or nearly sub-opposite; simple, exstipulate, may be cauline or radical, hairy having variously incised margins usually somewhat succulent, usually sessile and auriculate in the floral region.

Inflorescence

Inflorescence is mostly of racemose type and may be either a typical raceme, corymb or a corymbose raceme. Bracts and bracteoles are usually absent. Bracts may occasionally be present, and the presence of bracteoles can sometimes be traced in very young flowers.

Flowers

Flowers are mostly actinomorphic (regular), hermaphrodite, complete and hypogynous.

Calyx

It consists of typically four free arranged in two whorls of two each. The sepals are sometimes saccate at the base, especially the two inner

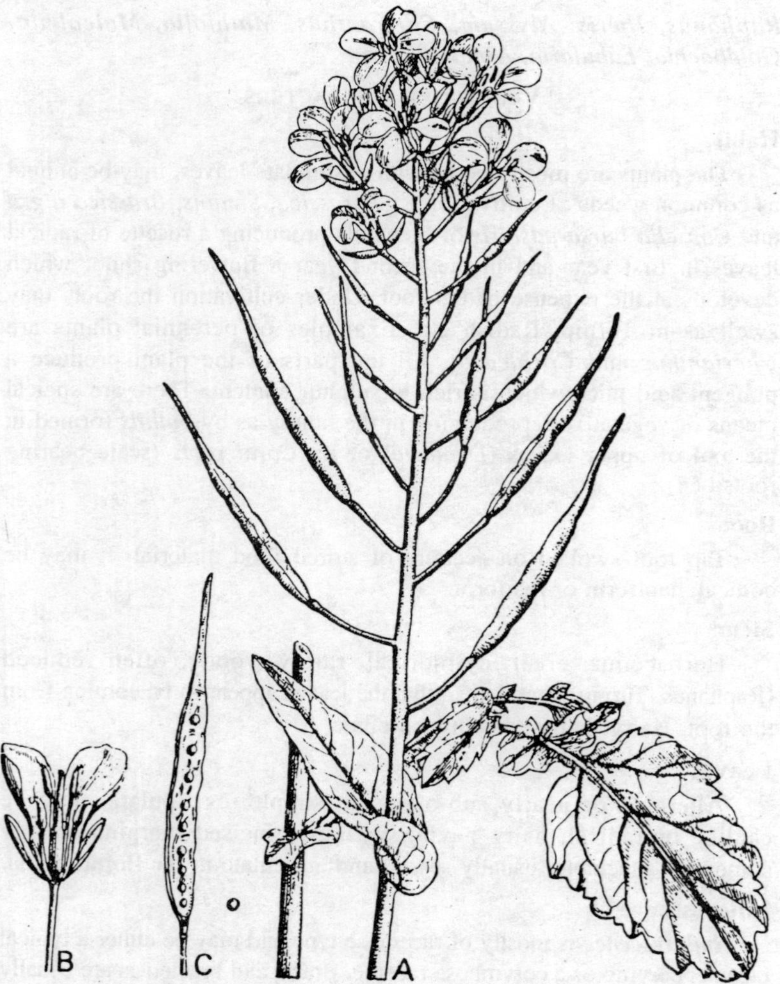

Fig. 12.1. Brassica campestris L. var. sarson Prain. A—Flowering and fruiting branch; B—Flower longitudinal section.

ones, and these sacs serve as nectar containing pouches. The aestivation of the calyx is imbricate.

Corolla

It consists of typically four petals arranged in one whorl only, and alternate with the sepals. They are polypetalous, often with long claws and spread out in the form of a Greek cross. This arrangement of the petals which is characteristic of the family is known as the *Cruciform arrangement* and the corolla is described as *Cruciform corolla*. Sometimes the flowers may become zygomorphic and depart from the normal type, by the enlargement of the two outer petals e.g., *Iberis* (Candytuft), *Teesdalia* etc. In certain cases the petals are extremely minute or may be altogether absent e.g., species of *Lepidium*, *Nasturtium* etc. In *Capsella bursa-pastoris* (Shepherd's purse) the stamens are often more or less aborted and the petals are sometimes replaced by four stamens.

Androecium

The *androecium* consists of usually six stamens which are tetradynamous, i.e., the two outer stamens are opposite the lateral sepals and the four inner ones are opposite the petals which have longer filaments than the outer ones. The filaments of two inner pairs of stamens are occasionally connate. Sometimes, as in *Alyssum*, the filaments are winged or with tooth-like appendages. The anthers are usually dithecous, introrse and opening lengthwise. Some species of *Nasturtium* have four stamens and in *Coronopus didymus* there are only two lateral stamens. A disc is often present at the bases of the stamens which has usually four basal glands opposite the sepals. The nectar secreted by these glands is poured into the pouches of the inner gibbous sepals.

Gynoecium

Bicarpellary syncarpous, ovary superior, unilocular but becomes bilocular due to formation of a false septum or replum. Placentation is prietal, ovules are many on each placenta (one in *Lepidium*). Style long, stigma bifid and glandular. May be tricarpellary (*Draba*, *Nasturtium*) and even multi-carpellary pistils have also been seen in some genera.

Fruit

Fruit is a capsule of pod like form (Capsular-pod). If it is much longer than board, it is called a *siliqua* and if much shorter with almost the same width as length it is called a *silicula*. The replum which is thin and membranous is always present in the fruit dividing it into two parts. The fruit may sometimes be a lomentum jointed and constricted between the seeds, characteristic of the order Leguminales e.g., *Raphanus*. Achene-like one-seeded fruits also occur in a few genera. The fruit dehisces usually by two valves which separate from below in an upward direction, leaving the seeds attached to the replum.

Seeds

Many, exalbuminous. At the base of the stamens are nectaries which vary in form and number in different genera. They occur on each side of the lateral stamen either as distinct outgrowths or joining or completely covering the base of the filament. Smaller glands may be between the median 4-filaments and may unite with the glands of the lateral stamens. *Lindley* and *Saunders* consider the commissures (vertical line of tissue at the junction of 2 carpels) and the replum, the number of carpels in the pistil being 4 instead of two. According to them, the commissure represent the position of 2 undeveloped carpels and the replum bearing seeds another pair of carpels.

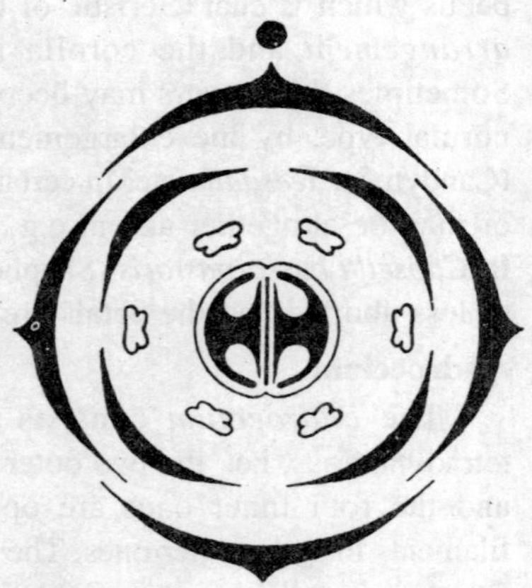

Fig. 12.2. Floral diagram of Brassica campestris var. sarson.

Pollination

The plants are self-pollinated and cross-pollinated. Many insects are not attracted by the flowers of the family. Anthers twist to felicitate the insects for self-pollination. *Pringlea* is, however, actinomorphic (It is thought that no suitable insect is available for its pollination). In fact, insects do not visit the flowers of this family as the nectar is not easily accessible to them.

Economic Importance

The family is economically very important as it includes many vegetables, plants of ornamental value and of medicinal importance. They may be described as follows:

1. *Brassica* has many important species: (i) *B. campestris* (Mustard); seeds yield oil and leaves used as vegetable. (ii) *B. nigra*, the black mustard. (iii) *B. alba*, the white mustard. (iv) *B. oleracea*, cabbage has varieties like bandh-gobhi (*B. oleracea var botrytis*) and phool gobhi (*B. oleracea var capitats*) which are well-known vegetables. In the former variety leaves serve as food, while in the latter it is the inflorescence. (v) *B. juncea*, Rye, used as condiment and also preserves pickles. (vi) *B. napus*, the Indian Rape.
2. *Matthiola*, *Alyssum* and *Iberis* (Candytuft) are cultivated for their ornamental value.
3. *Senebiera didyma,* the wart-cress is common during winter at moist places.

4. *Roripa officinale* (syn: *Nasturtium officinale*), the common watercress is cultivated in gardens. It is edible also. *R. indicum*, another species has laxative seeds.
5. *Lepidium sativum* is the common garden cress.
6. *Eruca sativa.* Seeds yield a crude oil, (resembling mustard oil) which is used in the lamps in the villages and is called *Duan* in vernacular.
7. *Cardamine hirsuta,* the bitter-cress is a weed found in winter season.
8. *Raphanus sativus* (Radish), fleshy fusiform roots make good food and are of medicinal importance.
9. *Cochlearia flava.* It is a branched glabrous annual on the banks of rivers or on damp soils.
10. *Cherianthus cheiri,* the wall flower is cultivated for its ornamental value.
11. *Sisymbrium thalianum.* The Thale-cress is an annual weed.

13

CUCURBITACEAE

This family contains about 100 genera and 850 species mostly distributed in tropical and subtropical regions of the world. In cold countries, the representation of the family is extremely poor. In this country it is very well represented by a large number of plants which are cultivated for the fruits, e.g., *Citrullus vulgaris* (Watermelon, Hindi—Tarbuz), *Curcurbita maxima* (Pumpkin, Hindi—Kaddu), *Cucumis sativus* (Cucumber, Hindi—Khira), *C. melo* (Melon, Hindi—Kharbuza), etc. The plants are mostly climbing or prostrate annual herbs. Sometimes they may be perennial and only rarely there may be shrubs or small trees.

All the plants named above are annual herbs, either climbing or prostrate. *Acanthosicyos horrida* growing on sanddunes (S. W. Africa) is a leafless spiny shrub 3—5 feet in height. The Plant possesses a very long and thick root which may go as deep as 40 feet or even more. Species of *Dendrosicyos* found in the island of Socotra are small trees.

Distribution

The plants are mainly tropical or subtropical, a few, however, occur in the temperate zones. *Ecballium* is mediterrenean. *Luffa*, *Lagenaria*, *Momardica*, *Cucurbita*, *Melothria*, *Bryonopsis*, *Trichosanthes*, *Cucumis*, *Citrullus* etc. grow throughout India. *Herpetospermum* occurs in temperate Himalayas from Simla to Bhutan. *Benincasa* grows in UP, Punjab and West Bengal. *Coccinia* is reported from the parts of Northern India. *Zanonia*, *Biswaria*, *Zehneria* have been reported from Eastern and North-Western Himalayas.

Indian Representatives

Cucurbita, *Cucumis*, *Coccinia*, *Trichosanthes*, *Herpetospermum*, *Luffa*, *Lagenaria*, *Benincasa*, *Mamordica*, *Malothria*, *Blastania*,

Corallocarpus, *Bryonopsis*, *Thladiantha*, *Actinostemma*, *Cyclanthera*, *Zanonia*, *Biswaria*, *Gymnospetalam*, *Zehneria*, *Citrullus* etc.

Vegetative Characters

Root

Tap and branched.

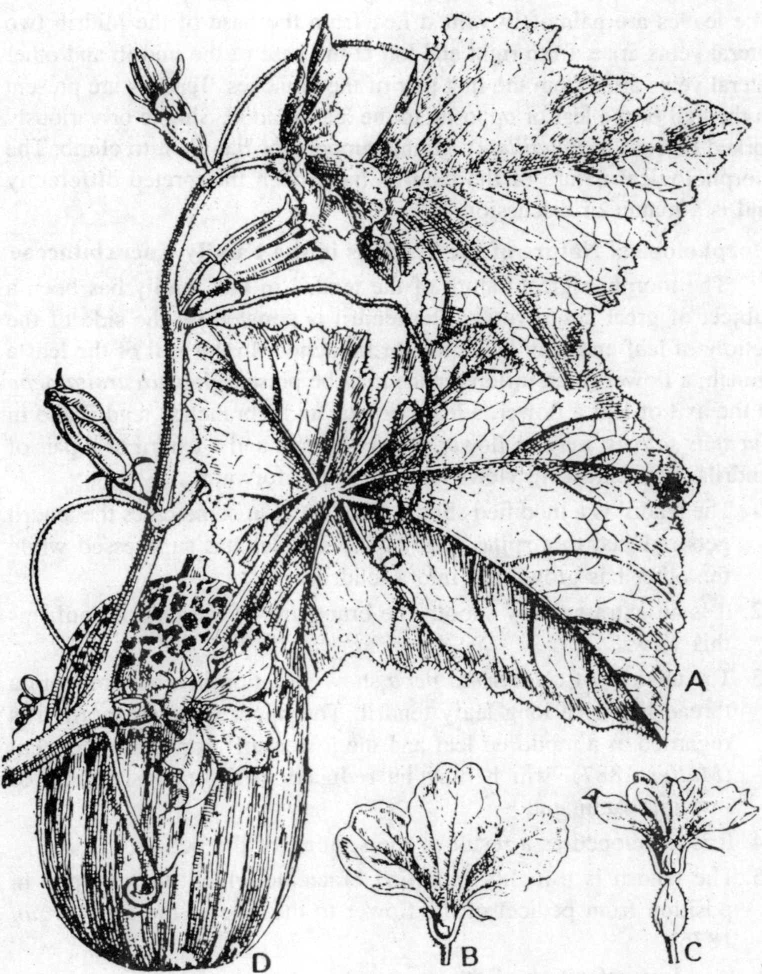

Fig. 13.1. Cucumis sativus L. A—Flowering branch; B—Staminate flower, partial longitudinal section; C—Pistillate flower; D—Fruit.

Stem

Herbaceous, climbing, hollow, often 5—angled and either glabrous, hairy or prickly (bi-collateral vascular bundles present).

Leaves

Alternate, the common arrangement being 2/5, simple, deeply lobed (palmately or pinnately), petiole long and hollow: stipules are absent. The leaves are palmately veined i.e., from the base of the midrib two lateral veins arise, from right and left at the base of the midrib and other lateral veins arise from the first pair of the branches. Tendrils are present in the axil of the leaf or opposite to the leaf at node. Simple or variously forked (absent in *Ecballium*), giving support for the plant to climb. The morphological nature of the tendrils have been interpreted differently and is a matter of discussion.

Morphological Nature of the Tendrils in the Family Cucurbitaceae

The morphological nature of the tendril in this family has been a subject of great controversy. The tendril is situated by the side of the petiole of leaf and may be simple or branched. In the axil of the leaf a branch, a flower or an inflorescence may be present. In *Curcurbita pepo* in the axil of leaf a flower, a leafy branch and a branched tendril and in *Cucumis sativus* may be flower, a leafy branch and a tendril or a pair of tendrils. The following views have been put forward:

1. The tendril is a modified stipule (*Engler*) as in some cases the tendril gets reduced to a spine. One of the stipules has suppressed while the other has grown out into a tendril.
2. It is an extra-axillary shoot. The branching of the tendril confirms this view.
3. The tendrils of *Curcurbita pepo* show variation in structure from a thread like to a long leafy tendril. The upper twinning portion is regarded as a modified leaf and the lower stiff portion as the stem (*Muller* 1887), which may be reduced to an inconspicuous or invisible rudiment.
4. It is developed as a result of the spliting of the petiole.
5. The tendril is a highly modified bracteole which has dropped in position from pedicel of the flower to the side of the leaf (*Braun* 1876).
6. It is a modification of the penduncle.

Out of these views, put forward by different authors *Muller's* view is currently the most accepted one and has also been supported by

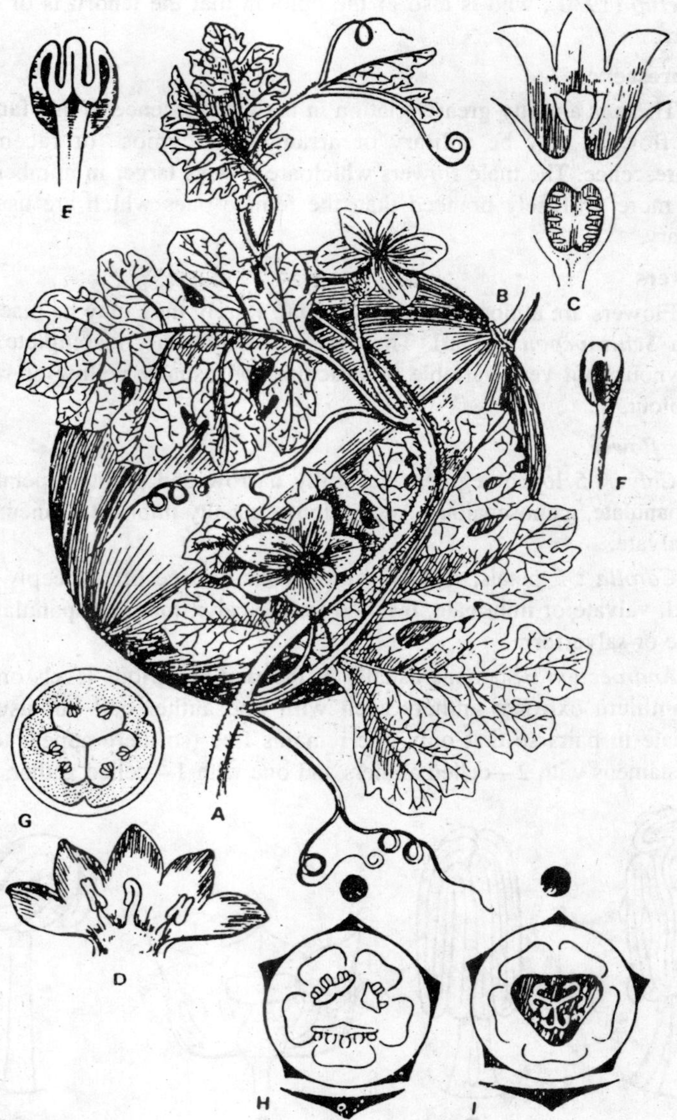

Fig. 13.2. Citrullus vulgaris. A—Twig; B—Fruit; C—L.S. of female flower; D—Splitted corolla showing stamens; E—Part of stamen; F—Loose stamen; G—T.S. of ovary, H—Floral diagram of male flower and I—Floral diagram of female flower.

Hagerup (1930), who is also of the opinion that the tendril is of dual nature.

Inflorescence

There is a pretty great variation in the inflorescence in this family. The flowers may be solitary or arranged in cymose or racemose inflorescence. The male flowers which are always larger in number are also more profusely branced than the female ones which are usually solitary.

Flowers

Flowers are almost always unisexual. Rarely they may be bisexual as in *Schizopepon* (japan). They are actinomorphic, incomplete and epigynous, but very variable in structure and usually yellow or white in colour.

Male flower

Calyx : 5 lobed segment generally narrow and pointed, petaloid, campanulate, gamosepalous. Aestivation is usually imbricate quncuncial or valvate.

Corolla : 5 petals, mostly sympetalous or free often deeply 5—lobed, valvate or imbricate, inserted on the calyx-tube, campanulate to rotate or salverform.

Androecium : Stamens usually 5, inserted at various levels on the hypanthium extrorse anthers each with one anther cell but usually connate in pairs so that only one remains free (so there appear to be two stamens with 2—celled anthers and one with 1—celled anthers and

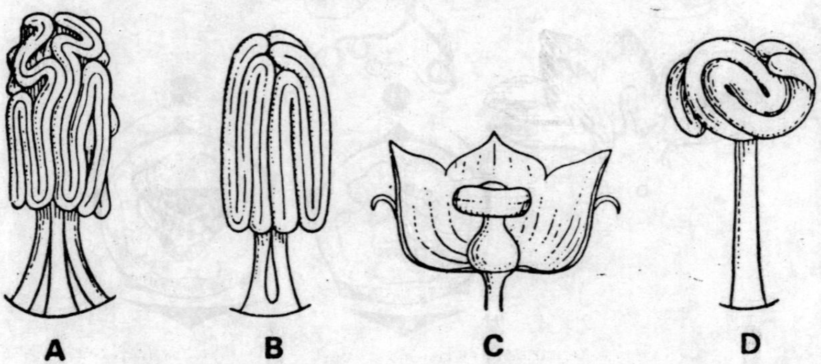

Fig. 13.3. Androecium types. A—Logenaria; B—Cucurbita, anthers united into a column; C—Cyclanthera, stamens are all united into a column; D—Sicyos, filaments as well as anthers united.

are described as three stamens), anthers often connate and their cells straight or variously curved or twisted (sigmoid). Stamens alternating with the petals. Anthers always dehisce by longitudinal slits.

Female flower

Calyx : As in male flower.

Corolla : Also as in male flower.

Androecium : Absent. Sometimes staminodes are present.

Gynoecium : 3 carpels (tricarpellary), syncarpous, ovary inferior, usually unilocular (one-celled), the three parietal placentas often however meeting and filling up the cavity of the ovary or ultimately 3—celled. Apparently the placentation looks like axile, but it is not so, the ovules do not born in the centre, because of the fusion and curving back of the placentas so, on reaching the carpellary walls the placentas fork into two and bear the ovules. The ovules usually numerous, anatropous, style short, stigma 3 to 5 usually forked.

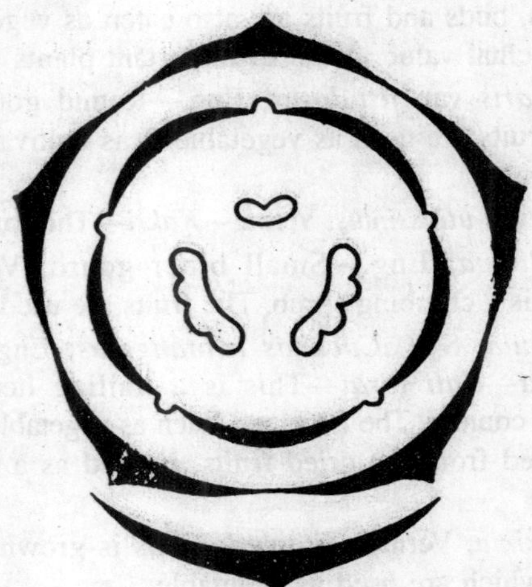

Fig. 13.4. Floral diagram of Cucumis sativus, staminate flower.

Fruits and Seeds

The *fruit* is a fleshy berry-like with soft or hard pericarp. This type of fruit is called pepo. It is usually indehiscent but rarely dehiscent as in *Herpetospermum*. The shape and size of fruit varies considerably. The *seeds* are often packed in a pulp or fibre. They are compressed and nonendospermic with a straight embryo and large and leafy cotyledons.

Pollination

The flowers are conspicuous by their colour and size (except some British species where the flowers are green) with abundant nectar secreted by the floor of the cup-shaped disc, which in male flower becomes roofed in by the stamens. The insects approach the netar through the 3 longitudinal apertures (present between the stamens) and through the top (between the upper ends of the stamens). As a result of the twisting of the anther cells, the pollengrains after dehiscence collect towards these apertures, so that, when the insect visits the flower in search of nectar, the pollengrains collect on its head and undersurface of its body. In the female flowers, the style rises in the centre of the cup and divides into three branches which are lobed broadly and the papillae of which collect the stamens brought by the insects. In this way cross-pollination takes place in Cucurbitaceae.

Economic Importance

Economically the family is fairly important. They possess edible fruits. The flowers, buds and fruits are also eaten as vegetables. Some plants are of medicinal value. A list of important plants is given here.

1. *Citrullus vulgaris* var. *fistulosus*; Eng.—Round gourd; Verna.—*Tinda*—The fruits are used as vegetable. It is cultivated chiefly in Northern India.
2. *Cucumis melo* var. *utilissimus*; Verna.—*Kakri*—The fruits are edible.
3. *Momordica dioica;* Eng.—Small bitter gourd; Verna—*Jangli Karela*—This is a climbing shrub. The fruits are used as vegetable.
4. *Luffa acutangula*; Syn. *Cucumis acutangulus*; Eng.—Vegetable sponge; Verna—*Kali torai*—This is a trailing herb cultivated throughout the country. The fruits are eaten as vegetable. The fibrous material obtined from the dried fruits, is used as a substitute for bath sponges.
5. *Cucurbita maxima*; Verna—*Sitaphal*—This is grown for its large edible fruits, which are used as vegetable.
6. *Coccinia cordifolia*; Syn. *C. indica*; Verna—*Kanduri*—This is a twining or spreading shrub. The fruits are used as vegetable. It is cultivated in Assam, Bihar, Bengal, Orissa, Maharashtra and Madras.
7. *Benincasa hispida*; syn. *B. cerifera*; Eng.—Ash gourd; Verna—*Petha*—This is a trailing or climbing herb. The fruits are edible. Sweets are prepared from the pulp of fruits.
8. *Momordica tuberosa*; Syn. *Luffa tuberosa*; Verna—*Kadavanchi*—This is a trailing or climbing herb, found mainly in Maharashtra and Madras. The fruits are eaten as vegetable.

9. *Trichosanthes cucumerina*; Verna—*Jangli chachinda, Rambel*—The fruits are edible and eaten as vegetable.
10. *Citrullus colocynthis*; Eng.—Colycynth; Bitter apple; Verna—*Tinda*—The fruits are used as vegetable. It is cultivated chiefly in Northern India.
11. *Trichosanthes anguina*; Eng.—Snake gourd; Verna—*Chichenda*—This is a climbing or trailing herb, cultivated throughout the country. The fruits are eaten as vegetable.
12. *Cucurbita moschata*; Eng.—Squash; Verna—*Mitha Kaddu*—The fruits are edible. They are eaten as vegetable.
13. *Cucumis melo*; var. *agrestis*; Eng.—small gourd; Verna—*Meki, Takmak*—The fruits are edible.
14. *Cucumis melo*; var. *momordica*; Verna.—*Phunt, Kachra*—The fruits are edible.
15. *Cucurbita pepo*; Eng.—Pumpkin; Verna—*Safed Kaddu Kumra*—The plant is commonly cultivated in Northern India. The fruits are edible.
16. *Lagenaria siceraria*; Syn. *L. vulgaris*; Eng.—Bottle gourd; Verna—*Lauki*—This is a climbing or trailing herb. It is cultivated throughout our country. The fruits are eaten as vegetable.
17. *Citrullus vulgaris*; Eng.—Water melon; Verna—*Tarbooz*—This is a native of tropical Africa but now cultivated thoughout our country for its large edible fruits.
18. *Luffa cylindrica*; Syn. *Luffa aegyptiaca*; *Momordica cylindrica*; Eng.—Vegetable sponge; Verna—*Ghia torai*—It is cultivated throughout the country for its fruits which are used as vegetable. The dried fruits yield a spongy substance which is used as a bath sponge.
19. *Cucumis anguria*; Eng.—Gherkin—This is a creeping herb, introduced from Canada. The fruits are edible.
20. *Momordica cochinchinensis*; Verna—*Bhat Karela*—This is cultivated throughout India for the fruits which are used as vegetable.
21. *Luffa echinata;* Verna—*Chagarabela*—It is found in Bihar, Bengal, Gujarat and Dehradun. The fruits possess medicinal properties and are used to cure dropsy. They are also given as purgative.
22. *Momordica charantia*; Eng.—Bitter gourd; Verna—*Karela*—This is climbing or trailing herb. This is cultivated throughout the country for its fruits which are eaten as vegetable.

23. *Momordica balsamina*; Eng.—Balsam apple; Verna—*Mokha*—This is climbing or spreading herb. The fruits are edible.
24. *Momordica subangulata*—This is a climbing shrub, grown as an ornamental.
25. *Cucumis sativus*; Eng.—Cucumber; Verna.—*Khira*—The fruits are edible. They are used in salad.
26. *Cucumis melo*; Eng.—Musk melon; Verna—*Kharbuza*—It is cultivated in Northern India. The fruits are edible.
27. *Trichosanthes dioica*; Verna—*Parwal*—It is cultivated commonly in Northern India, Assam and Bengal. The fruits are eaten as vegetable. The leaves are also consumed as vegetable.

14

MALVACEAE

The family includes 75 genera and 1000 species, mostly confined to the tropical and subtropical regions of the world. In India the family is represented by 22 genera and about 110 species occurring mostly in the warmer parts. The well known examples of the family are species of *Hibiscus* (Rose mallow), *Gossypium* (Cotton), *Malva* (Mallow) and *Abutilon* (Flowering Maple).

Salient Features

Plants mostly herbaceous, glandular and stellate hairs and mucilage cells or canals present on vegetative parts. Epicalyx present in most genera. Stamens monadelphous and epipetalous, anthers monothecous, pollen grains minute spiny and mostly multiporate. Placentation axile. Fruit a capsule or schizocarp.

Distribution

The members of the family are widely distributed over the globe. The number of the species increases as we approach tropics. In warmer parts the plants are often shrubby and large. The plants are less common in colder regions. In India the members of the family are found commonly as wild or cultivated. *Malva* (4 species) : *M. sylvestris* as an ornamental plant and *M. parviflora* and *M. verticellata* as a wild plant is South India, Bengal, Punjab and Rajasthan; *Althea* (2 species). *A. rosea* (Hollyhock) as an garden plant and *A. officinalis* a wild plant in Kashmere; *Malvestrum* (*M. Coromandalianum* = *M. tricuspidatum*) a wild undershrub in Northern India, Punjab, Haryana, Delhi, South India and Bengal; *Sida* (6 species) growing wildly in various parts of the country; *Abutilon* (5 species); *A. indicum* the most common wild representative occurring in different parts; *Hibiscus* (16 species); *H. rosasiensis* grown as an ornamental plant; *Gossypium* (6 species) grown as a chief fibre plant, are some commonly distributed members of this family.

Indian Representatives

Abutilon, Hibiscus, Malva, Malvestrum, Althea, Gossypium, Sida, Malachra, Pavonia, Urena, Thespesia, Decaschistia, Senra, Abelmoscus.

Vegetative Characters

Habit

The plants of this family are annual herbs (*Abutilon*, *Malva*, *Urena*) sometimes perennial shrubs or tree (*Levatera*) occurring both as wild and as cultivated; with a mucilaginous sap in all parts. Trees are usually with butresses. Species of *Sida*, *Hibiscus rososinesis* are undershrubs.

Root

Tap root system.

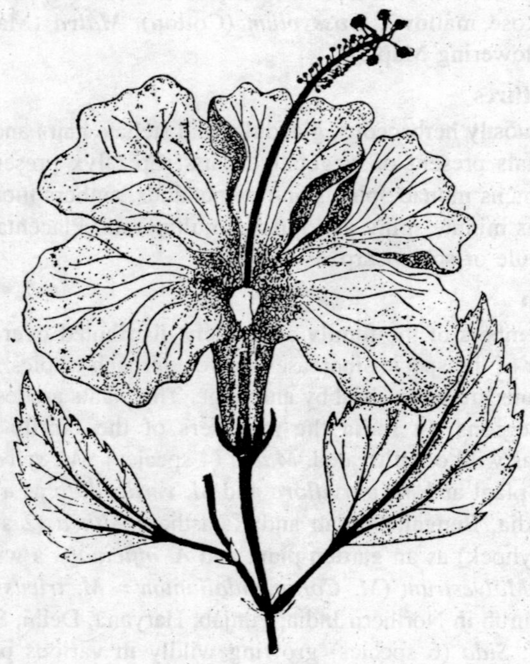

Fig. 14.1. A twig of Hibiscus rosa-sinensis.

Stem

Herbaceous (*Malachra*, *Malva*) or woody (species of *Hibiscus* and *Salmalia*). Erect or sometimes spreading (*Sida*, *Malva parviflora*), cylindrical, generally hairy, sometimes prickly (species of *Hibiscus*), pricks become hard and conical in *Salmalia*, branched and solid. Some plants, like the species of *Hibiscus* possess mucilage in young and tender regions.

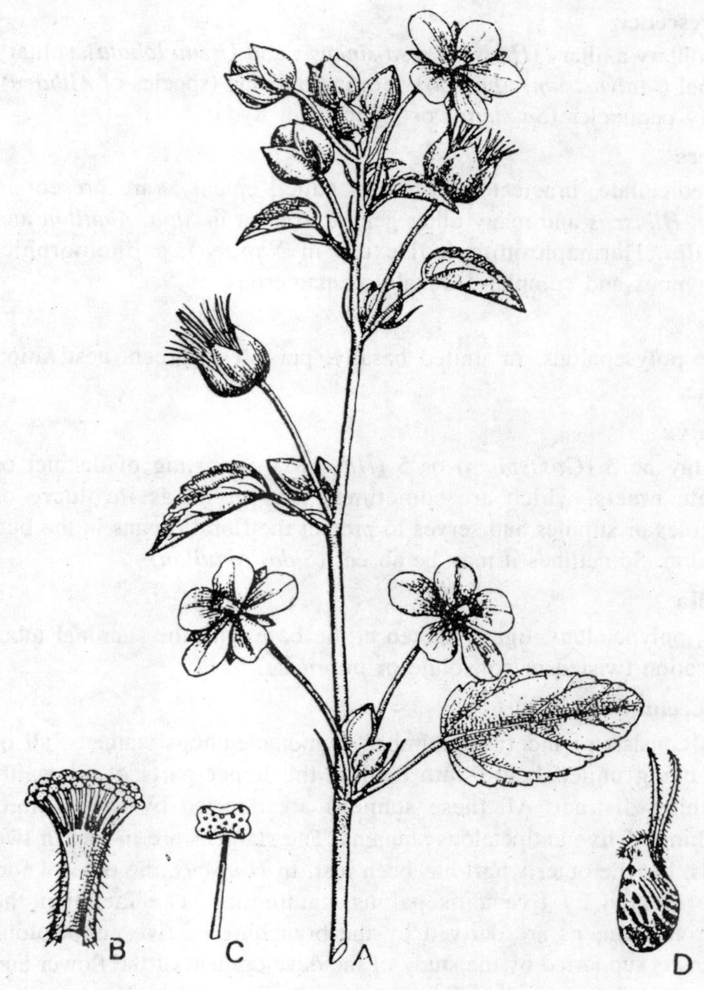

Fig. 14.2. Sida cordifolia L. A—Flowering and Fruiting branch; B—Androecium; C—Stamen; D—Schizocarp.

Leaves

The *leaves* are alternate, simple, either entire, palmately divided or variously lobed. They are usually palmately veined. Sometimes the lower leaves of a shoot are more or less rounded with entire margins while the upper leaves become palmately cut. The leaves are mostly stipulate, but the stipules are often caducous and fall off at an early stage of development.

Inflorescence

Solitary axillary (*Hibiscus rosa-sinensis* and *Urena lobata*), solitary terminal (*Malvastrum*, *Abutilon*), terminal raceme (species of *Althaea*), axillary peduncles (*Salmalia*) or panicled in *Kydia*.

Flowers

Pedicellate, bracteate, bracteoles called epicalyx are present in *Malva*, *Hibiscus* and many other genera. Absent in *Sida*, *Abutilon* and *Salmalia*. Hermaphrodite (unisexual in *Napaea*), actinomorphic, hypogynous and complete. Usually pentamerous.

Calyx

5, polysepalous, or united basally, persistent, green, aestivation valvate.

Epicalyx

May be 3 (*Gossypium*) or 5 (*Hibiscus*) consisting of distinct or connate bracts, which are sometimes interpreted as involucre or bracteoles or stipules and serves to protect the floral organs in the bud condition. Sometimes it may be absent (*Sida*, *Abutilon*).

Corolla

5, polypetalous, lightly united at the base with the staminal tube. Aestivation twisted or convolute or imbricate.

Androecium

It consists of indefinite number of monadelphous stamens, all of them being united below into a tube, the upper parts of filaments remaining distinct. All these stamens are derived by the copious branching of five antipetalous stamens. The stamens are in fact in two whorls, but the outer whorl has been lost. In *Hibiscus*, the outer whorl is represented by five antisepalous staminodes. The fact that the numerous stamens are derived by the branching of five antipetalous stamens is supported by the study of the development of the flower and by the fact that in some wild species only five antipetalous stamens have been observe. The staminal tube becomes united with the corolla

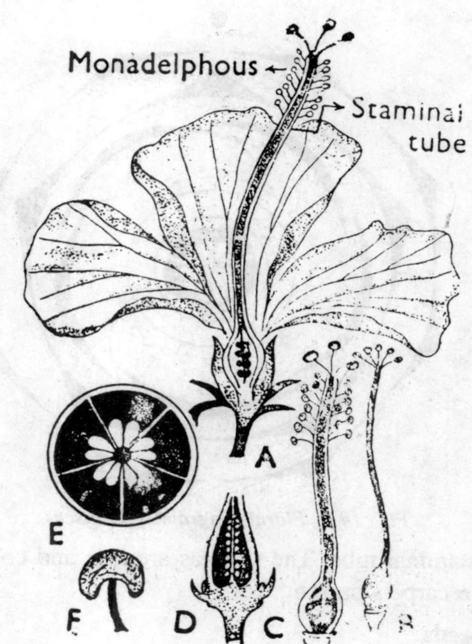

Fig. 14.3. Hibiscus rosa-sinensis. A—L.S. of the flower showing monadelphous condition and staminal tube; B—Carpel and five stigmas; C—L.S. of the carpel showing long style passing through the staminal tube; D—L.S. of the ovary; E—T.S. of the ovary; F—Filament and the anther.

(epipetalous), and make the petals to appear gamopetalous. The anthers are one-celled (monothecous), reniform (kidney shaped) and open by a slit which runs across the top dividing the anther into two valves. They are extrorse (turn outwards) at the time of dehiscence. The pollengrains are large, spherical and spiny.

Gynoecium

It consists of carpels the number of which varies greatly in different genera. Usually the number is either five or ten, but sometimes the carpels may be less than five or more than ten. In *Hibiscus* the carpels are five. In *Althaea rosea* (Hollyhock) they are ten. In *Abutilon* the number is indefinite (15-20 in *A. indicum*). In Kydia the number is reduced to three only. The gynoecium is syncarpous. The ovary is superior, penta or multilocular with axile placentation. The number of ovules varies from one to many in each loculus. They are anatropous and mostly ascending, but sometimes pendulous. The style passes

Fig. 14.4. Floral diagram of Hibiscus.

through the staminal tube. The stigmas are free and correspond to the number of the carpels present.

Fruits and Seeds

The *fruit* is a loculicidal capsule as in *Hibiscus* and *Gossypium* or more often it is dry and indehiscent or dehiscent. In *Sida, Malva* and *Abutilon* schizocarps separate from one another and from the persistent central axis and each is one seeded or occasionally two to many seeded as in some species of *Abutilon*. In *Sida* schizocarps dehisce irregularly or by a small chick. In some species of *Urenā* the mature carpels

Fig. 14.5. Floral diagram of Sida cordifolia.

are covered with hooked bristles. The fruit is a fleshy berry in *Malvaviscus*.

The *seeds* are reniform or obovoid with a scanty endosperm which is often mucilaginous, and a curved embryo. They are often pubescent or densely clothed with woolly hairs as in *Gossypium*.

Pollination and Dispersal

The flowers are mostly insect pollinated. The insects are attracted by the large size and bright colours of the flowers. The nectar is secreted at the bases of the sepals or petals. Extrafloral nectaries are present in many species of *Gossypium*. The humming birds are common pollinators in *Abutilon* and *Gossypium*. *Malva rotundifolia* is self pollinated.

The seeds of *Gossypium* and *Bombax*, which have a hairy covering are dispersed by wind. The seeds of several species of *Malva* are buoyant and this favours their dispersal by water. In some species such as *Urena lobata* the seeds have hooked spines and they are dispersed by adhesion to animals and humans. Ants help in dispersal of seeds in *Sida* and *Abutilon*.

Economic Importance

The family is of great economic value as mentioned below:

1. *Salmalia* and *Adinsonia* are huge trees grown in gardens and on roads. The former is popularly known as silk cotton tree and yields fine silken structures. These are, in fact, the outgrowths of the inner wall of the capsule and they surround the seeds. Sometimes pillows are stuffed by these fine and delicate outgrowths. The common species with palmately compound leaves and red beautiful flowers without epicalyx is *S. malabaricum* (Syn : *Bombax malabaricum*). *Adinsonia digitata* is a big sized tree, but both these genera are now put in a separate family.
2. *Hibiscus rosa–sinensis*, the popular shoe-flower is highly ornamental. The name shoe-flower is given to it because the mucilage from the petals was formerly used in place of shoe polish. When the latter was not known. *Abelmoschus esculentus* (syn: *H. esculentus*), commonly known as lady's finger (*Bhindi*) is cultivated for its fruits which form a good vegetable. Another species *H. cannabis* is cultivated for its fruits which form a good vegetable. Another species *H. cannabis* is cultivated widely for the popular fibre, hemp. The latter is used in making ropes.
3. *Althaea*, the common Hollyhock is ornamental and is, therefore, cultivated in gardens and houses. *A. rosea* and its other varieties flower from March to May. It is a pubescent herb or shrub.

4. *Gossypium.* This is the most important genus of the family. The cotton of the commerce is taken out from the capsules of this genus. The cotton is thehairy outgrowth of the seeds. Its common species are *G. herbaceum* and *G. neglectusm*. Other species cultivated for commercial purposes are *G. indicum*, *G. hirsutum* and *G. peruvianum*. The tree cotton is *G. arboreum*.
5. *Malachra*, *Urena lobata* and *Pavonia odorata* are perennial shrubs or undershrubs growing in forests and woods.
6. *Malva*, *Malvastrum* and *Sida* grow wild almost throughout the year on waste lands and fields.

15

LEGUMINOSAE

The plants are herbs, shrubs and trees. Flowers are bisexual, generally zygomorphic, sometimes actinomorphic. Sepals 5, odd sepal anterior or more or less united with imbricate or valvate aestivation (in regular flowers). Petals free, united in zygomorphic flowers, alternating with the sepals. Stamens 10, slightly perigynous; free or more or less united in a tube—one or more diadelphous. Monocarpellary, style is simple, ovule 1 per loc. or many on 2 alternating rows along ventral suture. Fruit is generally legume or pod.

Distribution

This is the third largest family of the flowering plants. It is world-wide in distribution. The family contains about 600 genera and 1200 species (Willis). It is represented in almost all parts of the world; the subfamilies Mimosoideae, and Caesalpinoideae are mostly represented in tropical parts while the sub-family Papilionateae is mostly restricted to temperate regions. Tropical rain forests are rich in trees of Mimosoideae and Caesalpinoideae while the papilionaceous herbs are present and the arboreal forms are few in sub-tropical forests. In the regions of deciduous forests, where growth is interrupted by cold winter, leguminous trees are rare.

VEGETATIVE CHARACTERS

Habit

The plants differ very much in habit, may be herbs, shrubs, climbers or trees. They may be inhabitants of dry and moist places. Some are aquatic, others are xerophytic and may be found in all types of climate.

Root

Tap root, branched. There are swellings or tubercle on the secondary roots which contain a large number of bacteria, fixing the atmospheric nitrogen into the soil in the form of organic nitrogenous compound. The nitrogen fixing bacteria as *Bacilus radicicola* enter the young roots and multiply rapidly so as to form a swelling on the root. In some cases the roots are tuberous (*Vigna vexillata*, *Dolichos* (some sp.) *falcatus*, *Lathyrus macrorhizus*).

Stem

It is usually erect, but may be climbing (*Bauhinia*) or twining (*Dolichos*) or creeping (*Parochetus communis*). The stem may climb by means of their tendrils (*Lathyrus aphaca*, *Pisum sativum* or hooks *Caesalpinia*).

Leaves

Mostly alternate, simple or compound, pinnate, bipinnate, stipulate (stipules leafy as in *Lathyrus aphaca*), or reduced to spines as in *Acacia*, *Mimosa* or small and toothed as in *Vicia sativa*. In some (*Acacia*) petiole of the leaf get flattened called *phyllode*. Some of the leaves are very sensitive to touch (*Mimosa pudica*) while others show sleeping movements (*Desmodium gyrans*, *Albizzia*); there is a pulvinus at the base of the leaves.

Floral Characters

Inflorescence

It varies very much in the family. It may be racemose (*Melilotus*, *Trigonella*), panicle, solitary axillary (*Cicer*), terminal (*Dalbergia*) and commonly cymose head (*Acacia* sp.).

Flowers

Actinomorphic, or zygomorphic, complete, bisexual, hypogynous or perigynous, generally bracteate.

Calyx

Calyx 4-5, gamosepalous, sometimes poly, odd sepal anterior, sometimes bilipped.

Corolla

Corolla 4-5 (fifth one supp-ressed), polypetalous, aestivation valvate (Mimosoidae), ascending imbricate (Caesalpinoidae) of descending imbricate (Papilionatae).

Androecium

10 rarely few (numerous in *Acacia* and *Albizzia*) free or variously united, filaments thread-like sometimes dilated toward tip, stamens 2-celled, uniform.

Gynoecium

Monocarpellary, ovary superior, unilocular, placentation marginal, ovules many in two alternating whorls on the ventral suture. Style simple, slender; stigma usually small, terminal.

Fruit

Typically a legume or a pod (about 3 m. long in *Entada*), flattened, round or like (*Cassia fistual*), coiled (*Medicago*). Sometimes a lomentum dehiscing both by dorsal and ventral suture; in a few genera breaking up into indehiscent one seeded joints. In *Dalbergia* fruits are indehiscent, one seeded and winged.

Seeds

Attached along the upper margin of the valves alternatley sometimes spearated from one another by partitions or by constrictions of the pod. Exalbuminous or little in Mimosoidae and Papilionatae. In *Caesalpinia* and *Bauhinia* it is present as a thin layer round the embryo. In some genera seed coat is copious and coloured red in *Abrus precatorius*).

PAPILIONACEAE (PEA FAMILY)

This is the largest family of order Leguminales and is spread over 375 genera. In India the family is represented by about 70 genera and 867 species.

Salient Features

Herbs, shrubs or trees, often climbing. Leaves alternate, stipulate, simple or compound. Flowers, bisexual, zygomorphic. Sepals more or less connate. Petals 5, papilionaceous. Stamens 10 or 9, mono- or diadelphous, rarely free. Carpel one, ovules many on ventral suture. Fruit a legume, sometimes indehiscent or lomentaceous.

Distribution

Plants are distributed mostly in temperate regions of both northern and southern hemisphere. The representative of this family grow all over India in plains as well as on hills upto 3000 to 8000 feet. They also grow as halophytes along sea shore. The family belongs predominantly to drier regions and its member show a decline as we proceed to the regions of heavy rainfall. It includes most of our vegetables and pulses etc. Some common genera found in India are : *Cratalaria* (18 species) growing throughout India, *Trigonella* (7 species), *Indigofera* (17 species), *Desmodium* (6 species), *Cicer* (1 species), *Glycine* (2 species), *Dolichos* (3 species), *Lens* (1 species), *Abrus* (2 species), etc.

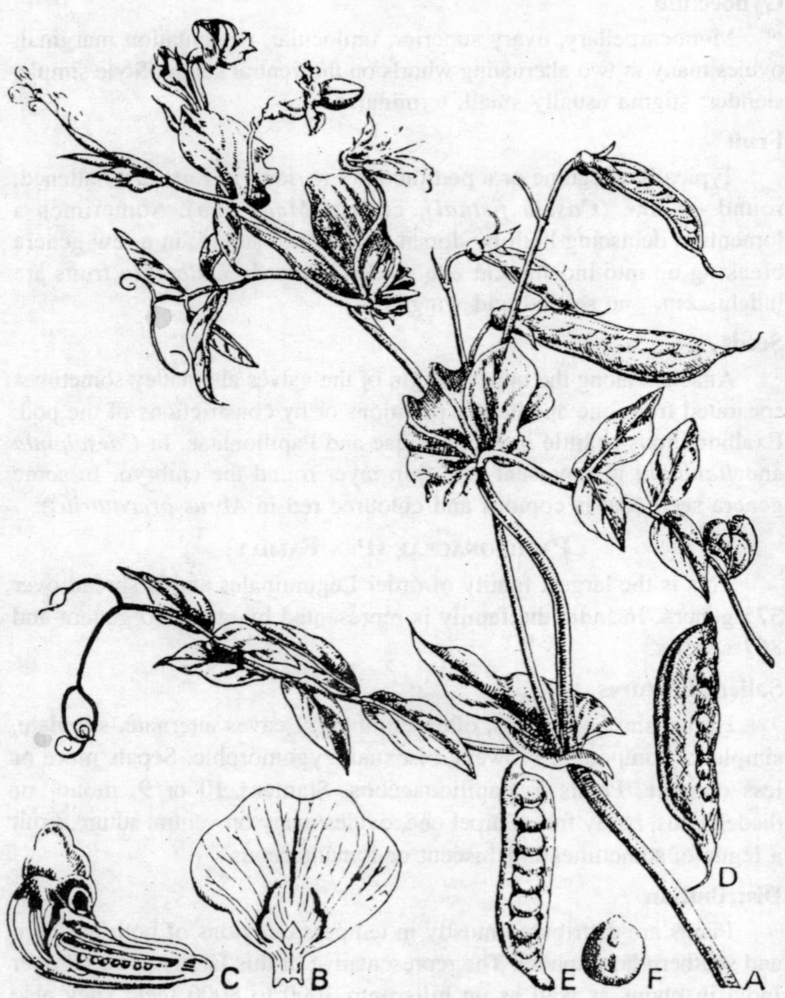

Fig. 15.1. Pisum sativum L. A—Flowering and fruiting branch; B—standard petal; C—flower longitudinal section; D—Pod; E—Pod longitudinal cut; F—Seed.

Indian Representatives

Arachis, Astragalus, Alysicarpus, Alhagi, Atylosia, Abrus, Butea, Crotalaria, Canavalia, Clitoria, Cajanus, Gyamopsis, Dolichos, Desmodium, Dalbergia, Derris, Durmasia, Erythrina, Indigofera, Lens, Lespedeza, Lathyrus, Melilotus, Medicago, Pterocarpus, Pisum, Phaseolus, Pongamia, Sesbaenia, Sophora, Stauteria, Smithiama, Trigonella, Tephrosia, Uraria, Vicia, Zorina etc.

VEGETATIVE CHARACTERS

Habit

Herbs (*Desmodium, Melilotus, Phaseolus*), shrubs (*Cajanus, Butea*) and trees (*Dalbergia, Sesbaenia, Erythrina, Pongamia*).

Root

Tap root system. The roots of the members of this family are characterized by the presence of nodules which make good home for nitrogen fixing bacteria. These bacteria fix the nitrogen of the soil and that is why the papilionaceous crops are sown in rotation with other crops. This is called "Rotation of Crops".

Stem

Herbaceous (*Pisum, Desmodium* and *Melilotus*) to woody (*Dalberia, Sesbania* and *Cajanus*). Sometimes twining (*Dolichos, Mucuna*, and *Vigna*), climbing (*Clitoria* and *Shuteria*) or erect, glabrous (*Trigonella*) or with prickly branches (*Erythrina*), cylindrical and branched.

Leaves

Alternate rarely opposite, petiolate, stipulate, stipules free, adnate (*Medicago* and *Trifolium*) or sometimes foliaceous as in *Pisum*, occasionally simple (*Heylandia* and *Crotalaria*) but generally compound, may be pinnate (*Melilotus, Tephrosia, Millettia, Cicer* and *Lens* etc.) or palmate (*Butea, Erythrina, Cajanus, Phaseolus* etc.), leaves or leaflets of various shapes, leaf base pulvinous, leaf or leaflets show unicostate reticulate venation. Leaves occasionally show characteristic, movement e.g., *Desmodium gyrans*.

Inflorescence

Inflorescence is usually of racemose type, but variation occur in different genera. In *Crotalaria*, the flowers are arranged in terminal or leaf opposed racemes. In *Melilotus, Cajanus, Trigonella* etc., also the flowers are arranged in racemes. In *Cajanus indicus* (Hindi—Arhar), the flowers are arranged in corymbose racemes or form a terminal panicle (Raceme or corymb with branching pedicles). In *Dolichos biflorus*, the inflorescence is axillary with 1—3 flowers in the axils of leaves. In *Cicer*

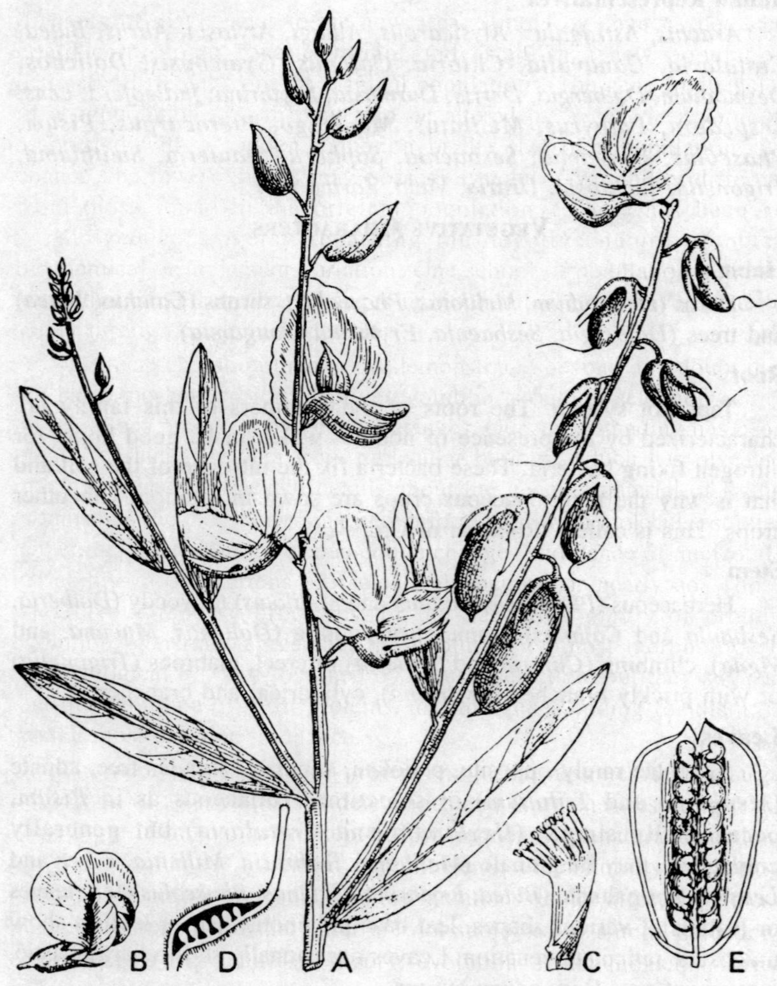

Fig. 15.2. Crotalaria juncea L. A—Flowering and fruiting branch; B—Flower; C—Expanded androcium; D—Gynoecium longitudinal section; E—Pod longitudinally cut.

(gram) the flowers are solitary axillary. In *Dalbergia*, the inflorescence is either a terminal or lateral panicle.

Flowers

The *flowers* are often bracteate as well as bracteolate. Sometimes the bracts are foliaceous and persistent as in *Flemingia*. They are complete, zygomorphic, hermaphrodite, pentamerous and hypogynous or perigynous.

Calax

The *calyx* is composed of usually five sepals which are more or less united into a tube. They show valvate or imbricate aestivation in bud. The odd sepal is usually anterior.

Corolla

The *corolla* is papilionaceous. The five petals are unequal and have a bilateral symmertry. The posterior (outer most) petal which is largest is called standard. The lateral pair or petals (one on either side of the standard) which are similar to each other and often clawed are called the wings. The two anterior petals are united to form the keel which encloses the stamens and carpel. The petals show descending imbricate aestivation. In *Amorpha* wings and keel are absent and in some species of *Lespedeza* the flowers are apetalous.

Androecium

The *androecium* is composed of ten stamens which are arranged in a single whorl in mature flower. The stamens are usually diadelphous. The filaments of nine stamens are united into a long tube which encloses the ovary and tenth posterior stamen remains free. Sometimes stamens are monadelphous as in *Crotalaria*. Anthers are usually uniform but sometimes, as in *Crotalaria*, they are dimorphous. They are dithecous, introrse and dehisce longitudinally.

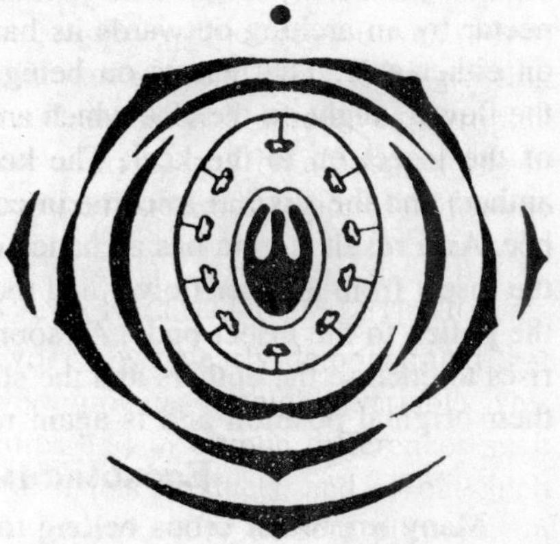

Fig. 15.3. Floral diagram of Crotalaria juncea.

Gynoecium

The *gynoecium* is monocarpellary with a superior or partly inferior unilocular ovary. The ovules are many to several on the ventral suture and the placentation is marginal. The style is simple and the stigma is capitate or terminal.

Fruits

The *fruit* is a legume or pod dehiscing by one or both sutures into two valves or it is indehiscent (e.g. *Dalbergia* and *Pongamia*) or sometimes it is jointed separating into one-seeded parts (*Alysicarpus*). The fruits of *Arachis hypogaea* develop undergound. After fertilization the flowers in the axils of lower leaves turn downward and get burried into the soil.

Seeds

The *seeds* are nonendospermic or with a scanty endosperm and with a large curved embryo.

Pollination

The flowers are specially adapted for cross-pollination by insects, the standard petal acts as a flag apparatus rendering the flowers conspicuous, the two lateral wings serve the purpose of platform for alighting the insects and the keel protects the anthers and gynoecium from rains. Besides that the standard also serves as a fulcrum against which the bee pushes its head whilst standing on wings. Nectar is secreted as the base of the filaments round the ovary and a passage is established through the slit of the staminal tube enclosing the ovary and other connate 9 stamens. The wings and standard also acts as mechanical device for closing and opening the passage slit.

Only the long tongued insects can enter and make a passage to the nectar by an arching outwards its base or of bases adjoining filaments on either side. The insects on being attracted by the bright colour of the flower, alight on the alae which are depressed and transfer the weight of the insect on to the keel. The keel on being depressed allows the anthers and the stigmas to come in contact with the undersurface of the bee. As a result stigma has a chance of receiving the pollen brought by the insect from another flower and the anther give over a new charge of the pollen to the insect body. As soon as the insect flies, the keel again rises to enclose the anthers and the stigmas and the various parts regain their original position and is again ready for another visit.

ECONOMIC IMPORTANCE

Many important crops belong to this sub-family and therefore its economic importance is too much. Besides, there are many plants which provide other materials of commerce which are highly useful. Some of them are described below:

1. *Clitoria tarnatea* is an ornamental climber, and *Erythrina suberosa* is a beautiful tree with scarlet red flowers. *Butea frondosa* besides being an ornamental perennial shrub yields a beautiful yellow dye from its flowers which is chiefly used during holi festival. The flowers are sold in the market under the name of '*Tesu*'.

2. *Arachis hypogaea*, the common ground nut is an important crops. Its seeds are eaten and yield vegetable oil which is of great use in the industry of vegetable oils.
3. *Lathyrus sativus* (*Matar*). The plants are used as fodder while seeds form an important pulse.
4. *Cicer arietinum* or gram is one of the most important and widely cultivated crops of the country.
5. *Crotolaria.* It includes about 250 species and is an important genus of the sub-family. Some species are ornamental, some are grown for fodder and few are medicinally important. *C. medicaginea* is a species which is medicinally useful. *C. juncea* is of special mention because it yields sun hemp, a very important fibre. Some other species also yield fibre.
6. *Dlabergia latifolia* and *D. sissoo* are the well known timber trees and are largely cultivated for their valuable wood.
7. *Desmodium* with its 150 species is another important plant. *D. gyrans*, the well known telegraph plant shows characteristic movement.
8. *Melilotus.* It includes two important species *M. indica* and *M. alba* which make good fodder for the cattle.
9. *Trigonella* is also an important fodder. Its leaves are used as vegetable (methi. *T. foenum-graecum*).
10. *Cajanus indicus* commonly known as pigeon-pea (arhar) is an important crop and yields the pulse which is very much eaten throughout the country.
11. *Aeschynomene aspera*, the sola plant is important as it is used in the manufacture of pith-hats or sola-hats.
12. *Phaseolus.* It is an important genus and includes many useful pulses like *P. lunatus* (*Lima beam*), *P. vulgaris* (*French Bean*), *P. aconitifolius* (*moth*), *P. radiatus* (*moong*) and *P. mugo* Roxburghii (*Urd*) which are used commonly in the different parts of the country.
13. *Vigna* (*lobia*), *Dolichos* (*sem*) yield nutritive pods which are used as vegetables.
14. *Indigofera* with its 250 species is another important plant of the family. Of special mention are two species *I. tinctoria* and *I. articulata* which yield '*Neel*'.
15. *Sesbania aegyptiaca* forms a good hedge because its growth is very quick. *S. grandiflora* is a beautiful tree with large white flowers. The latter are offered to the Deity and are supposed to be sacred.

16. *Alhagi camelorum* is a wild herb given to the camels as fodder.
17. *Abrus precatorius* is a wild herb and its seeds, called '*Ratti*' are used by the jewellers as weights. Seeds are also medicinally important.
18. *Lens esculenta,* the lentil (*masur*) is an important pulse. *Vicia faba* (Broad bean) forms an important vegetable.
19. *Pisum sativum*, the garden pea is ornamental and forms an important crop.

MIMOSACEAE (ACACIA FAMILY)

This is the smallest family amongst families of the order Leguminales. The family is spread over 40 genera (Rendle, 1959) and 2,000 species. In India the family is represented by 12 genera and about 90 species.

Salient Features

Trees or shrubs. Leaves pinnate. Flowers arranged in spherical heads or spikes in lax receme or umbel. Flower small 4 or 5 merous, actinomorphic. Sepals fused, valvate. Petals valvate. Stamens equal to the number of petals or indefinite, free, pollen grains united in packets. Carpel 1, superior, unilocular, placentation marginal. Fruit a legume or pod.

Distribution

The plants are distributed in tropical and sub-tropical parts of the world. Many of them are found in dry regions. In India, following genera are very commonly found. *Acacia* with its 25 species is perhaps the most common and widely distributed genus. *Acacia arabica* (Hindi — Babul) is the most common species. Some common species of *Albizzia* are : *A lucida*, *A. lebbek*, *A. procera*, *A. odoratissima*, *A. thompsoni*, *A. amara*, *A. stipulata* etc. *Ademanthera pavonia* grows wild in sub-Himalyan tract from Gorakhpur eastwards. Western Ghats and East and West Khandesh, *Calliandra* is a popular genus of Khasia hills. *Entada scandens* commonly occurs in sub-Himalyan tract from Nepal eastwards, in Western Ghats and in Andaman and Nicobar islands. *Mimosa* (*M. pudica*) is cultivated throughout the country. *Neptunia* grows in swamps and slow flowing streams throughout India. *Prosopis* is very common in Delhi, Punjab, Western U.P., Rajasthan, Gujarat, Central and South India. *Xylia* occurs very commonly in Orissa and parts of Central India.

Indian Representatives

Adenanthera, *Acacia*, *Albizzia*, *Calliandra*, *Dichrostachys*, *Entada*, *Mimosa*, *Neptunia*, *Piptandaenia*, *Prosopis*, *Pithecolabium*, *Xylia*.

Vegetative Characters

Habit

Herbs (*Mimosa pudica* and *Neptunia oleracea*), shrubs (*Acacia cancinna* and *Dichrostachys cinera*) and trees (*Acacia arabica* and *Albizzia*). Rarely climber (*Entada gigas*) is also found.

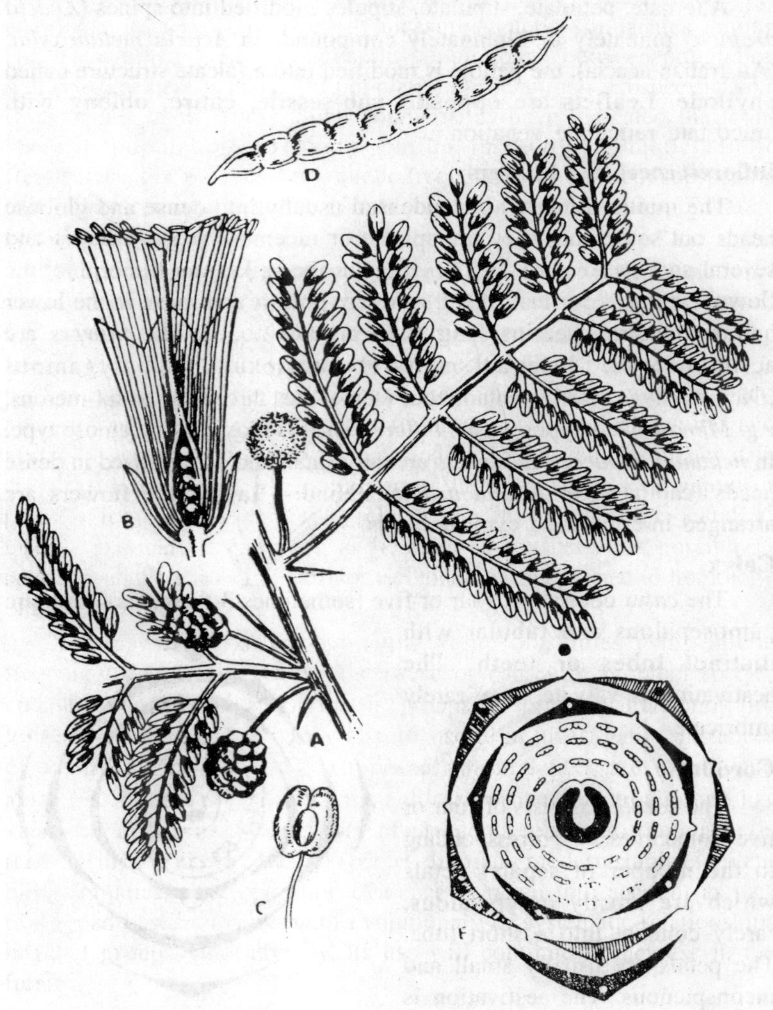

Fig. 15.4. Acacia arabica. A—Twig; B—L.S. of flower; C—Stamen; D—Fruit; and E—Floral diagram.

Stem

Woody, erect, rarely climbing, cylindrical, branched and solid. In some plants like *Acacia arabica*, gum passages are met with in the pith and medullary rays.

Leaves

Alternate, petiolate, stipulate, stipules modified into spines (*Acacia arabica*) pinnately or bipinnately compound. In *Acacia melanoxylon* (Australian acacia), the petiole is modified into a falcate structure called phyllode. Leaflets are opposite, sub-sessile, entire, oblong with unicostate reticulate venation.

Inflorescences and Flowers

The minute flowers are condensed usually into dense and globose heads but sometimes they are spicate or racemose (e.g. *Prosopis* and several species *Acacia*). In *Dicrostachys* the spikes are dimorphic, the flowers in the upper half of the spike are perfect and those in the lower half are neutral, bearing long filiform staminodes. The *flowers* are actinomorphic, bisexual or rarely unisexual or polygamous (*Dicrostachys*), mostly pentamerous (sometimes three-four-or six-merous, e.g. *Mimosa*) and hypogynous. *Inflorescence* is mostly of racemose type. In *neptunia oleracea* the flowers are very small and are arranged in dense heads (capitate). In *Acacia arabica* (Hindi—Babul), the flowers are arranged in compound cymose heads.

Calyx

The *calyx* consists of four or five (sometimes 3-6) sepals which are gamosepalous and tubular with distinct lobes or teeth. The aestivation is valvate, very rarely imbricate.

Corolla

The *corolla* consists of four or five (sometimes 3-6 corressponding to the number of sepals) petals which are mostly polypetalous, rarely connate into a short tube. The petals are usually small and inconspicuous. The aestivation is valvate. The number and cohesion of *stamens* show much variation. In

Fig. 15.5. Floral diagram of Pisum sativum.

Acacia there are numerous stamens which are free, whereas in *Albizzia*, *Pithecolobium* and other related genera indefinite stamens are monadelphous at the base. In several other genera (e.g. *Prosopis* and *Leucaena*) the stamens are twice the number of the petals and are free and diplostemonous. Sometimes, as in *Acrocarpus* and some species of *Mimosa* the stamens are as many as the petals. The filaments are long, filiform and often exerted. The anthers are small, dithecous, introrse and dehiscing longitudinally. The pollen grains are often collected into two to six masses in each anther chamber.

Gynoecium

The *gynoecium* is of a single carpel. The ovary is superior and unilocular with usually several ovules along the ventral suture. The placentation is marginal. The style and stigma are one.

Fruits and Seeds

The *fruit* is a legume or indehiscent. In several species of *Acacia* the fruit is a lomentum. It is considered between the seeds and breaks up into one-seeded segments. The fruit of *Entada* is very large and woody, composed of many one-seeded discoid joints.

The *seeds* are with a scanty endosperm or without any endosperm. Sometimes, as in *Pithecolobium dulce* the seeds are half enveloped in a pulpy aril. In *Inga* naked embryo falls from the fruit and it directly germinates.

Pollination and Dispersal

The flowers are pollinated by insects which are attracted by long exerted and beautifully coloured stamens or in some cases they visit the flowers for nectar. The seeds are dispersed by wind, birds or animals.

Important Genera

Acacia

Acacia nilotica ssp. *indica* (*Acacia arabica*). Common tree in dry places in the plains of U.P., Punjab, Haryana and Rajasthan. The bark yields gum and is used for tanning leather. The twigs are used as brushes for cleaning teeth.

Acacia catechu (Vern. Khair). Common tree in the Siwalik forests of Saharanpur mixed with Shisham (*Shorea robusta*). The leaflets are 6 to 100 in each pinna. Flowers are pale in colour. Corolla 2-3 times longer than sepals. The bark yield catechu so sommonly used as an ingredient with betel leaves.

Acacia pseudo-eburena (Vern. Pahari Kikar). A small deciduous tree commonly associated with Khair.

Acacia farnesiana (Vern. Wilayati Kikar). A thorny shrub with bright yellow flowers having sweet scent. Wild in Dehra Dun and Sharanpur districts.

Acacia Senegal (Vern. Kumat). A medium-sized deciduous tree. Common on hills in Rajasthan. Source of gum arabic.

Acacia modesta (Vern. Phulai). A small spiny tree. The branches are used as tooth brush.

Acacia leucophloea (Vern. Safed-Kikar), A small deciduous tree. Flowers yellow in heads. Common in dry parts.

Other species of *Acacia* commonly found in Siwaliks are *A. concinna*, *A gageana*, *A. pennata*. In many Australian species (*A. melanoxylon*) the petiole gets modified into a leaf-like structure called the phyllode; in some the thorns are hollow (modified stipules) and swollen, affording shelter for ants which get food from the nectary. Various exotic species are cultivated in Nilgiri and Palni hills in South India and in arid Rajasthan.

Albizzia

Albizzia lebbek (Vern. Siris) Large deciduous tree common in low valleys and along bank of streams. $K_{(5)}C_{(5)}A_{(8)}G_1$.

Albizzia procera (Vern Safed Siris). Large tree with white scented flowers. The trees are cultivated and are also found wild in most localities in mixed forests of low lying hills.

Albizzia jullibrissin (*A moltis*). A sparingly branched tree with rose coloured flowers. Wild in hills, up to 1500 m.

Other species of *Albizzia* found in the hills are *A chinensis* (*A. stipulata*), *A. lucida* etc.

Mimosa

Mimosa pudica (Vern. Lajwanti). Described as type 2. Cultivated and also found as wild. Roots and leaves are used medicinally.

Mimosa rubicaulis (=*M. himalayana*). A straggling prickly shrub, found commonly in grasslands and in valleys in the hills. It is excellent for hedges. The wood yields gunpowder charcoal. Leaves and seeds are used medicinally.

Parkia

Parkia roxyburghiana. A deciduous tree with bipinnate compound leaves. Flowers are white. *Parkia biglandulosa* is cultivated in the gardens.

Prosopis

Prosopis cineraria (= *Prosopis spicigera*, Vern. Khejri Jund). A small to large tree. Flowering in spring. Common in arid regions. Considered sacred. Pods used as vegetable, leaves as fodder.

Neptunia

Neptunia oleracea. An acquatic plant sensitive to touch. Flowers white.

Pithecolobium

Pithecolobium dulce. A middle-sized tree with stipular spines. Stamens monadelphous.

SUB-FAMILY CAESALPINOIDEAE (CASSIA FAMILY)

Trees or shrubs, more rarely herbs, with pinnate or bipinnate, more rarely simple leaves; stipules absent. The flowers are zygomorphic. Calyx is free or the 2 upper sepals may be united; the aestivation is imbricate. Corolla is ascending imbricate. Stamens are 10 or few, free, or united in one or two bundles.

Distribution

Plants are cosmpolitan, distributed throughout the world but more commonly in tropical regions. The representatives of the family occur throughout India. *Caesalpinia* with 7 species, *Cassia* with about a dozen species *Bauhinia* with about 30 species and other genera like *Delonix*, *Parkinsonia*, *Saraca*, *Tamarindus* occur throughout the country.

Indian Representatives

Bauhinia, Cassia, Caesalpinia, Tamarindus, Delonix, Saraca, Parkinsonia, Haematoxylon, Peltaphorum, Mezoneurum, Pterolobium, Pongamia, Acrocarpus, Hardwickia, Intsia, Humboldtia etc.

VEGETATIVE CHARACTERS

General Habit

Majority are trees or shrubs, some are woody climbers (*Bauhinia vahlii*), rarely herbs (*Cassia*). *Cassia obtusifolia* and *C. obtusa* are annual herbs (2-7 feet tall), *C. occidentalis* is an undershrub, *C. fistula* is a medium-sized tree, *C. sophera* is a diffuse shrub 2.5 to 3 metre high. *Parkinsonia* is a xerophytic plant, either shrub or a tree. *Tamarindus indica* is a very large-sized tree.

Roots

The roots in a large number of plants of the order Leguminales as a whole show special swellings or tubercles which are simply metamorphosed lateral roots containing a large number of bacteria. These bacteria are *Pseudomonas radicicola*, the nitrogen fixing bacteria, and fix the atmospheric nitrogen in the form of organic nitrogen compounds. The plants bearing these root tubercles absorb and use these organic nitrogen compounds, while the bacteria inside the root obtain nourishment from the protoplasm of the roots of these plants (symbiosis). These nitrogen fixing bacteria enter the roots when they are quite young and tender, and once they enter these roots, multiply very rapidly and their presence results in the development of tubercles in the soil especially when the roots are allowed to decay in the soil, enrich it with nitrogen compounds. These leguminous plants are thus used in the rotation of crops, especially when the soil is poor in nitrogen compounds.

Stem

The stem is usually erect. Some plants are also climbing some species of Bauhinia climb by means of tendrils. B anguina possesses flat ribbon-like stem which is twisted to form depressions on alternate sides thus giving the appearance of a snake. The stem pieces which look like snakes are supposed to pup off serpents.

Leaves

The *leaves* are usually alternate and rearely simple as in *Bauhinia* but more commonly they are pinnate or bipinnate. The stipules are present and they are free and of various types. The stipels are mostly absent. The leaf base is often swollen. In *Parkinsonia* the leaf rachis is flat and ends in a sharp point. When the minute leaflets are fallen or do not develop, the rachis resembles the phyllode of an acacia.

Inflorescence

The flowers are arranged usually in racemes or panicles or rarely the *inflorescence* is a cymose. In some species of *Cassia* there are only one or two flowers in a leaf axil.

Flower

Medianly zygomorphic, zygomorphy due to unequal insertion of petals on thalamus and not to their size and form. One half on the thalamus is occupied by two petals and the other half bear three petals, bisexual, perigynous; generally 5-merous rarely 4-merous.

Calyx

5, poly or gamosepalous (two upper sepals united in *Tamarindus indica*), aestivation imbricate, rarely valvate (*Cercis*).

Corolla

5, polypetalous, aestivation is ascending imbricate. In *Cercis* the corolla is papilionaceous, in *Copaifera* and *Ceratoma* petals are absent, while in *Krameria* anterior petals are represented by glandular scales and absent in *Tamarindus* sp.

Androecium

The *androecium* consists of mostly ten stamens which are all free or sometimes variously connate. Sometimes they may be fewer in number or rarely numerous. In *Cassia* all the stamens are only very rarely perfect, and some of them usually 3-5 are almost always reduced to staminodes. Sometimes a few may be entirely absent. In species of *Bauhinia* the stamens may be ten in number or they may be reduced to five, three or even one. The missing stamens may be represented by staminodes or may be entirely absent. In *Tamarindus* only three stamens are developed and they are monadelphous. The missing stamens are only represented by bristles, which may be regarded as extremely reduced staminodes. Anthers are various, sometimes opening by terminal pores.

Gynoecium

The *gynoecium* is monocarpellary with a superior unilocular ovary. There are usually two rows of ovules on the marginal placentation. The style is simple with a terminal, oblique or capitate stigma.

Fruits and Seeds

The fruit is, a legume or indehiscent. It often becomes transversely septate in *Cassia*. The *seeds* are endospermic or nonendospermic with a large embryo.

Pollination and Dispersal

The flowers are often large and showy and the stamens and *stigma* are freely exposed. They are pollinated by insects. Some species of *Cassia* are visited by sun birds. The dispersal of seeds takes place by wind or by animals.

Economic Importance

The family is of fairly great importance. The plants are either ornamental or of medicinal importance. A few plants have food and other values.

1. Some common ornamental plants grown in the gardens and place of botanical importance are *Amherstia nobilis* (native of Burma but also found in Bengal), *Bauhinia* (*B. varietata*, Hindi—Kachnar, *B. acuminata*, *B. corymbosa*), *Cassia* (*C. artimisiodes*, *C. fisutla*,

Hindi—Amaltas, *C. javanica*, *C. nodosa*, *C. siamea*, *C. renigera*, *C. sophera* etc.), *Caesalpinia* (C. giliesii), Colvillea (C. racemosa), *Delonix* (= *Poenciana*) (*D. elata*, *D. regia*), *Parkinsonia* (*P. aculeata*), *Peltophorus* (*P. pterocarpum*), *Saraca* (*S. declinata*, *S. indica*) etc.

2. The bark of *Bauhinia perpurea*, *Tamarindus indica* and roots and fruits of *Caesalpinia digyna* are used in tanning. wood of economic importance is obtained from *Acrocarpus fraxnifolius*, *Afzelia* (*A. palembanica* and *A. bijuga*—(both grown in Andaman), *Hardwickia binata* (grown in Andhra, Maharashtra and Coromandal Coast) and *Tamarindus indica*. Haematoxylon campachianum (a native Haematoxylon and is used as a staining agent in biological preparations. *Afzelia palembanica* is a source of yellow dye used for colouring clothes and mats. Hard wood of *Caesalpinia sappan* yields a red dye. A resin tapped from the heart wood of *Hardwickia* is used to prepare varnish.
3. Seeds of *Tamarindus indica* yields a jelly-like substance which is used as a sizing material in cotton and jute industry. A charcoal is prepared from the wood of *Parkinsonia*. The bark of *Hardwickia* and species of *Bauhinia* (*B. racemosa*, *B. tomentosa* and *B. Vahlii*) are source of fibre that is spun into ropes and cordage.
4. The leaves of *Cassia alata* (found in Bengal and Madras) are used to cure ringworm and skin diseases. The leaves and fruits of *C. angustifolia* (found in Madras) and *C. obovata* are used as laxative. The fruit pulp of *C. fistula* is used as purgative. The bark and leaves of *C. glauca* are used in diabetes and gonorrohea. The roots of *C. mimosoides* are given in spasms of stomach. The roots, leaves and seeds of *C. occidentalis* and *C. pumila* are used as purgative. The leaf juice of *C. tora* is an excellent cure of malaria. The bark of *Caesalpinia curiaria* is used in chronic fevers. Tender leaves of *C. crista* are used in liver disorders. Dried leaves and flower buds of *Bauhinia tomentosa* are given in dysentery. Oil extracted from seeds of *Cynometra cauliflora* is applied externally in leprosy and skin diseases. The bark of *Saraca indica* (Hindi—Ashok) is used in uterine disorders. The fruits of *Tamarindus indica* possess carminative and laxative properties.
5. The leaves and flower buds of *Bauhinia variegata* are used as vegetable. The young leaves of *Cassia tora* are made into vegetables and seeds into sweets. The acidic fruits of *Tamarindus indica* are edible and are rich in tartaric acid.

16

UMBELLIFERAE

A family of 275 genera and 2,850 species widely distributed throughout the world but they are most abundant in north temperate zone. In India the family is represented by 53 genera and about 200 species occurring chiefly in the temperate and alpine Himalayas but some species are also cultivated in warmer parts. The common examples are Coriander (*Coriandrum sativum* L.), Carrot (*Daucus carota* L.), Fennel (*Foeniculum vulgare Mill.*) and Brahami (*Centella asiatica* L. Urban).

Salient Features

Plants mostly herbs: annual, biennial or perennial. Stem herbaceous, with hollow internodes. Leaves generally compound, alternate with a broad a sheathing leaf base, exstipulate. Flowers arranged in simple or compound umbels, pentamerous except gynoecium, actinomorphic. Ovary inferior with an epigynous disc. Fruit a schizocarp, dehiscing into two dry indehiscent mericarps. Seeds with minute embryo and large oily endosperm. Plant organs characteristically have etherial oil or resin containing canals.

Distribution

The plants are cosmopolitan mostly occuring in north temperate regions extending between subarctic and the subtropical regions and also in the tropical mountains. In India the family is represented by a number of important plants which are cultivated during winters. *Centella* grows throughout India is marshy places up to 2100 metres.

Indian Representatives

Anethum, Angelica, Anthriscus, Apium, Bupleurum, Carum, Centelia, Cicuta, Coriandrum, Cumium, Daucus, Eryngium, Ferula,

Foeniculum, Heracleum, Opopanax, Pteroselinum, Peucedanum, Pimpinella, Psammogeton, Prangos, Seseli, Trachyspermum, Zosimia etc.

VEGETATIVE CHARACTERS

General Habit

Mostly annual (*Coriandrum*) or perennial (*Caraway*) sometimes biannial (*Daucus*) herbs, rarely shrubs (*Ferula communis*), *Pseudocarum* an East African plant is a climber, *Azorella* with about 100 species growing in Antarctic islands and Andes is an evergreen cushion like plant aging more than 100 years, usually the plants have an aromatic smell.

Leaves

The *leaves* are alternate. Rarely they may be opposite e.g., *Apiastrum*. They are usually pinnately compound and in several cases these pinnate divisions continue to fifth or sixth order. Sometimes palmate divisions also occur. Only rarely, the leaves are simple e.g., *Hydrocotyle*, *Bupleurum* etc. Some of the genera e.g., *Eryngium* (South America) and *Aciphylla* (Australia) possess monocot-like leaves which are narrow with parallel venation and sheathing bases. In most of the cases the leaves are exstipulate with sheathing bases.

The chief anatomical feature of the family is the presence of ethereal and often scented oils, balsam or resin-ducts in all the organs of plants except roots. Collenchymatous strands are found in the primary cortex, corresponding to the ribs of the stem. The pith is absent except at the nodes. Cork formation, where it occurs, is usually superficial. Branched hairs of various forms occur in several genera.

Inflorescence

As the name of the family suggests, an umbel *inflorescence* is the characteristic of the family. Rarely, the umbel is reduced to a single flower as in some species of *Hydrocotyle* or forms a compact head (*Eryngiunm*). The compound umbel is most common, although simple umbel also occurs in some genera (e.g. *Hydrocotyle* and *Bupleurum*). The primary umbel is surrounded at the base by an involucre of bracts and each secondary umbel is subtended by an involucel of bractlets.

Flower

The flowers are usually hermaphrodite but sometimes unisexual flowers are also found. In several genera the outer flowers of the inflorescence are male only. In *Astransia* sp. majority of the flowers of the inflorescence are bisexual and some are only male flowers. In

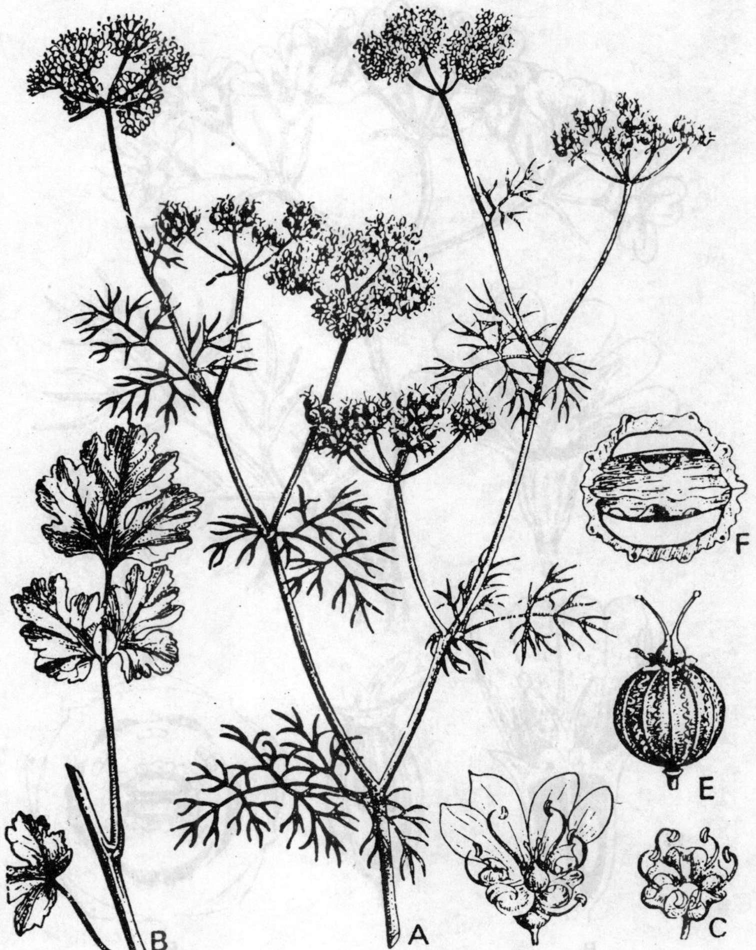

Fig. 16.1. Coriandrum sativum L. A—Flowering branch; B—basal leaf; C—Central flower; D—Peripheral flower of inflorescence with two enlarged anterior petals; E—Fruit; F—Fruit transverse section.

Echinophora sp. the monoecious plants bear the unisexual flowers. The plants of *Arctopus* sp. are dioecious and bear unisexual flowers. In majority of the inflorescences the flowers are regular (actinomorphic) but sometimes the outer flowers of the umbles are zygomorphic and irregular e.g. *Coriandrum sativum*. The flowers are usually pedicellate, breacteate complete and epigynous.

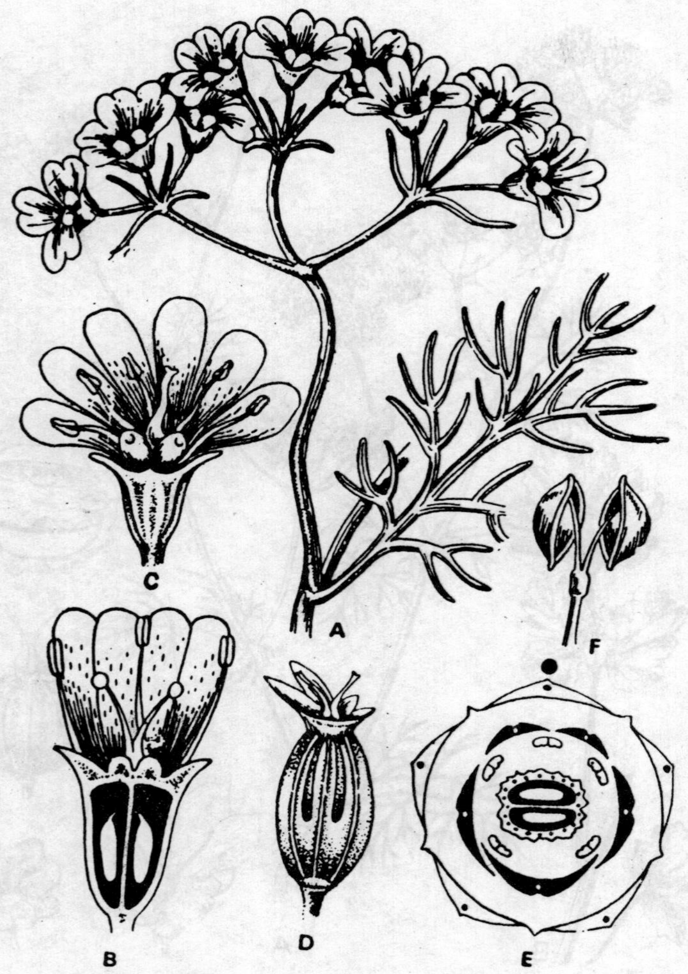

Fig. 16.2. Coriandum sativum. A—Twig; B—L.S. of flower; C—Dissected flower; D—Fruit; E—Dehisced fruit and F—Floral diagram.

Calyx

It consists of five sepals. The sepals are very minute in structure. The odd sepal being posterior. In many the calyx is absent, e.g., *Foeniculum vulgare*.

Corolla

The *corolla* consists of five petals, polypetalous and usually apically inflexed. The outer flowers of the umbels may have zygomorphic corolla. The petals are mostly white, some times becoming pink, yellow or rarely blue. They are only rarely scented, but the entire plant smells of ethereal oils which serve the purpose of attraction for the insects. The aestivation may be valvate or imbricate.

Androecium

Androecium consists of five stamens alternating with the petals and arising from an epigynous disk. Anthers are introrse, 2-celled, basi or dorsi fixed, sometimes versatile, dehiscing longitudinally.

Gynoecium

It consists of two carpels, syncarpous, the ovary being inferior, bilocular with axile placentation, having a single ovule in each loculus. The ovules are anatropous and pendulous. The carpels are being placed anterioposteriorly in the flower. On the top of the ovary is an epigynous disk, which is continued upwards into two, usually short styles. The stigmatic tip of each style is not well differentiated.

Fruit

Schizocarpic cremocarp, crowded by the rim of calyx; splitting into two one seeded mericarps. The mericarps are first attached to a stalk called the carpophore. Each mericarp have five longitudinal ribs (costae) and between these ridges are furrows (valeculae) under, which oil ducts are present (*vittae*). Seeds albuminous.

Pollination

The flowers are rendered conspicuous by aggreation of a large number of flowers into dense inflorescence. The outer flowers being often sterile or male, recalling the development of ray florets in Compositae. They are generally white or yellow, rarely blue and scented, but having ethereal odour which probably serves to attract the insects. The nectar is secreted in the disc, surrounding the stigma, and is easily accessible to insects having short tongue. Cross-pollination is effected by well-marked protandry.

Economic Importance

The family is fairy important form the economic point of view. The fruits of some plants are used as condiments whereas the other plants are of medicinal value.

1. *Ferula assafoetida*; (Verna.—*Hing*)—A perennial herb, commonly grown in the Punjab and Kashmir. The gumres in, obtained from the roots, is used in perfumery andfor flavouring food products. It

is also used in medicines in the treatment of asthama, cough and indigestion.

2. *Foeniculum vulgare*; (Verna.—*Saunf*)—An aromatic herb, native of the Mediterranean region, now cultivated mainly in the Punjab, Assam, Uttar Pradesh and Maharashtra. The fruits are used as spice and condiment. Also used medicinally as carminative and stimulant. The residue, left after the distillation of an essential oil, makes a valuable fodder.
3. *Anethum graveolens*; Syn. *Peucedanum graveolens*; (Verna.—*Soya*)—Native of Eurasia, now cultivated on a small scale in Jammu and Kashmir. The fruits are used as spice and condiment and are also used as carminative.
4. *Seseli indicum*; (Verna.—*Banajowan*)—A herb, commonly found in Uttar Pradesh, Bengal and Assam. The fruits are used as stimulant, stomachic, and are also used in expelling roundworms.
5. *Daucus carota* var. *sativa* (Eng.—Carrot, Verna.—*Gajar*)—An annual or biennial herb, grown chiefly in the Punjab, Uttar Pradesh and Madhya Pradesh. The roots are edible.
6. *Apium graveolens* var. *duluce* (Eng.—Garden celery; Verna.—*Ajmud, Karas*)—Native of Europe, now cultivaed in the North-Western Himalayas and in the hills of Uttar Pradesh, the Punjab and South India. Used as a vegetable. The roots and seeds are used medicinally.
7. *Ferula galbaniflua* (Eng.—Galbanum; Verna—*Gandhabiroza*)—Found in North Western India. The oleo-gum resin, obtained from the stem, used in perfumery, and is also used in the treatment of chronic bronchites and asthma.
8. *Apium graveolens* var. *rapaceum* : Syn. *A. rapaceum* (Eng—Celeriac; Verna.—*Ajmud, Salari*)—The roots are eaten as vegetable. The fruits are used for extracting an oil, which is much valued, and the seeds as spice. Cultivated in the hills of Uttar Pradesh the Punjab and South India.
9. *Pastinaca sativa*; Syn. *Peucedanum sativum*; (Verna.—*Gujur*)—A herb, native of Europe. The roots are eaten as vegetable.
10. *Corandrum Sativum* (Eng.—Coriander; Verna.—*Dhania*)—An aromatic herb, native of the Mediterranean region, now cultivated chiefly in Madhya Pradesh, Maharashtra, Mysore, Bihar and Uttar Pradesh for its fruits and leaves, which are used as condiment and spice. The fruits are also used as stimulant, carminative, stomachic and tonic.

11. *Ferula narthex*; (Verna.—*Hing*)—Commonly found in Kashmir. It produces an oleo-gum resin, which is used for flavouring food products.
12. *Anethum sowa*; (Verna.—*Sowa*)—Cultivated all over India. The leaves are used for flavouring and the fruits are used as carminative, stomachic and stimulant.
13. *Cuminum cyminum*; (Verna.—*Jira*)—A herb, native of the Mediterranean region but commonly grown in the Punjab and Uttar Pradesh, for the aromatic fruits, which are used for flavouring purposes.
14. *Centella asiatica*; Syn. *Hydrocotyle asiatica* (Eng.—Asiatic pennywort; Verna.—*Brahmi*)—A creeping herb, used as an antidote against cholera and also to cure madness. The plants are used as remedy of certain forms of leprosy.
15. *Carum carvi*; (Eng.—Caraway; Verna.—*Shiajira*)—A herb, native of Europe and West Asia, now cultivated in Bihar, Orissa, Punjab, Bengal and Andhra Pradesh. The fruits are used as condiment,and are also used medicinally as stomachic and carminative.
16. *Ferula jaeschkena*—Grows in Kashmir. The roots and seeds produce an essential oil.
17. *Bumium persicum*; Syn. *Carum bulbocastanum*; (Verna.—*Kala zira*)—A perennial herb, found in Kashmir. The starchy tubers are eaten as vegetable and the seeds are used as spice.
18. *Trachyspermum ammi*; Syn. *Carum copticum*; (Verna.—*Ajwain*)—A herb, cultivated throughout India. The fruits are used as spice, and are also used medicinally as carminative stimulant, tonic and in indigestion.
19. *Trachyspermum roxburghianum*; Syn. *Carum roxburghianum*; (Eng.—Ajmud; Verna.—*Ajmuda Radhuni*).

17

SOLANACEAE

The *Solanaceae* are a large family containing 90 genera and over 2,000 species distributed in the tropical and temperate regions of the world but the chief centre of their distribution is Central and South America. In India the family is represented by 15 genera and 88 species.

Many species are cultivated throughout India whereas others occurring chiefly in the Himalayas and Southern and Eastern parts of India. The well known examples of the family are mostly cultivated species, such as Potato (*Solanum tuberosum* L.), Egg plant (*Solanum melongena* L.), Tomato (*Lycopersicon esculentum* L. Karsten), Raspberry (*Physalis peruviana* L.), Chillies (*Capsicum annuum* L.) and Tobacco (*Nicotiana tabacum* L.), and Dhatura (*Datura stramonium* L,) which occurs , in the temperate Himalayas from Kashmir to Sikkim.

Salient Features

Plants are generally herbs or shrubs. Stem possesses bicollateral vascular bundles. Leaves simple, alternate in vegetative parts and opposite in or near the inflorescence due to congenital union of leaf bases with the axis i.e., adnate leaf bases. Inflorescence usually on axillary cyme.

Flowers bisexual, hypogynous, actinomorphic (slight zygomorphy due to oblique placenta). Calyx 5-partite or lobed, persistent and usually emerged in the fruit. Corolla 5-lobed, plicate, twisted. Stamens generally 5, epipetalous with connivent anthers. Gynoecium bicarpellary, carpels placed obliquely in relation to mother axis, placentation axile, placenta swollen, ovules numerous. Fruit a berry or a capsule. Seeds smooth or pitted.

Distribution

The plants are distributed in the tropical and temperate regions of the world with chief centres of distribution in Central and South America. Several plants are cultivated all over the world for their great economic use, e.g., *Solanum*. (*S. tuberosum*, *S. melongena*), *Lycopersicum*, *Capsicum*, *Atropha*, *Nicotiana* etc. The plants grow wild or are cultivated all over India.

Fig. 17.1. Solanum nigrum L. A—Flowering and fruiting branch; B—Flower; C—Stamen; D—Fruit.

Indian Representatives

Atropha, *Browallia*, *Capsicum*, *Cestrum*, *Cyphomandra*, *Datura*, *Hyoscymus*, *Lycium*, *Mandragora*, *Nicotiana*, *Nicandra*, *Petunia*, *Physoclaina*, *Solanum*, *Withania* etc.

VEGETATIVE CHARACTERS

The plants are usually annual or perennial herbs (*Solanum nigrum*, *Physalis* sp. *Nicotiana* sp.) or shrubs (*Datura*, *Solanum surattense*, *S.*

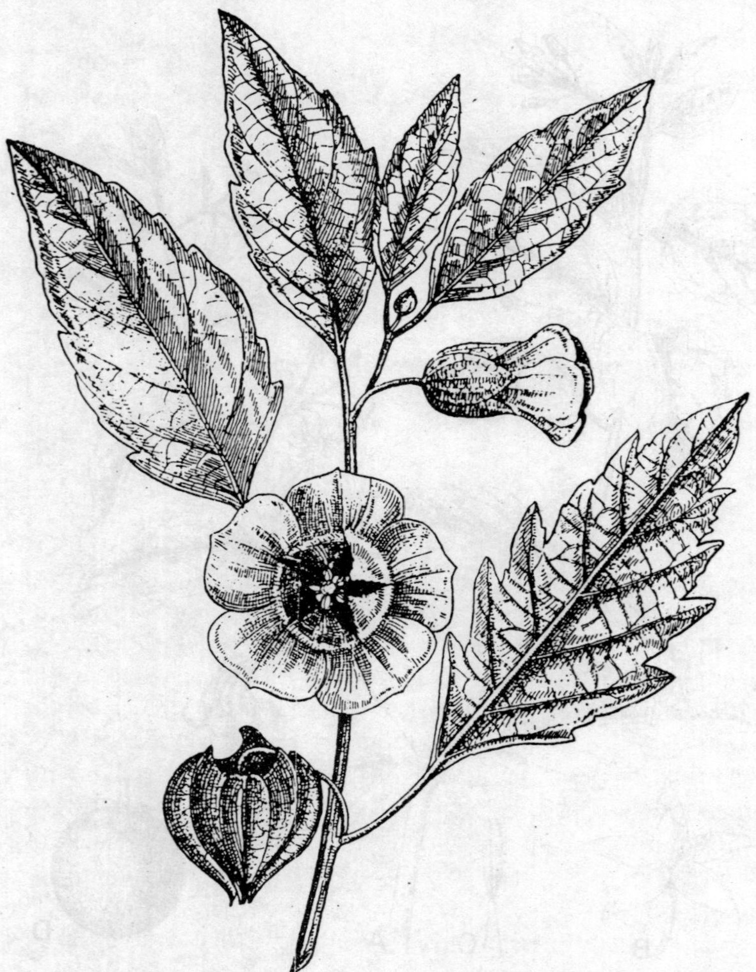

Fig. 17.2. Nicandra physaloides Gaertn. flowering and fruiting branch.

indicum, *Cestrum noctrurnum*), rarely soft wooded trees. Some of the plants may be climbing (*Solanum dulcamara*). Plants of various habits are found in the single genus *Solanum*. The production of underground tubers in *S. tuberosum* is exceptional.

Root

Tap root system. In case of modified stems (Tubers) adeventitious roots are also present.

Stem

Generally herbaceous, erect or climbing (*Solanum dulcamara*). The genus *Solanum* is interesting in view of the presence of herbs, shrubs, trees, creeping, erect and climbing plants. Cylindrical, branched and solid. *Solanum xanthocarpum* is prickly. *S. tuberosum* has underground stem, the tuber (Potato).

Leaves

Alternate, may be opposite in the floral region, exstipulate, simple, shape of the lamina variable, sometime hairy as in *Datura* and *Nicotiana*, unicostate reticulate venation.

Inflorescence

Inflorescence is usually of cymose type. The flowers are arranged in either axillary or extra-axillary cymes. Sometimes they are solitary or clustered together. In *Solanum nigrum*, the small white drooping flowers are arranged in extra-axillary subumbellate cymes. In this species there is an adnation or union of the flower-shoot for some distance above the position of the bract with the axis, so the inflorescence appears to spring from the middle of an internode. Hence it is described as extra-axillary. In *S. xanthocarpum*, the flowers are arranged in extra axillary, shortly peduncled cymes. Sometimes they become solitary. In *Physalis* the flowers are solitary axillary. In *Withania somnifera* (Hindi—Askand) the flowers are arranged in subsessile umbelliform cymes. In *Datura*, the flowers are large solitary.

Flowers

Regular (strictly speaking the flowers are rarely actinomorphic due to the oblique position of carpels), sometimes zygomorphic (*Salpiglossis*), hermaphrodite, complete, usually pentamerous (cleistogamous in *Salpiglossis*).

Calyx

5, gamosepalous, 5-lobed (may be 4-6 lobed) or toothed, persistent in fruit (*S. melongena*, *Physalis*). In *Physalis* the calyx develops a bladdery husk round the fruit.

Corolla

5, gamosepalous, may be rotate (*Solanum*), campanulate (*Atropa*), infundibuliform or lobed, often plaited. Limbs 5 or more, lobed or entire, sometimes zygomorphic, bilabiate (*Schizanthus*). The aestivation is contorted, plicate or convolute, rarely valvate. The *stamens* are usually five, epipetalous on the corolla tube and alternate with the lobes. They are commonly of unequal height. In zygomorphic forms there are only four (*Salpiglossis*) or two (*Schizanthus*) fertile stamens and the remaining are represented by the staminodes. The anthers are ovate or oblong, sometimes connivent into a cone as in *Solanum*, dithecous, introrse and dehiscing by longitudinal slits or by apical pores (*Solanum*).

Gynoecium

Bicarpellary syncarpous, carpels situated obliquely in the flower. The ovary is superior, bilocular may be 4 (*Datura*) or 3-5-loculed (*Nicandra*) due to formation of false septa. The placentation is axile, ovules many in each locule placed on a swollen oblique placenta. Rarely the ovary may be unilocular (*Henoonia*) with the ovule. In *Capsicum* the ovary may be 1-loculed apically. Style is single, stigma may be bilobed. There is a honey secreting disc present below the ovary.

Fruits and Seeds

The *fruit* is a berry which is sometimes (e.g. *Physalis*) enclosed within an inflated bladder-like calyx or it is a capsule which dehisces by valves (*Datura*) or circumcise above the middle (*Hyoscyamus*). The seeds are very numerous, compressed, discoid or subreniform, endospermic and with a curved or straight embryo.

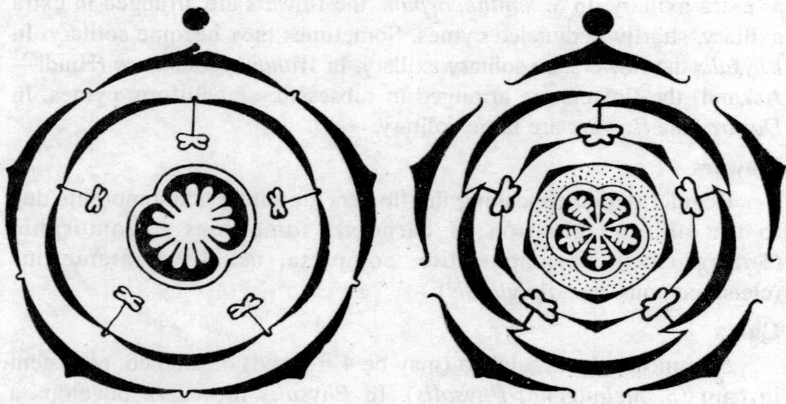

Fig. 17.3. Floral diagram of Solanum nigrum.

Fig. 17.4. Floral diagram of Nicandria physaloides.

Pollination and Dispersal

Usually conspicuous flowers of the Solanaceae are visited by insects for nectar secreted by a hypogynous disc. In *Hyoscyamus* cross pollination is insured by the prominent position of stigma. In *Schizanthus* the two stamens are enclosed by the lower lip of the corolla and they spring up when an insect alights on the lower lip. The stigma is at first far below the stamens but after the dehiscence of anthers it lengthens and projects beyond the stamens. It is now the first part to be touched by an insect visitor. *Solanum tuberosum* is devoid of nectar and is scarcely visited by insects. Here usually self pollination occurs by the style curving backwards to touch the anthers. The seeds are dispersed usually by birds and animals. Species of *Datura*, *Atropa* and *Hyoscyamus* are dispersed by water.

Economic Importance

1. The members of the family Solanaceae are of great economic importance. It includes many plants that furnish foods, medicines and ornamentals.
2. Some of the plants are cultivated in the gardens for their beautiful flowers. These include *Petunia*, *Nicotiana*, *Cestrum*, *Brunfelsia*, *Salpiglossis*, *Datura*, *Schizanthus* etc.
3. *Atropha belladona* provides "belladona" rich in an alkaloid "atropine" which is used against severe colds. *Nicotiana tabacum* (Hindi—Tambaku) and *Datura stramonium* (Hindi—Datura) contains alkaloids, nicotine and daturin respectively, which are used medicinally. *Hyoscyamus niger* is used as a narcotic. *Withania somanifera* (Hindi—Asgranth) is useful as a nervine totic. Plants of minor medicinal importance include *S. nigrum*, *S. surattense* etc.
4. The leaves of *Nicotiana tabacum* (tobacco) in various forms are used as intoxicant. 'Bidis' and 'Cigarettes' are filled with the dried leaves of tobacco.
5. A number of plants are used as vegetables, e.g., stem tubers of *Solanum tuberosum* (Hindi—Aloo) and fruits of *S. melongena*, (Hindi—Baingan), *Lycopersicum esculentum* (Hindi—Tamatar), *Capsicum annum* (Hindi—Mirach) etc. The fruits of *Capsicum annum*, *C. fatigiantum* and *C. frutenscens* are used as condiment. Fruits of *Physalis* (Hindi—Rasbhari) are edible and those of *Solanum quitoense* for a refreshing beverage.

18

CONVOLVULACEAE

This is a small family containing 50 genera and about 1,200 species. It is a most cosmopolitan being distributed throughout the world except the coldest regions. The major development of the family is however, in the warmer parts of the world (tropical and warm temperate regions). The family is represented by a number of important genera in this country e.g., species of *Ipomoea*, *Convolvulus*, *Cuscutta* etc.

Salient Features

Plants generally herbs, often twinning, milky sap present. Stem has bicollateral vascular bundles. Leaves simple, entire or variously lobed, exstipulate, inflorescence dichasial cyme or solitary. Flowers with jointed pedicels, pentamerous, bisexual, actinomorphic. Sepals 5, free. Petals 5, fused and forming a funnel-shaped and convolute corolla. Stamens 5, epipetalous, intrastaminal disc present. Pistil bicarpellary, bilocular, placentation axile with two ovules in each loculus. Fruit a loculicidal capsule. Seeds with folded or bilobed cotyledons, endosperm hard and cartilaginous.

Distribution

The plants are distributed in tropics—tropical Asia and America, but are well known thoroughout the world except for the coldest regions.

In India the plants grow all over but are more common in the Eastern and Western Himalayas and the plains of north India. Some of these also grow in south India.

Indian Representatives

Convolvulus, Evolvulus, Argyeria, Ipomoea, Exycibe, Quamoclit, Colonyction, Merremia, Exogonium, Lettsomia, Porana, Brewaria, Aganosoma, Cressa etc.

Vegetative Characters

Habit

Usually twining herbs or shrubs sometimes erect rarely trees. The plants may be xerophytes, hydrophytes and sometimes parasites. Usually the species of *Ipomoea* are twining or prostrate herbs, rarely shrubby or erect, the most popular species *Cuscuta reflexa* (*Amar bel*) is leafless twining parasitic herb. The species of *Evolvulus* are small usually silky pubescent, prostrate or erect herbs sometimes undershrubs. *Ipomoea reptans* (*Nari*) is an aquatic herb. The family is very much varied in its habit. The plants usually contain some milky sap. Sometimes when the plants are shrubby they prodcue thorns.

Root

The root system is quite variable from plant to plant. Sometimes as in *Convolvulus arvensis* the roots are long and slender, where they serve the purpose of vegetative propagation. Sometimes they are thick, fleshy and tuberous as in *Ipomoea batatas* where they store great quantity of food. In *Convolvulus scammonia*, thick tuberous rhizomes are found. In *Cuscuta reflexa* (*Amar bel*), they are represented by small haustoria, through which they absorb their nutrition from the host.

Leaves

Simple exstipulate, usually petiolate and alternately arranged. On the dodders (*Cuscuta* sp.) they are reduced to scales or altogether absent. Usually entire but sometimes pinnately divided.

Inflorescence

Inflorescence is mostly a dichasial cyme with tendency to become cincinnus Sometimes the flowers are solitary. In *Cuscuta*, the flowers may be either solitary or in clusters or shortly racemose. In *Evolvulus*, the flowers are either solitary axillary or in a few flowered cymes. In *Convolvulus* also the flowers are either solitary axillary or in a few flowered cymes. In *Porana paniculata* (Bridal creeper), the flowers are in large drooping panicles. *Bracts* and *bracteoles* are present, the former mostly in pairs and sometimes very close to the calyx forming a sort of involucre.

Calyx

5, mostly free and distinct but sometimes more or less united, sepals persistent. Aestivation imbricate (quincuncial) odd sepals being posterior. The sepals may be smooth, warty or hairy and may vary in size. The sepals may be orbicular or elongated determining the shape of the bud; 2 or 3 outer sepals are sometimes larger than the inner ones.

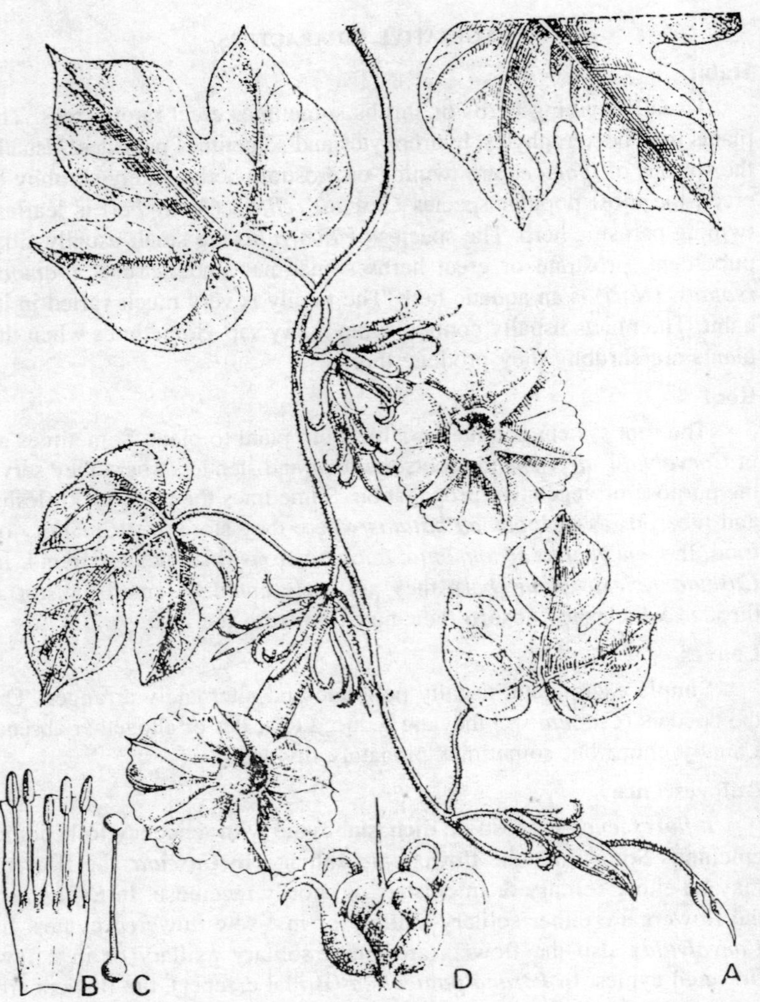

Fig. 18.1. Ipomoea nil Roth. A—Flowering branch; B—Corolla expanded with stamens; C—Gynoecium; D—Fruit enclosed in persistent calyx.

Corolla

5, gamopetalous, often entire or lobed, infundibuliform or campanulate, rarely tubular or salver form. Aestivation is induplicate valvate (plaited or twisted in bud). In *Cuscuta* the petals are tubular and corolla is appendaged within.

Androecium

5, distinct, polyandrous, stamens epipetalous at the base of corolla tube alternating with the corolla lobes. The length of the filament is variable even in the same flower. The anthers are 2-celled, introrose, dehiscing longitudinally. The pollengrains are of two types: (i) *Convolvulus type*. Ellipsoidal with longitudinal bands and (ii) *Ipomoea type*. Spherical with warty exine. Intrastaminal disc is usually present; ring or cup-shaped, usually lobed.

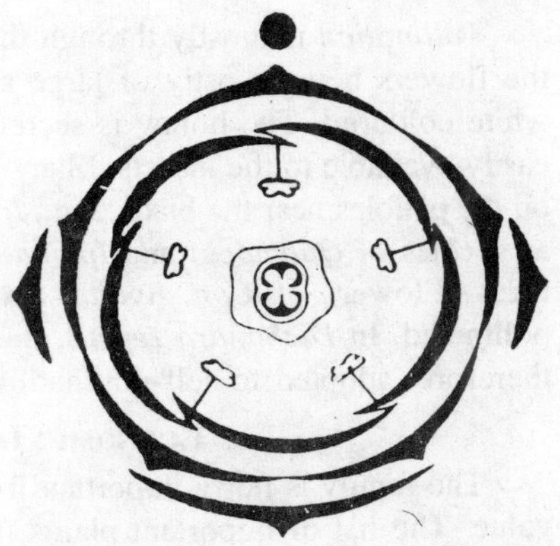

Fig. 18.2. Floral diagram of Ipomoea nil.

Gynoecium

Bicarpellary, syncarpous. Ovary is superior, bilocular, sometimes 1-4 celled due to false septum. Carpels may be sometimes 3-5; but vary rarely. Each is 1-2 ovuled, borne on axile placenta. Style is single, filiform, may be two (*Cuscuta*) or 3 or absent and the stigma is capitate and terminal (2-lobed in *Convolvulus*). The ovary is often covered with hairs and situated on a hypogynous, cup shaped, nectar secreting disc.

Fruits and Seeds

The *fruit* is usually a two or four valved capsule. In *Cuscuta* the fruit is a dry or fleshy capsule, dehiscing transversely or irregularly. Sometimes it is fleshy and indehiscent (*Argyreia*).

The *seeds* are smooth or hairy, non-endospermic or with a scanty endosperm. The embryo is usually large and the cotyledons are often plaited. The seeds of *Cuscuta* germinate in the ground and produce a very short anchorage root and a filiform leafless stem. The stem nutates and as it comes in contact of a host it develops suckers for deriving nourishment from the host and the root dies away.

Embryo is usually large with folded or bilobed cotyledons. In *Cuscuta*, the embryo is filiform and spirally twisted with usually no trace of cotyledons. Some authors place *Cuscuta* in an entirely different family called *Cuscutaceae* on account of its *parasitic habit, imbricate corolla* with a whorl of scales on the inside of the corolla tube and more abundant endosperm with filiform, spirally coiled embryo with almost no trace of cotyledons. The floral structure, however, resembles that of the family *Convolvulaceae*. It is always better to regard this genus as a reduced parasitic form of the family *Convolvulaceae*, the presence of such a simple embryo is associated with its parastic mode of life.

Pollination

Pollination is mostly through the agency of insects, on account of the flowers being mostly of large size and bright red, violet, blue or white coloured. The honey is secreted of the hypogynous disk and is easily available to the insects. Many plants possess extra floral nectaries on the petioles near the blade e. g., *Ipomoea batatas*. Some flowers such as species of *Quamoclit* and *Ipomoea* are pollinated by honey-seeking birds. Flowers of *Convolvulus* arvensis with short styles are self-pollinated. In *Dichondra repens*, the flowers are cleistogamic and are, therefore, adopted to self-pollination.

Economic Importance

The family is fairly important from the stand point of its economic value. The list of important plants is given below:

1. *Argyreia nervosa*; Syn. *Convolvulus nervosus*; Verna.—*Ghabel.*—This is a climbing shrub; grown as an ornamental.
2. *Operculina turpethum*; Syn. *Ipomoea turpethum*; Eng—India Jalap Verna—*Nisoth.*—This is a common twining herb. A resinous substance known as turpethin, obtained from the root bark, is used as a purgative and act as a substitute for jalap.
3. *Ipomoea cairica*, Syn. *I. pabnata*; *Convolvulus cairicus*; Eng.—Railway creeper—This is climbing shrub, grown as an ornamental. The flowers are campanulate and violet coloured.
4. *Cuscuta reflexa*; Eng.—Dodder; Verna—*Amarbel*. Akasbel—This is total stem parasite. It is a twining leafless herb. The seeds are carminative.
5. *Ipomoea aquatica*; syn. *I. reptans*; convolvulus reptans; Eng-Swamp cabbage; Verna—*Kalmisag, Ganthian*—This is an aquatic herb, commonly found in Maharashtra, Bihar, orissa and South India. The leaves and shoots are eaten as vegetable.
6. *Ipomoea batatas*; syn. *Convolvulus batatas*; *Batatas edulis*; Eng.—Sweet potato: Verna—*Shakarkand*—This is a prostrate herb, cultivated throughout our country for the edible tuberous roots, which are also in the manufacture of industrial alcohol, syrup and starch. It is native of tropical America.
7. *Ipomoea leari*; Eng.—Blue dawn flower.—This is an ornamental climbing shrub. It is native of tropical America.

8. *Ipomoea alba*; syn. *I. bonanox*; Eng.—Moon flower Verna.—*Dudhiakalmi*.—This is an ornamental herb. The leaves and fleshy calyx are eaten as vegetable.
9. *Exogonium purga*; Syn. *Ipomoea purga*; Eng.—Jalap. This is a twining herb, cultivated in the Nilgiris and Poona. The roots are the source of a resinous drug, which is used as a purgative.
10. *Ipomoea hispida*; syn. *I. eriocarpa*; *Convolvulus hispidus* Verna—*Boota*.—The leaves and stems are used as vegetable. It is found throughout India.
11. *Ipomoea quamoclit*; Syn. *Quamoclit pinnata*; Eng.—Indian pink; Verna.—Kamalata—This is a twining shrub; grown as an ornamental. It is native of tropical America.
12. *Ipomoea maxima*; syn. *I. sipiaria*; Verna.—*Bankalmi*.—This is a twining herb, used as a vegetable.
13. *Ipomoea muricata*; syn. *Calonyction muricatum*; Verna.—*Michai*—This is a herb, commonly found in Uttar Pradesh, Madhya Pradesh, Bengal and Bihar. The flowers and pedicels are eaten as vegetable.
14. *Argyereia splendens*; Syn. *Convolvulus splendens*; Eng—Silver morning glory. This is grown as an ornamental in the gardens.
15. *Evolvulus alsinoides*; Verna—Sankhpushpi.—It is used to cure bronchites and asthma. It is also used in dysentery.
16. *Porana paniculata*; Eng.—Christamas vine; Verna—Dela—This is an ornamental shrub, grown in the gerdens.
17. *Ipomoea tricolor*; Syn. *I. rubrocaerulea*; Eng,—Morning glory.—This is grown as an ornamental in the gardens.
18. *Ipomoea angulata*; Eng.—Star Ipomoea—This is a climbing shrub, grown as an ornamental in the gardens for its scarlet flower. It is native of north Mexico.
19. *Ipomoea crassicaulis*; Syn. *I. carnea*—This is a stout straggling shrub grown as a hedge plant.
20. *Ipomoea nil*; syn. *I. hederacea*; verna—*Kaladana*—It is found throughout our country. The seeds are used as purgative and as a substitute for jalap.
21. *Ipomoea digitata*; verna—*Bilaikand*—This is a common climbing herb, used as a fodder.

19

RUBIACEAE

Herbs, shrubs or trees. Leaves opposite or whorled, with inter-petiolar or intra-petiolar (sometimes foliaceous) stipules. Flowers 4-5-merous, bisexual (rarely unisexual), regular, sometimes irregular. Corolla tubular, rotate or funnel-shaped, lobes sometimes valvate. Staments 4-5, inserted on the corolla tube and alternating with the corolla lobes, introrse, 2-celled with longitudinal dehiscence. Ovary generally 2 celled, less often 3-5 celled and rarely one-celled, crowned by a more or less developed fleshy disc; inferior. Ovules 1 or more in each cell. Fruit usually a capsule.

Distribution

The family is cosmopolitan in distribution, but the great majority of the species are found in tropical regions. The species of *Galium*, *Oldenlandia* etc., are frequently found in the temperate regions. Certain species of *Galium* are found even in the Arctic zone or high elevations on tropical mountain ranges. The species of *Nertera* are found to be spread along the Andes. The genus *Comprosma* is commonly found in the south temperate and Antarctic regions, *Guettarda speciosa* is quite common round the Indian ocean from Africa to Malacca. The family is represented by several genera in our country, e.g., *Cinchona*, *Coffea*, *Adina*, *Hamelia*, *Ixora*, *Galium*, *Gardenia*, *Mussaenda*, *Rubia*, *Morinda* and many others.

VEGETATIVE CHARACTERS

Habit

The plants show great variations in habit., Majority of plants are either trees or shrubs. The herbaceous plants are also common and mainly they are confined to certain tribes. The members of the tribe *Galieae* are exclusively herbaceous. The tendency towards the formation of herbaceous forms is widespread, as certain herbaceous genera, such as, *Bouvardia* sp. is found in the typically woody tribe *Cinchoneae*. The climbing habit is also found in the family. Some climbers are herbaceous, whereas certain others are shrubby twiners, e.g. *Manettia*. *Uncaria* is a typical example of hook climbers. However, *Galium*

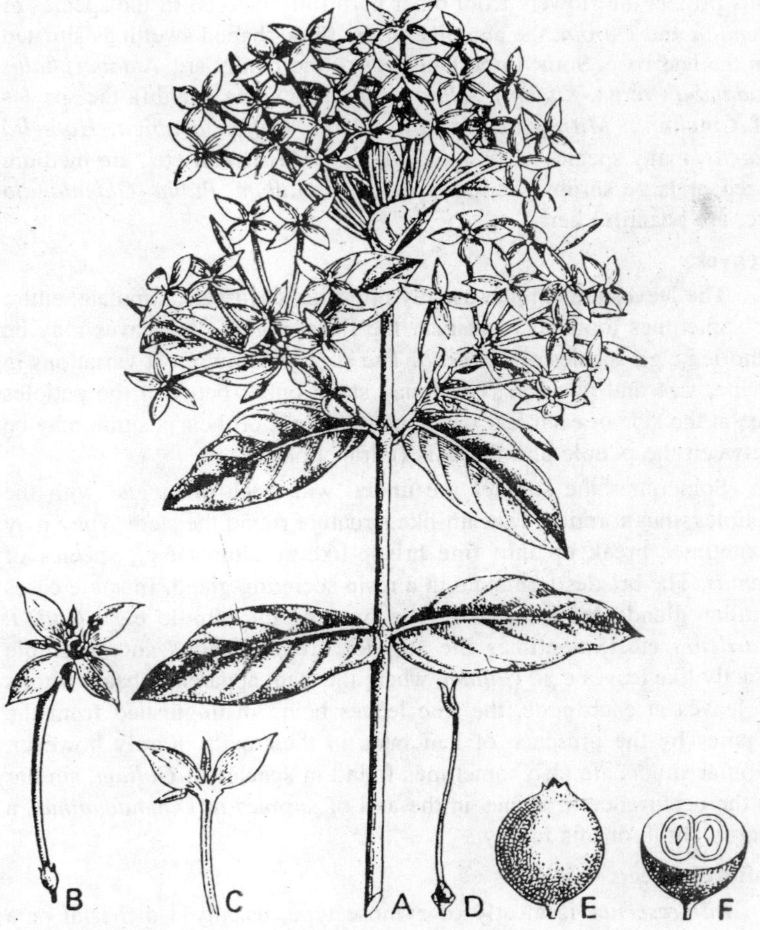

Fig. 19.1. Ixora coccinea L. A—Flowering branch; B—Flower; C—Flower longitudinal section in part; D—Gynoecium; E—Fruit; F—Fruit transversally cut.

aparine possesses stiff recurving hairs, which help in climbing on the supporting vegetation. The epiphytes also occur, e.g., *Myrmecodia.*

Myrmecodia possesses a tuber-like stem developed from the swollen hypocotyl. This tuber-like stem contains a number of communicating galleries which harbour the ants. Such type of plants are known as *myrmecophilous* but the exact relation of ants to plant is not clear. In the species of *Nauclea*, *Duroia* etc., the ants live just beneath the inflorescence in the swollen hollow internodes. It is thought that the ants protect the flowers from other harmfully insects. In the species of *Remijia* and *Duroia*, the ants live in the flask-shaped swellings situated on the leaf base. Some important trees of the family are, *Anthocephalus cadamba* (Verna.-*Kadam*). *Adina cordifolia* (Verna.-Haldu), the species of *Cinchona*. *Mitragyna parviflora* etc., *Coffea arabica*, *Hamelia patens*, many species of *Ixora*, *Gardenia*, *Mussaenda* etc., are medium sized or large shrubs, Certain species of *Galium*, *Rubia*. *Oldenlandia* etc, are beautiful herbs.

Leaves

The leaves are simple, usually opposite, decussate, stipulate, entire or sometimes toothed. In or near the floral region, the leaves may be whorled e.g., species of *Hamelia*. The stipules show great variations in shape, size and position. They may stand either between the petioles i.e., at the side of each leaf base (interpetiolar), or their position may be between the petiole and the axis (intrapetiolar).

Sometimes the stipules are united with each other and with the petioles, thus forming a sheath-like structure round the stem. They may sometimes break up into fine bristle-like structures e.g., species of *Pentas*. The bristles terminate in a resin secreting gland. In some cases similar glands are situated at the base of the stipule e.g., *Coffea*, *Gardenia* etc. Sometimes the stipules are foliaceous and resemble exactly like leaves e.g., *Galium*, where the stem apparently bears whorls of leaves at each node, the true leaves being distinguished from the stipules by the presence of leaf buds in their axils. Rarely however, stipular shoots are also sometimes found in species of *Galium*, similar to the occurrence of spines in the axil of stipules in *Damnacanthus*, a thorny shrub of this family.

Inflorescence

Inflorescence is mostly of cymose type, usually a dichasial or a panicled cyme. Sometimes the dichasial cymes are condensed or aggregated together into globose heads. Solitary flowers also occur. In

Mussaenda glabra the flowers are arranged in dense terminal cymes. In *Ixora subsessilis* the flowers are in sub-sessile corymbose cymes. In *Oldenlandia* the flowers are arranged in panicled cymes.

In *Coffea arabica* the flowers are in axillary cymes. In *Anthocephalus cadamba*, *Adina codrifolia* etc., the flowers are aggregated together and condensed into globose heads. In *Randia tetrasperma*, *Dentella* etc., the flowers are solitary. In *Coffea bengalensis*, the flowers are either solitary or 1—3 together in the axil of leaves. In cases where the flowers are aggregated together in heads, it often happens that the ovaries of the flowers which are inferior get completely united e.g., *Sarcocephalus*, *Morinda* etc. Sometimes, the union goes still ahead and the calyx of two paired flowers may unite into one as in the Australian genus *Pomax*.

Flower

Actinomorphic, rarely zygomorphic (somewhat bilabiate in *Henriquezia*), mostly hermaphrodite, rarely unisexual, epigynous.

Calyx

Usually 5, sometimes 4, gamosepalous. Sometimes from the inflorescence one or more sepals of few flowers get enlarged and leaflike; coloured variously for attracting the insects. Sometimes the calyx becomes enlarged in the fruit (*Nematostylis*) or in certain cases may be inconspicuous or entirely absent. In some cases enlargement of the sepals occur after the fertilization has taken place and serve as a means for the dispersal of the fruits (*Albertia*). Aestivation is usually valvate.

Corolla

Usually 5, sometimes 4, gamopetalous. The petals may be campanulate, tubular funnel-shaped or rotate. The aestivation is various and may be valvate, imbricate or contorted. Sometimes the flowers are zygomorphic showing bilabiate characters. Aestivation imbricate (*Henriquezia*).

Androecium

5, may be 4 or as many as the corolla lobes, epipetalous. Anthers introrse, dithecous, dehiscing longitudinally, sometimes by apical pores, when the filaments are united. There is heterostyly in *Oldenlandia*.

Gynoecium

Carpels 2 or more, usually bicarpellary syncarpous. Ovary inferior (may be superior in *Pogamea*, or half inferior in *Synaptanthera*). The number of locule is generally 2 (one chambered in *Gardenia* with 2-

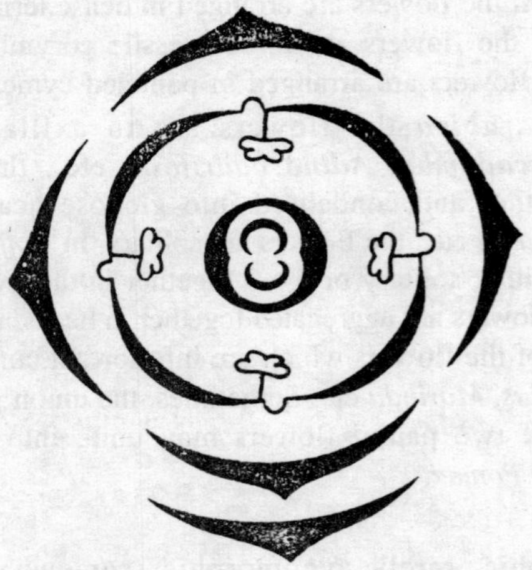

Fig. 19.2. Floral diagram of Ixora coccinea.

many parietal placenta). One to many ovules in each locule; placentation is axile. The style is single, often branched with usually linear stigmas on each style or 2-lobed on a single style.

Fruit

Capsular (*Anotis*, *Silivianthus*) loculicidal or septicidals berry (*Mussaenda*) and fleshy (*Coffea*). In *Galium* the fruit separate, into one seeded segments that are indehiscent. Seeds endospermic sometimes winged.

Pollination

It is thorugh the agency of insects. In some cases the flowers are long tubed, white or bright coloured while in others the flowers are aggregated into conspicuous inflorescence. In the flowers there is an epigynous disc in the form of a cushion. Ring or the cup above the ovary secrete nectar and the heterostyly is well seen in the tribe Galieae.

The flowers are protandrous and as soon as the flower opens, stamens shed their pollen and bend outwards before the two styles arms separate. Nectar is slipped by small-tongued insects which carry the pollengrains in passing from one flower to another. The long-tongued flowers are visited by insects having long proboscis and contain enough of nectar. When the insect after visiting the flower withdraws its proboscis, pollengrains covering the mouth or the corolla tube adhere to its body and when the insect visits a second flower, some of the pollengrains are deposited on the stigmas which stand close together in the tube, below the anther, thus affecting cross pollination. And if, by chance, the insects do not visit the flower self-pollination occur. In zygomorphic flowers (*Posqueria*) the pollination mechanism is very much like that in Papilionaceous plants. The pollen dehisce into a chamber, formed by close approximation of the anthers which are in a state of tension. As soon as the proboscis of the insect touch the anthers, the pollens are scattered over the body of the insect and are carried away to the stigma of another flower, along with the visit of the insect.

Economic Importance

From the point of view of its economic importance, the family is of great value. A list of few important plants is given below:

1. *Gardenia gummier*; Gummy cape jasmine; Verna.—Dikamli—The plant yields a gum-resin. It is used as a carminative and stimulant, and also given in dyspepsia. The fruits are edible.
2. *Randia uliginosa*; Syn. *Gardenia uliginosa*; Verna—*Pindalu*—The fruits are edible and used for any purposes. The leaves are used as fodder. The unripe fruits are used as a remedy for: dysentery and diarrhoea. The plants are commonly found throughout the country except in the Northern India.
3. *Gardenia jasminoides*; Verna.—*Gandharaj*—This is an ornamental shrub grown in the gardens. A yellow dye is obtained from the pulp of its, fruits. An essential oil is extracted from its; flowers, which is used in making of perfumes.
4. *Mitragyna parviflora*; Verna—*Kadam*—This is a tree found throughout our country. A fibre is obtained from the bark which is used for cordage. The furniture, agriculture implements and carvings are prepared from its wood. The leaves are used as a fodder.
5. *Oldenlandia umbellata*; Eng.—Indian madder Verna—*Saya*—This is a herb, commonly found in Bengal, Orissa and South India. A red dye is obtained from the bark of its root.
6. *Randia dumetorum*; Syn. *Gardenia dametorum*; Verna.—*Mainphal*—This is a large shrub or a small-tree. The fruits are edible. The pulp of the fruits is used to cure dysentery. The plants are commonly found in Madras, Maharashtra, Gujrat and Uttar Pradesh.
7. *Hymeodictyon* excelsum; Verna.—*Bhaulan.*—This is a large tree. The bark is used in tanning. Match splints, tea boxes and packing cases are made from its wood. The leaves are used as fodder.

17. *Cinchona officinalis*. The tree yields 'quinine' used for - malarial fevers. It is grown in South India and Ootacmund.

9. *Gardenia campanulata*; Verna.—*Bitmara*—A large shrub. The leaves and fruits are edible. The plants; are found in Assam and Bizarre.
10. *Morinda angustifolia*; Verna.—*Banhardi*—This is large climbing shrub. The roots yield a yellow dye. The plants are found in the Khasia hills, Assam and Sikkim.
11. *Cinchona calisaya*; Eng.—Quinine; Verna.—*Cinchona.*—This is a tree. The famous medicine of malaria known as quinine is obtained from the bark of this tree. The tree is native of South America, but found in the Nilgiris and Sikkim.
12. *Oldenlandia corymbosa*; Verna.—*Daman paper*—The decoction is used as a remedy for dysentery and cholera.
13. *Anthocephalus indicus*; Syn. *A. cadamba*; Eng.—Kadam: Verna.—Kadamba.—This is a tree grown as an ornamental for its golden coloured flowers. The fruits are edible. The bark is used as a tonic. It is also given as antidote in snake-bite.
14. *Rubia tinctrum*; Verna.—*Bacho*—The roots of this plant also yield a red dye which is used for dyeing cloth. The plant is also used as fodder plant.
15. *Cinchona ledgeriana*. The medicine known as quinine, which is used for malarial fevers is also extracted from the bark of the tree. This is grown in Bengal the Khasia hills and South India.
16. *Cinchona succirubra*. The quinine is also extracted from the bark of this tree. The medicine is used for the treatment of malarial fevers. It is grown in South India, Sikkim and Madhya Pradesh.
17. *Mussaenda glabrata*; Verna—Bedina—This is a small tree, grown as an ornamental in the gardens. The plant is used for asthma, fevers and dropsy.
18. *Hamelia patens*. This is beautiful shrub with red flowers, grown in the gardens as an ornamental. The fruits are edible. The syrup of berries is given in dysentery.
19. *Rubia sikkimensis*. A red dye is obtained from its roots. It is used for dyeing both cotton and hair.
20. *Rubia cordifolia*; Verna.—*Manjit*—This is a herb. A dye is extracted from its roots.
21. *Coffea arabica*; Eng.—*Arabian coffee*; Verna.—*Kafi.*—This is a large shrub. The famous non-alcoholic beverage ‘coffee’ is obtained from its seeds. This is a native of Abyssinia, but now cultivated in the Nilgiris, Mysore, Coorg and Travancore:

22. *Coffea liberica*; Eng.—Liberian coffee.—A non-alcoholic beverage is obtain-from its seeds. The bush is native of West coast of Africa, but now also cultivated is, Mysore and Travancore.
23. *Coffea robusta*; Eng.—Congo coffee—This is also a source of commercial coffee. It is native of Congo but now grown in Travancore, Mysore and Madras.
24. *Ixora chinensis*; Eng.—Chinese Ixora—This is a beautiful small shrub grown in the gardens as an ornamental.
25. *Ixora coccinea*; Eng.—Jungle flame Ixora Verna.—*Rangan*—This is, a shrub, grown as an ornamental. This is a good hedge plant.
26. *Ixora fulgens.* This is a shrub grown as an ornamental.
27. *Ixora lutea.* A shrub, grown as an ornamental.
28. *Ixora nudulata.* A shrub; grown as an ornamental, for its white fragrant flowers.
29. *Adina cordifolia*; Eng,—Saffron teak; Verna—*Haldu.*—This is a beautiful shade given tree. The wood is used for carvings, construction work, flooring and for railway carriages. The plant possesses antiseptic properties.
30. *Rubia khasiana.* A red dye is obtained from its roots.
31. *Gardenia latifolia*; Verna.—*Papra*—This is a small tree, usually grown as hedge plant.
32. *Gardenia resinifera*; Verna.—*Dikamli*—This is a large ornamental shrub with fragrant flowers.
33. *Mussaenda luteola.* This is shrub of ornamental value.
34. *Galium* sp. The plants are very ornamental, and grown in the gardens.
35. *Ixora arborea*; Syn. *Ixora parviflora*; Eng.—Torch tree Verna.—Jilpai—This is a small tree. The torches are made from its branches.

20

Compositae

Plants generally herbs. Leaves alternate or opposite, exstipulate, simple or variously lobed and divided. Inflorescence head (capitulum with an involucre of bracts). Flowers bisexual or unisexual, epigynous, usually of two types, the central "disc florets" are actinomorphic and the peripheral or "ray florets" are zygomorphic. Calyx modified into pappus. Corolla 5-lobed; tubular or ligulate. Stamens 5, syngenesious, epipetalous. Gynoecium bicarpellary, syncarpous, unilocular, inferior, ovule one, basal placentation, style one with bifid curled stigmas whose receptive surface is on the inside. Fruit a one-seeded cypsella crowned with pappus. Seed fills the fruit, large straight embryo without much endosperm.

Distribution

The plants are widely distributed and cosmopolitan occurring in almost all habitats. They are most abundant in the tropical and temperate lands but are also found in the arctic and alpine regions. Aparts from mesophytes, some occur as xerophytes, aquatic or marsh plants and epiphytes. In India the plants occur in all possible climates and places but are less common in areas under rain forests. They grow both in hills and plains and play an important role in the vegetation.

Indian Representaives

Ageratum, *Artemisia*, *Vernonia*, *Inula*, *Plucnea*, *Blumea*, *Helichrysum*, *Senecio*, *Helianthus*, *Eclipta*, *Tagetes*, *Calendula*, *Sonchus*, *Tridax*, *Dahlia*, *Zinnia*, *Chrysanthemum*, *Guizotia*, *Cosmos*, *Coreopsis*, *Erigeron*, *Cnichus*, *Centaurea*, *Dimorphotheca*, *Gnaphalium*, *Xanthium*, *Mikania*, *Gerbera*, *Taraxacus*, *Lactuca*, *Bellis* etc.

Vegetative Characters

Habit

Most of the members of this family are herbs, either annual (*Helianthus*, *Tagetes*, *Calandual*, *Lactuca*) or perennial; but a small portion are shrubs and a few reach the habit of tree (*Vernonia arborea* and *Leucomeris* in Nepal). Many of the plants are xerophytes (*Proustia*, *Baccharis*), hydrophytes (*Bidens*, *Cotula*) some are semi-aquatic (*Caseulia axilaris*, *Sphaeranthus indicus*) found in the rice fields. The plants may be twining (*Mikania*) or climbers (*Mutisea*).

Root

Normally a tap root (Dandelion bearing adventitious buds), branched and fibrous. In a few species the roots get thickened and form tubers (*Taraxacum*, *Dahlia*).

Stem

Erect or prostrate, herbaceous, may be woody below (*Artemisia Tenacetum*), usually hairy with latex. On the rhizome, of the perennial species, stem tubers are present (*Helianthus* sp.) and are formed for carrying on the vegetative reproduction. The tubers are eaten like potato (*Helianthus tuberosum*, Jerusalem artichoke and *Helianthus maximiliani*, wild artichoke). In xerophytes (*Proustia*) the branches are converted into thorns. The stem may be winged, leaf-like (*Baccharis*) and the leaves are more or less suppressed. In some cases the vegetative production is by means of runners (*Achillea millaefolium*) and in some the stem is undeveloped and gives a scapigerous habit (Dandelion).

Leaf

Alternate, sometimes opposite (*Helianthus*) rarely whorled (*Zinnia verticillata*, *Eupatorium*). Frequently the leaves are radicle being crowded in rosettes. Although the leaves are simple, exstipulate but show a great diversity in size form and depth of incision. The leaves may be compound, pinnately or palmately lobed. In xerophytic species the leaves are small, scale-like or needle-like or may be large (in some species of *Senecio*). The base of the leaves is frequently decurrent (*Senecio*) while in some cases it may be auriculate. In some genera the leaves are narrow and show parallel venation like the leaf of a monocot (*Corymbium*).

Covering of the leaf with hairs may be present or absent depending on the nature of the habitat. In Dandelion, Daisy and others, there is complete absence of hairs, while alpine and xerophytic, plants (*Gnaphalium*, *Filago*, *Senecio* sp.) have a cottony covering. The cottony

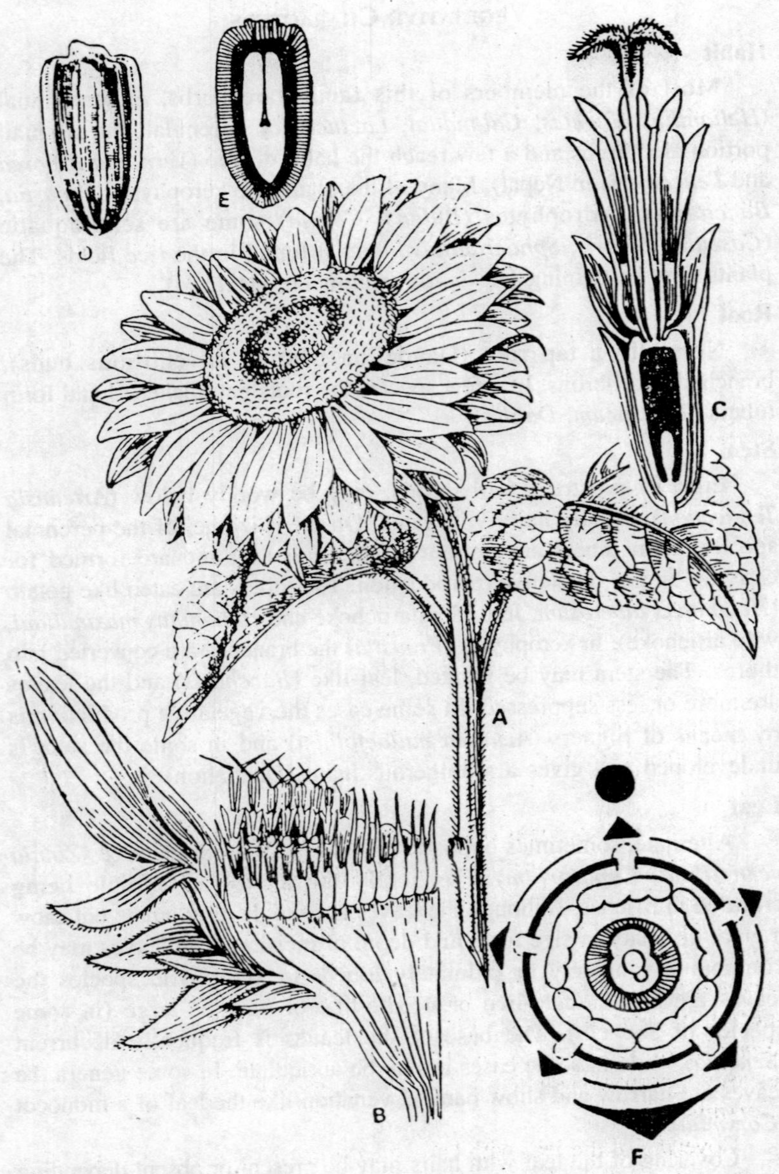

Fig. 20.1. Helianthus annus. A—Twig; B—Part of head in V.S.; C—V.S. of flower; D—Fruit; E—L.S. of fruit and F—Floral diagram.

hairs may be stellate or surrounded at the base by silicified cells which render the surface rough, after the hairs have fallen. In some cases laticiferous vessels are present in the sieve tubes (Cichoreae) or in other plants, oil containing passages occur in the cortex of the vegetative parts. In the cellsap of roots and tubers (of *Helianthus tuberosus*, *Inula* and others) Inulin is present dissolved in the sap.

Inflorescence

Capitulum or head surrounded by an invlucre of bracts. The inflorescence may be heterogamous (when flowers are differentiated into ray-florets and disc-florets) or homogamous (when no such distinction exists). The arrangements of flowers in an inflorescence may be of three types: (a) Flowers differentiated into peripherals ray-florets which may be neuter or pistillate and zygomorphic and central disc-florets which are tubular, hermaphrodite and actinomorphic (*Helianthus*, *Cosmos*, *Aster* etc.). (b) All flowers ligulate, hermaphrodite and zygomorphic (*Launaea* and *Sonchus*). (c) All flowers tubular, hermaphrodite and actinomorphic (*Ageratum* and *Vernonia* species).

Flowers

The flowers are completely epigynous, and usually pentamerous with reduction in certain whorls. They may be hermaphrodite or unisexual, complete or incomplete, tubular (actinomorphic) or ligulate (zygomorphic), and bracteate or ebracteate. When bracteate, the bracts are usually scarious, either deciduous or persistent.

Calyx

It may be either very rudimentary or entirely absent, its function being mostly performed by the involucre. In *Ambrosia*, *Siegesbeckia* etc., it is entirely suppressed. In certain cases it is represented by a slightly five-lobed rim-like structure at the top of the ovary. In most of the cases, however, it is represented by hair or bristle-like structures called the *pappus* which enlarges in the fruit and acts like a parachute helping in the dispersal e.g., *Dandelion*. In *Bidens* there are hooked bristles which also perform a similar function.

Corolla

The *corolla* consists of mostly five petals which are gamopetalous and are variously coloured. It may be tubular (actinomorphic) or zygomorphic. In the latter case there may be two conditions or varieties, and either the corolla is bilabiate (2-lipped) or strap-shaped (ligulate). The number of the teeth or the segments present in the ligulate or the bilabiate corolla correspond to the number of the petals present.

The *stamens* are five, epipetalous and alternating with the corolla lobes. The filaments are free but anthers are usually connate into a tube around the style (syngenesious). The anthers are dithecous, introrse and opening by the longitudinal slits. The stamens are usually included in the corolla tube.

Gynoecium

The *gynoecium* is bicarpellary and syncarpous. The ovary is inferior and unilocular with a single basal anatropous ovule. The style is mostly bifid or bilobed with the stigmas of various forms.

Fruits and Seeds

The *fruit* is an achene which is often compressed and crowned by a pappus comprised of hairs. plumes, barbs or scales. The *seeds* are non-endospermic and with a straight embryo.

Pollination and Dispersal

The Asteraceae are one of the most widely spread families of the flowering plants. Their success lies in their adaptation—to cross pollination by a large variety of insects. Small flowers are rendered conspicuous by aggregation into heads. The heads become further attractive by the development of the ray florets. A single visit of insect may pollinate large number of flowers. The nectar is secreted by a ring shaped disc around the base of the style and collects in corolla tube which is accessible to a wide variety of insects. In the absence of cross-pollination, self-pollination may occur in several genera in late stages of flowering. The stigma lobes come in contact with the style and brush the pollens which are retained by the stylar hairs.

Those fruits which have a pappus of plumes or hairs are adapted for wind dispersal. The pappus of *Taraxacum* enlarges into a parachute. Hooked bristles on the pappus of *Bidens* and the development of hooked spines on the fruiting receptacle in *Xanthium* favour their distribution by birds and animals to the feather or fur of which they cling.

Economic Importance

Though the family is very big, its economic importance is not that much. Most of the plants are ornamental and only few are used as food. However some important plants are described as follows:

1. There are many other plants besides the ones already described above which are cultivated in the gardens for their ornamental beauty. These are *Dahlia*, *Zinnia*, *Cosmos*, *Calendula*, *Aster*, *Tagetes*, *Gaillardia*, *Cynara* and *Centaurea* etc.

2. *Lactuca sativa* is cultivated for the leaves which make good salad on the dining table.
3. *Bidens* is an annual or perennial herb, *B. pilosa* is a common species.
4. *Chrysanthemum* is the common ornamental plant, cultivated for its beautiful flowers.
5. *Artemisia nelagarica* is an aromatic shrub and is quite common on hills. It is used as nermifuge.
6. *Tridax procumbens* is also a wild herb, growing near the banks of small streams.
7. *Sonchus* and *Launaea* are two annual herbs with bright yellow ligulate bisexual flowers.
8. *Helianthus annuus* is the common sunflower grown for its beautiful flowers.
9. *Vernonia* includes herb, shrubs, climbers or small trees and is a wild as well as a cultivated genus.
10. *Pluchea* is a tomentose xerophyte. It is a very troublesome weed because it has very deep roots and once it grows, it becomes difficult to eradicate it.
11. *Blumea* is a genus which is wild. Of special mention are the two species—*B. aromatica* and *B. lacera* the latter being of medicinal value.
12. *Echinops echinatus* is a tomantos thistle like herb.
13. *Eclipta erecta* is a small weed in the moist places.
14. *Ageratum.* These are herbs or shrubs which are wild during winter *A. conyzoides* is a common hairy annual.
15. *Gnaphalium.* It has got nearly half a dozen species which are wild herbs.
16. *Grangea maderaspatana* is a common shrub during summer.
17. *Xanthium* is a wild annual with 3-fid spines.
18. *Erigeron* is an annual or perennial herb.

21

LABIATAE

Plants generally herbs. Stem quadrangular with collenchyma at the corners. Leaves opposite and decussate, glands containing aromatic oil occur in the epidermal cells. Inflorescence verticellaster. Flowers bisexual, zygomorphic. Calyx 5-lobed, persistent. Corolla 5-lobed, tubular and often bilabiate (2/3), imbricate. Stamens 4, epipetalous, didynamous. Gynoecium bicarpellary, syncarpous, superior, tetralocular due to formation of a false septum; style gynobasic.Fruit four (1-seeded) nutlets surrounded by persistent calyx. Seeds with thin testa and scanty endosperm.

Distinguishing Characters

Mostly aromatic, annual or perennial herbs, rarely shrubs; glandular and hairy. Leaves rarely alternate, simple, exstipulate with hairs. Flowers in verticillaster inflorescence. Bracts and calyx persistent, bell-shaped or tubular. Corolla tubular, bilabiate, aestivation imbricate. Stamens 4, didynamous, epipetalous, alternating with the lobes of the corolla. Ovary 4-celled. Style is gynobasic, simple with bifid apex. Fruit a 1-4 seeded nutlet. Hypogynous disc is fleshy.

Distribution

This is a large family containing about 170 genera and 3000 species (Rendle) of world-wide distribution growing under great variety of soil and climate, but abundant in Mediterranean and in the hills.

Indian Representatives

Acrocephalus, *Ajuga*, *Anisochilus*, *Anisomeles*, *Calamintha*, *Colebrookea*, *Coleus*, *Dracocephalum*, *Dysophylla*, *Elscholtzia*, *Galeopsis*, *Hyssopus*, *Lamium*, *Leonotus*, *Leucas*, *Lycopus*, *Majorana*,

Mentha, Micromeria, Nepeta, Ocimum, Platystoma, Pogostemon, Prunella, Phlomis, Roylea, Salvia, Rosmarinus, Thymus, Stachys, Scutellaria, Zataria, Ziziphora etc.

Vegetative Characters

Habit

Majority of the plants are annual or perennial herbs, inhabiting the temperate regions. In warmer climates, the plants become shrubby in nature. Trees are rare. Certain plants are xerophytes with extremely reduced leaves. A few species are climbers, but the climbing habit is very rare. Some important herbs are, *Coleus aromaticus, C. forskohlii, Leuas cephalotes, Mentha arvensis, Ocimum sanctum, O. basilicum*, etc., *Salvia aegyptiaca, Coleus rotundifolius* are undershrubs, *Oscimum grotissimum, Meriandra bengalensis* are shrubs.

Certain species of Brazilian genus *Hyptis* and Indian genus *Leucosceptrum* are trees. The certain species of *Mentha* and *Lycopus* are found in marshy places. A few American species of *Scutellaria* are climbers. Certain plants are extreme xerophytes, e.g. *Rosmarinus*. In this plant the leaves are rolled back and stomata are found among the hairs in the grooves on the underside of the leaf. The propagation of many plants of this family takes place by means of new shoots which remain attached to the plant, e.g. in Salvia, *Ballota*. In *Ajuga reptans* the new shoots form aerial runners with somewhat reduced leaves; in *Lycopus*, the subterranean stolons (suckers) are found with short internodes and scaly colourless leaves. The suckers are found in the species of *Mentha*.

Stem

Angular, usually squares, herbaceous, hairy, hairs glandular. Sometimes the stem is underground in the form of sucker Mint. Propagation of the plants is by new shoots which remain attached to the plant (*Salvia*) or the latter dies off (*Mentha, Stachys palustris*) or by runners (*Ajuga*) or subterranean stolon.

Leaves

Usually the leaves are simple, opposite, decussate and exstipulate. They show many variations from an entire blade to toothed, lobed, cut or finely dissected as in many species of *Salvia*. A whorled leaf arrangement of 3 to 8 leaves is found in some genera, e.g., *Ocimum*. All parts of the plants, such as stem, leaves and inflorescence are more or less hairy and possess glandular hairs, which secrete characteristic scent of the genus or species. Sessile scented oil secreting glands are also found frequently on the epidermis.

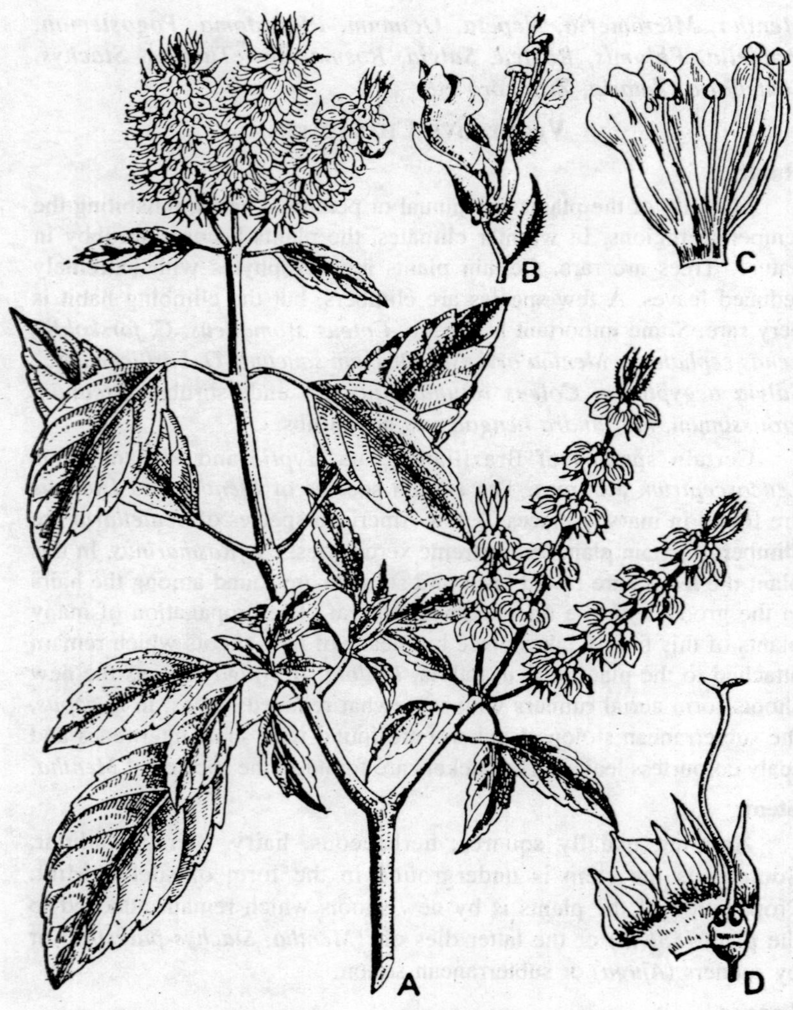

Fig. 21.1. Ocimum basilicum L. A—Flowering branch; B—Flower; C—Expanded corolla with stamens; D—Gynoecium with expanded calyx.

Inflorescence

Inflorescence in this family is usually of a condensed from, where a number of flowers, sometimes very large in number, form apparent whorls which are situated at each node or are axillary position. When examined closely, the apparent whorls are found to be consisting of two cymose inflorescences each forming usually a simple dichasium of three flowers. Such an inflorescence is termed as *verticillaster*, and is found in a large number of genera of this family such as *Salvia*, *Prunella* etc. Sometimes instead of a pair of simple dichasial cymes, there may be a pair of *cincinnus cymes*, where the simple dichasial cyme branches further and ends in a monochasial cyme e.g., species of *Lamium*, *Ballota*, *Nepeta* etc. In either case, the axis may be reduced, forming a dense sessile inflorescence or the main axis of both the main and the lateral ones may be more or less developed. Sometimes a number of whorls are aggregated together towards the apical portion of the stem, the subtending leaves being reduced to bract-like structure. This results in a raceme (if pedicels are present) or a spike-like inflorescence is (if flowers are sessile) e.g., *Prunella*, *Teucrium* etc. Sometimes a head-like inflorescence is also developed e.g., *Hyptis*, *Monarda* etc. The flowers may also be sometimes solitary in leaf axils e.g., *Scutellaria*.

Flowers

The plan of construction of the flowers is uniform. They are hermaphrodite, zygomorphic, rarely actinomorphic, e.g., in Mentha, Elshotzia, complete; hypogynous.

Calyx

The calyx consists of five sepals, gamosepalous. The sepals are inferior, persistent, campanulate or tubular, with free teeth or lobes, sometimes bilabiate (two-lipped), e.g., in *Thymus*, *Salvia* etc. In Origanum, the calyx is two lipped in *Hoslundia*, the persistent sepals become fleshy in the fruit. The aestivation may be valvate, imbricate or rarely quincuncial.

Corolla

It consists of five petals; pamopetalous, tubular and limb variously bilabiate. The corolla consists of two parts viz., tube and limb. The tube is straight or bent and wide towards the mouth. The limb is rarely equally fine toothed. In *Mentha*, two upper teeth become united forming an almost regular tetramerous corolla. The arrangement of the upper and lower lip of the limb is as follows. In majority of cases 2/3 arrangement it, found here the posterior pair of petals forms the upper lip which may be flat, e.g., in *Thymus*, or concave, e.g., in *Salvia*, *Stachys* etc. In the species of Ocimum, Hyptis etc., 4/1 arrangement of the petals is found. In such cases four upper petals form a developed upper lip while the anterior petal forms the lower lip. The 0/5 arrangement of petals is found in *Teucrium*. Here all the lobes being pushed forward. The aestivation is imbricate.

Androecium

Fig. 21.2. Floral diagram of Ocimum basilicum.

The stamens are typically four, didynamous, sometimes reduced to two. They are epipetalous and alternate with the corolla lobes. The fifth posterior stamen is usually absent and rarely developed. In majority of cases it is completely suppressed but in certain cases it is represented by a staminode. In *Salvia*, *Lycopus* etc., the two upper stamens are very often represented by staminodes or altogether suppressed. In *Coleus*, the stamens are monadelphous. Usually the anterior pair of the stamens is longer than the posterior pair, but in *Nepeta* and other allied genera the posterior pair is longer than the anterior pair. The anthers are twocelled; dehisce by longitudinal slits. In between the two cells of the anther, a connective is found. In *Salvia* this connective becomes filiform and articulated with the filament. In this case, the anterior cell of the anther is being reduced or sterile and modified. Generally a four lobed hypogynous disc is present. The anterior lobes of the disc secrete nectar.

Gynoecium

It consists of two median carpels, syncarpous, found to be seated on a hypogynous, nectar secreting disc. In very early stages, the ovary is two-celled (bilocular) but later on a constriction develops which divides the two carpels into four oneovuled segments. The ovules are anatropous. The ovules are attached at the inner corner of each ovary segment. The style is very characteristic of the family. It arises from the base of the ovary in between the four loculi, and is known as *gynobasic style*. The stigmatic papillae of bilobed stigma are situated at the tip of the style arms. The placentation is axile. The ovary is superior.

Fig. 21.3. Floral diagram of Lamium.

Fruits and Seeds

The *fruit* is of four one-seeded nutlets enclosed by the persistent calyx. The *seeds* are either non-endospermic or with a scanty endosperm and the embryo is usually straight.

Pollination and Dispersal

The brightly coloured flowers with nectar secreted by the posterior lobes of a hypogynous disc are adapted for insect pollination. They are mostly visited by moths and butterflies. Dichogamy is of universal occurrence.

The species of *Salvia* show an interesting mode of pollination. It has only two stamens (anterior pair). The two anther cells of each anther become widely separated by the development of the connective which becomes filiform and articulated with the filament forming a sort of lever mechanism. The anterior lobe is usually modified to a knob-like sterile structure and the posterior lobe only is fertile. The sterile lobes block the mouth of the corolla tube. When a bee thrust its proboscis in search of nectar, it pushes the sterile lobes.

A lever mechanism comes in action and the fertile lobes strike at the back of the insect and dust it with pollen. The flowers are . protandrous. As the stigma is ready to receive pollen, the style bends down and places the stigma in such a position to be touched first by a visiting insect. The persistent calyx sometimes forms a swollen bladder or develops barbed or thorny outgrowth which help in the distribution of nutlets.

Economic Importance

The family is of little economic value. Some plants are grown in the gardens as ornamentals, some plants yield essential oil while certain possess medicinal properties. A list of few important plants is given below:

1. *Pogostemon perilloides*; Verna.—Pacholi—This is a herb. The leaves yield an essential oil, which is used for making soaps and perfumes. The dried leaves are used for scenting clothes to keep off insects. It is commonly cultivated in the Western Ghats and Nilgiris.
2. *Anisomeles indica*. Syn. *A. ovata*; Verna.—Gopali, *Kalabhangra*—This is a woody herb, found throughout our country used as fodder and for flavouring purposes.
3. *Ocimum basilicum*; Eng.—Basil; Verna.—Bantulsi—This is an aromatic herb cultivated throughout our country. The leaves are used for flavouring purposes. The seeds are used medicinally as a remedy for dysentery and diarrhoea.

4. *Mentha longifolia*; Syn. *M. sylvestris*; Eng.—Mint; Verna—Pudina—This is an aromatic herb. The leaves are used for flavouring purposes. The dried leaves are used as stimulant and carminative. Usually cultivated in Uttar Pradesh, Kashmir, Maharashtra and the Punjab.
5. *Mentha pulegium*; Eng.—European pennyroyal—An essential oil is extracted from the leaves and tops of the plants, which is used in perfumery and cosmetics. It is a herb, cultivated in Jammu and Kashmir.
6. *Mentha arvensis var piperascens*; Eng.—Japanese peppermint. The leaves are the source of an essential oil, from which menthol is prepared. It is used in the treatment of colds. The plant is a herb and commonly cultivated in Jammu and Kashmir.
7. *Coleus sp*; Several species of *Coleus* are grown as ornamentals in the gardens for their variously coloured beautiful foliage.
8. *Mentha piperita*; Eng.—Pippermint; Verna.—Vilayatipudina—The leaves are used for flavouring purposes. An essential oil is obtained from the leaves, which is used in perfumery and soap making industries. The oil is also used medicinally as stimulant, carminative and for sickness and vomitting. This is a herb, cultivated in the Punjab, Kashmir and Maharashtra.
9. *Ocimum sanctum*; Eng.—Holy basil; Verna.—Tulsi—This is a small shrub and treated as holy plant by Hindus. Usually grown in the courtyards of Hindus and by the side of temples. This plant possesses several medicinal properties. The juice of the leaves is used as a stimulant and as a remedy for bronchitis. The seeds are used as a remedy for urino-genital troubles. The plant is used as an antidote for snake-bite.
10. *Coleus amboinicus*; Syn. *C. aromaticus*; Eng.—Indian borage; Verna-Pathorchur—This is a common herb. Its leaves are used for flavouring food products.
11. *Ocimum gratissimum*; Eng.—Shrubby basil; Verna.—Ranitulsi—This is a common ornamental, aromatic shrub. The plant possesses medicinal properties. The seeds are given in headache. This is also used as a mosquito repellant.
12. *Mentha arvensis*; Eng.—Field mint; Verna.—Pudina—This is a perennial herb. The leaves yield an essential oil which is used in the manufacture of certain kinds of cigarettes and pharmaceuticals. It is also used as a carminative, stimulant and refrigerant. Usually grown in the Punjab, Western Himalayas, Kashmir and Kumaon hills.

13. *Coleus forskohlii*; Syn. *C. barbatus*; This is a herb cultivated in South India. The roots are used as spice and condiment.
14. *Coleus rotundifolius*; Eng.—Madagascar potato; Verna—*Koorkan*—This is a herb or shrub commonly grown in South India. The underground tuberous stems are eaten as vegetable.
15. *Colebrookea oppositifolia*; Verna—*Binda*—The leaves are applied to wounds. The wood is used for making gun powder.
16. *Salvia officinalis*; Eng.—Sage; Verna.—*Salbia sefakuss Seesti*-This is a small shrub, grown as an ornamental. This is native of Mediterranean region.
17. *Satureia hortensis*. This is a herb found in Kashmir. The aromatic stem and leaves are used for flavouring food products.
18. *Mentha spicata*; Syn. *M. viridis*; Eng.—Spearmint; Verna.—Pahari pudina—The leaves yield an essential oil, known as spearmint oil, which is used for flavouring food products. This herb is cultivated in the Punjab, Uttar Pradesh and Maharashtra.
19. *Salvia splendens*; Syn. *S. colorans*; Eng.—Scarlet sage—This is grown as an ornamental.
20. *Ocimum kilimandscharicum*; Eng.—Camphor basil. The camphor is obtained from the leaves. It is cultivated on a very small scale in West Bengal, Assam, Madras and Dehradun.
21. *Leucas cephalotes*; Verna—Goma—This is a common herb. The flowers are used medicinally for cough and cold.
22. *Lavandula officinalis*; Syn. *L. vera*; *L spica*; Eng.—Lavender—This is a small shrub. It is native of Europe but now grown in Jamma and Kashmir. An essential oil is extracted from its flowers which is used in perfumery.
23. *Meriandra bengalensis*; Verna—Kafur ka pat—This is a shrub grown as an ornamental. The leaves possess medicinal properties. They emit a camphor-like scent. The leaves are used to preserve clothes against insects.
24. *Rosmarinus offcinalis*; Verna.—Rusmori—This is a herb grown as an ornamental. It yields an essential oil which is used in perfumery.
25. *Roylea calycina*; Verna—Patkarru—The plant possesses medicinal properties and used as a bitter tonic.
26. *Leucas zeylanica*; Verna—Gattatumba—The fresh juice of leaves is used as a remedy for headache and cold. It is also applied to skin diseases. The flowers are used for cough and cold.

22

APOCYNACEAE

A family of 180 genera and 1,500 species distributed mainly in the warmer parts of the world but a few are temperate. In India the family is represented by 29 genera and 60 species occurring chiefly in the Eastern Himalayas and Southern and the Peninsular India.

Trees or shrubs, generally woody climbers, more rarely erect or perennial herbs. Leaves opposite or whorled, very rarely alternate, simple, entire, rarely with small interpetiolar stipules. Flower hermaphrodite, 5-merous. Calyx 5, generally quincuncial. Corolla rotate, funnel-shaped or salver-shaped, lobes twistedly imbricate (contorted), throat with hairs. Stamens 5, on the corolla tube alternating with its segments sometimes adhering to the stigma by connectives. Anthers sometimes spurred downwards. Pollens granular. Ovary of two distinct or connate carpels. Fruit of 2 distinct or connate follicles or a drupe or berry.

VEGETATIVE CHARACTERS

Habit

There is a great variation in the habit of the plants of this family. They may be herbs, erect or twining shrubs or trees. *Vinca rosea* (Verna—*Sadabahar*) is a perennial herb; *Vallaris solanacea* (Verna-*Ramsar*) is a large twining shrub; *Nerium indicum* (*N. odorum*) is a large shrub with beautiful red or white flowers; *Thevetia peruviana* or *Thevetia neriifolia* (Verna—*Pili kaner*) is a large shrub or a small tree; *Plumeria acutifolia* is a small sized tree and *Alstonia scholaris* is a medium sized tree. In some genera, the stem becomes tuber like, e.g., *Adenium*. The species of *Landolphia* and *Clitandra* are climbing shrubs.

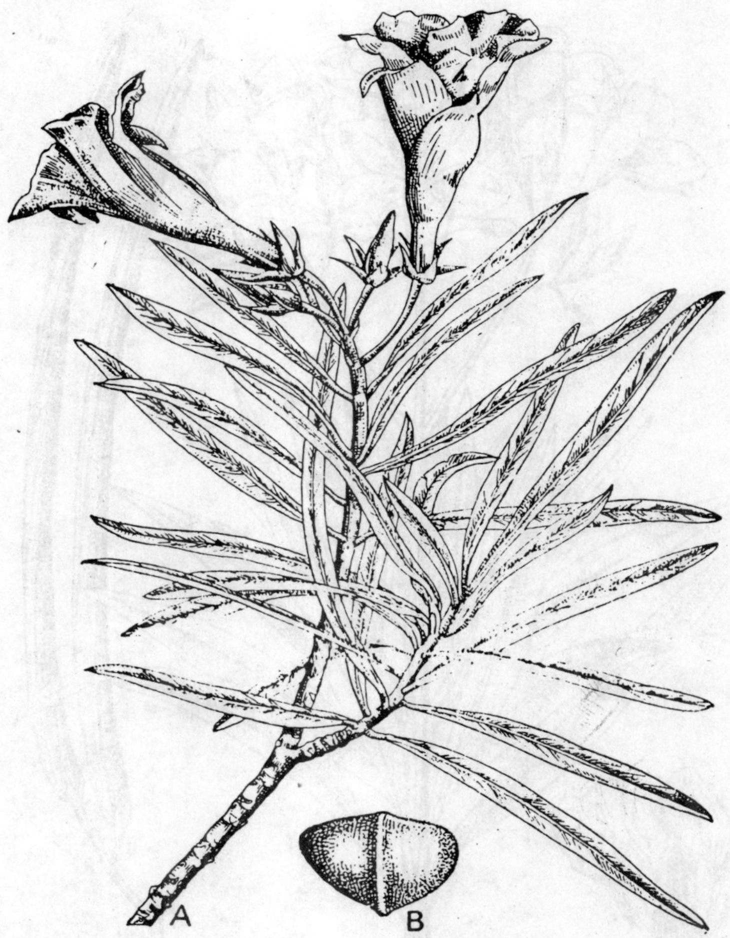

Fig. 22.1. Thevetia peruviana Schum. A—Flowering branch; B—Fruit.

Stem

Herbaceous, woody or climbing with collateral vascular bundles. Some stems have a tendency to become succulent by the development of thick parenchymatous cortex.

Leaves

The *leaves* are simple, mostly opposite and decussate. Sometimes they may be alternate or even whorled. They are exstipulate with entire margins. Only very rarely small interpetiolar stipules may be present.

Fig. 22.2. Nerium indicum Mill. A—Flowering branch; B--Corolla expanded.

Inflorescence

Inflorescence may be solitary e.g., *Lochnera* Syn. *Vinca* where sometimes the flowers are also in axillary pairs. Cymose type of inflorescence is however more common, and it may be either a terminal or an axillary cyme. In some cases it is a penicle or a cincinnus. In *Willughbeia edulis* the flowers are arranged in axillary cymes. In *Allamanda*, the flowers are arranged in axillary penicled cymes. In

Plumeria, the flowers are arranged in terminal cymes. In species of *Melodinus*, the flowers are arranged in either terminal or axillary cymes. In *Carissa* (Hindi—*Karaunda*), the flowers are arranged in corymbose cymes. In *Rauwolfia*, the flowers are arranged in either corymbose or umbellate cymes. In *Alstonia*, they are arranged in umbellately branched panicled cymes.

Flowers

The flowers are pedicellate, bracteate, bracteolate, hermaphrodite, actinomorphic, regular, sometimes slightly, zygomorphic, complete, hypogynous and pentamerous. In rare cases the flowers are tetramerous with reduction to two in the pistil.

Calyx

Usually it consists of five sepals, gamosepalous. The calyx is generally divided almost to the base. The aestivation is quincuncial.

Corolla

The *corolla* is gamopetalous, five-(rarely four) lobed and usually salver shaped but sometimes campanulate (as in *Apocynum*), urceolate (as in *Urceola*) or bell or funnel-shaped (as in *Beaumontia*). The aestivation of the corolla lobes is often twisted in bud or sometimes valvate. Some corona-like outgrowths are present at the mouth or within the corolla tube.

Androecium

Stamens 5-4 corresponding to the number of corolla lobes, alternating with the petals, inserted on the corolla tube with short filaments. The anthers are oblong linear or arrowshaped, more or less cohering in a cone, free or adherent to the stigma and apparently adnate to it. Anthers introrse opening by longitudinal slits. Sometimes spurred downwards with a ring-like or glandular disc at the base. The pollens are granular. Anther lobes are full of pollen up to base or they remain empty below and are prolonged into rigid spines. In *Condylocarpus* the pollengrains are united into tetrads.

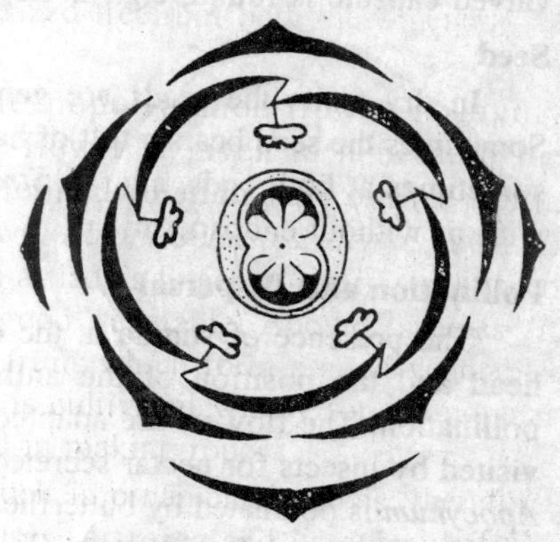

Fig. 22.3. Floral diagram of Nerium indicum.

Gynoecium

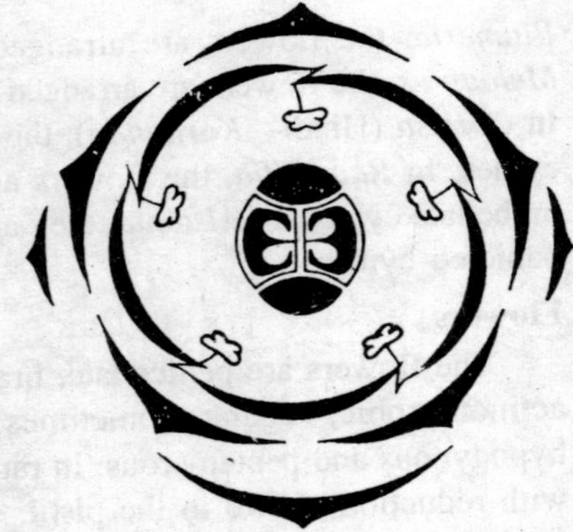

Fig. 22.4. Floral diagram of Catharanthus roseus.

Bicarpellary syncarpous or free, rarely the number of carpels may be 3-5. The ovary is superior or partly inferior (*Plumeria*). When the carpels are united the ovary may be unilocular with parietal placentation or bilocular with axile placentation. When the carpels are free, each ovary is one loculed and the placentation is marginal. The style is simple, stigma thickened, head-like (characteristic of the family is disc-like, or otherwise shaped, enlargement of stigamatic head with which anthers are closely related). There is a nectar secreting disc present below the ovary. Stigma is bilobed but the receptive portion is below the division on the median-like or at the base of the head.

Fruit

In the case of free ovaries, the fruit is a pair of follicles. Sometimes the fruits of separate ovaries are fleshy and indehiscent, or may be one seeded, e.g., *Cameraria*. In the case of syncarpous ovary, usually the fruit is indehiscent, fleshy and berry-like, e.g., in *Landolphia*. In *Cerbera*, it may be a drupe. This fruit is coconut like and distributed by means of water currents. In certain genera, possessing syncarpous ovaries a two-valved capsule is found, e.g., in *Aspidosperma* and *Allamanda*.

Seed

In dry fruits the seeds are generally winged, e g., in *Plumeria*. Sometimes the seed bears a tuft of hairs at the base. e.g., in *Kickxia*, and sometimes at both ends, e.g., in *Strophanthus*. The embryo is straight, with or without endosperm.

Pollination and Dispersal

The presence of stigma at the edge or under surface of the stylar head and the position of the anthers rules out the chances of self pollination. The flowers are adapted for insect pollination and they are visited by insects for nectar secreted by the hypogynous disc. Whereas *Apocynum* is pollinated by butterflies, the flowers of *Nerium* are adapted for long tongued Lepidoptera. In *Vinca*, when the insect enters its proboscis into the corolla tube it is smeared with adhesive matter and when withdrawn, it is cemented to the pollen. The winged seeds and presence of crown of hairs favour distribution by wind.

ECONOMIC IMPORTANCE

The family is of little economic value. Some plants are grown as ornamentals, while some possess medicinal properties. A list of some important plants is given below:

1. *Wrightia tomentosa*; Syn. *W. mollissima*; *Nerium tomentosum*; Verna.—*Dharauli*—The seeds and roots yield a yellow dye. The leaves and fruits are edible. Its soft wood is used for carvings. The bark and roots are used as an antidote for snake-bite.
2. *Rauvolfia serpentina*; Syn. *Ophioxylon serpentinum*; Verna.—*Chhotachand*—This is a small shrub found in Assam, Dehradun, Bihar, the Western Ghats and Bengal, the roots possess medicinal properties and are used in the treatment of hypertension mental disorders and related ailments.
3. *Carissa arduina*; Syn. *C. bispinosa*; Eng.—Natal plum.—This is a thorny shrub grown for its edible fruits.
4. *Wrightia tinctoria*; Syn. *W. rothii*; *Nerium tinctorum*; Versa;—*Dudhi*—A tree, found in Rajasthan, Madhya Pradesh and Madras. A blue dye is obtained from its flowers and fruits. The fruits are edible. The bark and seeds possess medicinal properties.
5. *Allamanda cathartica*; Eng.—Allamanda—This is a beautiful climbing shrub, grown as an ornamental in the gardens. It is native of Central America and Brazil.
6. *Ichnocarpus frutescens*; Eng.—blackcreeper; Verna.—*Dudhilata*, *Siamalata*—This is a twining ornamental shrub. It is found in Uttar Pradeṣh, Madhya Pradesh, Bihar, Assam and the Sundarbans. The stems are used for making ropes, baskets and fishing traps. The leaves possess medicinal properties.
7. *Landolphia kirkii*—The rubber is prepared from its latex. They have leaves with hook tendrils.
8. *Alstonia schoiaris*; Eng.—Dita bark; Verna—*Satwin*—This is a small tree grown as an ornamental. Its wood is quite light and used for carvings. In Burma, the black boards are prepared from its wood. The bark possesses medicinal properties, which is used for diarrhoea and dysentery. Its latex is applied to ulcers.
9. *Nerium indicum*; Syn, *N. odorum*; *N. oleander*; Eng—Oleander — Verna.—Kaner.—It is a shrub. They are grown as hedge plants. The plants possess medicinal properties.

10. *Beaumontia grandiflora*; Eng.—Nepal trumpet flower—It is a climbing shrub, usually grown as an ornamental for its large, white fragrant flowers. It is native of the Eastern Himalayas.
11. *Aganosma dichotoma*; Syn. *Echites dichotoma*; Verna.—*Malati*—This is a climbing shrub, grown as an ornamental in the gardens.
12. *Thevetia peruviana*; Syn. *Thevetia neriifolia*; Eng.—Yellow oleander; Verna.—*Pilikaner*—It is a shrub. The plants are grown as ornamental. They are also grown as hedge plants. The latex is highly poisonous.
13. *Carissa carandas*; Eng.—Karanda; Verna.—*Karaunda*—This is a spiny shrub grown throughout India for its sour edible fruits. The fruits are used as vegetable and pickle is prepared from them. The plant makes a good hedge.
14. *Carissa grandiflora*; Syn. *Arduina grandiflora*; Eng.—Natal plum—This is a large spiny shrub usually grown in Maharashtra and Baroda for its edible fruits.
15. *Beaumontia Jerdoniana*—This is also grown as an ornamental.
16. *Anodendron paniculatum*— Its leaves and roots possess medicinal properties.
17. *Plumeria alba*—A small tree, grown as an ornamental. The latex is applied to ulcers.
18. *Plumeria rubraforma acutifolia*; Syn. *P. acutifolia*; Verna.—*Goburchampa*—It is grown as an ornamental. It possesses several medicinal properties.
19. *Carissa spinarum*—This is a shrub or a small tree cultivated throughout India for fragrant flowers and hedge plants.

23

ASCLEPIADACEAE

The family is spread over 320 genera and 1,800 species. Depending on the variation in the number of genera (Woodson-100, Willis 130, de Wit-240, Rendle-280) the taxonomic status of the family needs revision. About 35 genera and 234 species (172 endemic species) are found. The name of the family Asclepiadaceae has been conserved (Lindley, 1847) over other names as Stapeliaceae.

Salient Features

Perennial erect herbs or shrubs with milky juice. Leaves opposite. Flowers generally regular, bisexual, pentamerous in terminal or axillary umbels or cymose. Sepals 5, free, imbricate. Petals 5, fused. Corona arising as outgrowths from petals or stamens. Stamens 5, filaments fused round the ovary and anthers adhering to the stigma by broad connectives forming gynostegium, pollen in 1 or 2 grandular or waxy masses in each cell at the angles of stigma. Carpels 2, free, placentation marginal, styles 2, connate, cohering above and dialated to form gynostegium. Fruit normally of 2 follicles. Seeds light and winged.

Distribution

The members of the family are mainly distributed in the tropics of the Old World, although some are found in tropical America. Some species are also represented in temperate zones. The plants chiefly inhabit arid soil. In India the plants chiefly occur in Himalayan and sub-Himalayan tracts. They also grow in Nilgiri hills and plains of UP, Bihar, MP, West Bengal and Punjab. *Calotropis* grows as a weed in dry soils in UP and Bihar. *Cryptostegia* is a climber very commonly occuring throughout India. *Hemidesmus* grows in Bihar. *Periploca* is found in outer Himalayas. *Pentanura khasiana* occurs in Khasia hills.

Indian Representatives

Calotropis, *Cryptostegia*, *Periploca*, *Myriopetron*, *Asclepias*, *Hemidesmus*, *Decalepis*, *Holostemma*, *Sarcostemma*, *Dregea*, *Cosmostigora*, *Finlaysonia*, *Streptocaulon*, *Brachylepis*, *Ulteria*, *Daemia*, *Marsdenia*, *Gymnema*, *Raphistemma*, *Taxocarpus*, *Sarcolobus*, *Congronema* etc.

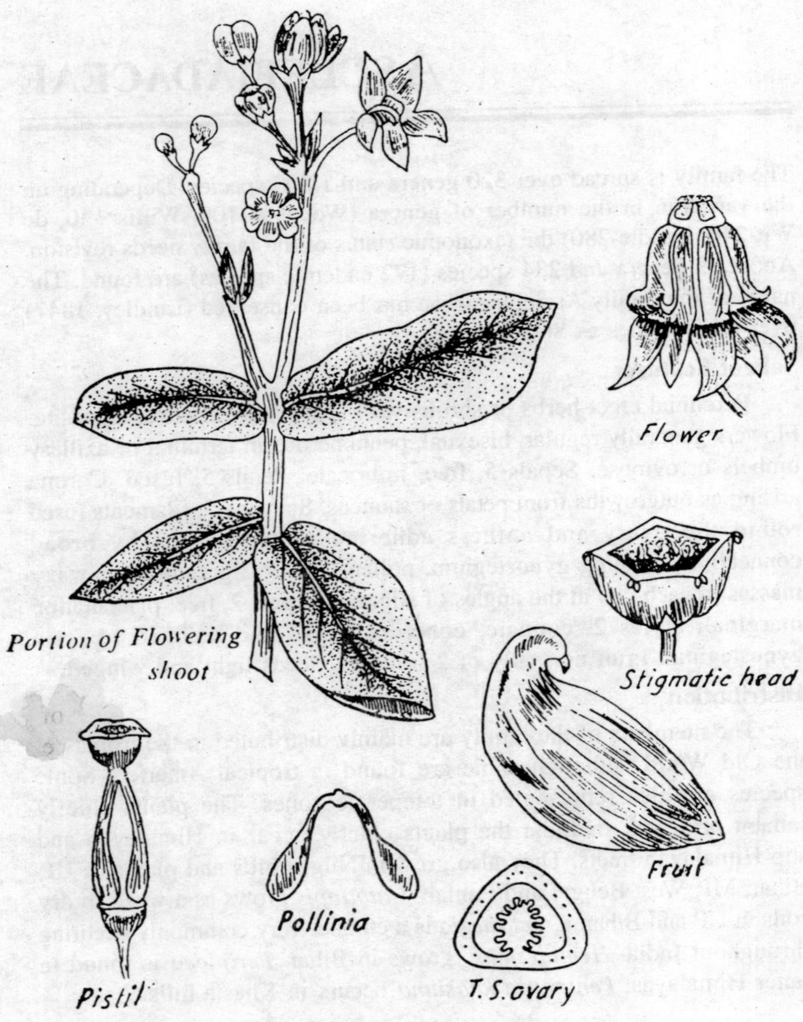

Fig. 23.1. Calotropsis gigantea.

Vegetative Characters

Habit

The plants of this family are mostly erect herbs or woody climbers but some are succulent. The perennial herbs are quite common in the dry regions of South and Central Africa. Some plants perennate by means of tuber-like rhizomes. Large shrubs and trees are rarely found. *Hoya* is a plant of succulent habit, it possesses fleshy stem and leaves. *Stapelia*, is a native of South Africa. It is cactus like in habit possessing thick and fleshy stems; the leaves are reduced to thorns or scales. *Dischidia* is an epiphytic plant; it climbs upon its host by means of adventitious roots, the leaves are succulent and covered with a coating of wax. *Cryptostegia grandiflora* is a climbing shrub. The species of Hemidesmus, Gymnema, Pergularia etc., are also climbing shrubs. *Leptadenia spartium* is a shrub. *Calotropis procera* (Verna-Ak) is a xerophytic shrub and *Calotropis gigantea* (Verna-Safet Ak) is a large shrub. *Asclepias curassavica* is a perennial herb. The species of *Ceropegia*, *Pentatropis* etc., possess underground tuber like parts. *Tylophora indica* is a herb. Most of the plants possess xerophytic characters and certain are extreme xerophytes.

Anatomically the plants of this family resemble with that of Apocynaceae. They also possess bicollateral bundles and the latex. According to Busich (1913), the endophytic mycorhiza is frequently found in the roots of succulent plants of this family.

Roots

A deep tap root, the root stalk may be tuberous and fleshy (*Brachystelma*, *Ceropegia bulbosa*) or food storage, adventitious (*Dischidia*) and epiphyte, climbing by means of these roots. Endotropic micorhiza in the roots in succulent species is frequent.

Stem

Woody below, herbaceous above, covered in many cases with wax (*Calotropis*) and show typical xerophytic characters. Intra-axillary phloem, bicollateral vascular bundles and laticiferous tubes in the anatomy of stem are common characters.

Leaves

Opposite, sometimes alternate or whorled, in come cases caducous and vestigeal, simple, entire and waxy. Sometimes the leaves are modified into pitchers (*Dischidia rafflesiana*) or they may be succulent (*Hoya*) or reduced to spines or scales (*Stapelia*) like that in cactus.

Inflorescence

Inflorescence is mostly of cymose type. It is a dichasial cyme, but one of the two branches of the dichasium grows faster than the other, and after a few branchings, the weaker one gets completely supperessed. This type of inflorescence which is similar to that found in the family *Caryophyllacae* is known as a *cincinnus*. (A dichasial cyme ending in a monochasial cyme). Sometimes it is of racemose type, either a *raceme* or an *umbel*; or the flowers may be arranged in umbelliform cymes. The inflorescence may also be axillary and solitary, but there is only one flower present at each node. The other leaf of that node remains sterile, with either only a vegetative shoot or absolutely nothing. In *Hemidesmus*, the flowers are arranged, in axillary cymes. In *Calotropis*, the flowers are arranged in umbellate cymes. In *Oxystelma*, the flowers are arranged in either umbelliform cymes or they are solitary. In *Brachystelma*, the flowers are few and are arranged in either axillary sessile umbels or they are solitary.

Flowers

The flowers are pedicellate, bracteate, hermaphrodite, actinomorphic, rarely zygomorphic, e.g., in *Ceropegia*, complete. The general plan of the flower is pentamerous with three regularly alternating pentamerous whorls of calyx, corolla and androecium, however, the number of carpels is reduced to two in the gynoecium. Usually the flowers are small in size, but the flowers of *Ceropegia*, *Stapelia* and *Stephanotis* are quite large in size.

Calyx

It consists of five sepals, which are either free (polysepalous) or somewhat connate at the base; with the odd sepal posterior. The aestivation is quincuncial or rarely valvate.

Corolla

It consists of five united petals (i.e., it is gamopetalous). The petals are spreading (i.e., rotate), but in *Stephanotis* the corolla tube is long, forming a salver-shaped corolla. In *Ceropegia* the corolla is pitcher-like in appearance (zygomorphic). The aestivation is contorted and rarely valvate. Sometimes the petaloid appendages arise either from the corolla or from the back of the stamens. Very often these hairs or appendages are found inside or at the mouth of the corolla forming the *corolline corona*.

Androecium

5, or as many as corolla lobes, free rarely, generally united into a short fleshy column (round the ovary) which usually bears a ring or series of processes that are attached to the filaments or to the back of the anthers; this called *staminal corona* (only in rare cases both coronas are present or absent). The apex of the staminal tube is united with the margins of the pentangular stigmatic head to which anthers are coherent by connectives. The structure thus formed by union of the anthers and the stigmatic head is known as the *Gynostegium*. The anthers are broad tipped two-celled (4-celled in *Secamone*).

In the sub-family *Cynanchoideae* the pollengrains at maturity are agglutinated (adhering together) into a small ovoid pendulous mass of pollen—the *Pollinium*. Thus in such stamen there are two such pollinia. The pollinia of two adjacent anther halves are united by means of short stalks or caudicles to from a *translator* or rider. This translator consists of two parts, the *Corpusculum* or gland like-body, which often gets attached to the margins of stigmatic surface between two anthers and a pair of arms the *retinaculi* or the connective by means of which the pollinia of adjacent anthers halves are attached to the corpusculum.

In the sub-family *Periplocoideae* the stamens are stalked and the pollengrains are granular in tetrads. The translator is spoon or funnel-shaped between each anthers and terminating below in an adhesive disc. The pollengrains of one half of each of the two adjacent anthers, when mature, fall into the concave receptacle of the spoon or funnel.

Gynoecium

Bicarpellary syncarpous, ovaries two, distinct, superior (in some cases slightly inferior during development). Each ovary unilocular, unicarpellate with several ovules on ventral marginal parietal placenta. Styles 2, free or may unite and in a dilated pentangular or lobed stigmatic head with which 5 epipetalous stamens are coherent. The 5 longitudinal strips of glandular stigmatic surface on the thickened edge or the lower side exposed between contiguous anthers is receptive.

Fruits and Seeds

The *fruit* is of two follicles which are close together or divergent. They vary in shape and are membranous to woody. The *seeds* are flattened and commonly bear a terminal tuft of long white silky hairs. The endosperm is dense and copious and the embryo is large.

Pollination and Dispersal

The flowers are perfectly adapted for insect pollination but the pollination mechanism is unique. In the subfamily Periploceae the pollen is transferred on to the spoon-shaped translators, each of which has a sticky basal disc. The insect visitor on retiring, carries the whole translator as the basal adheres to its head. The pollen contents of the

Fig. 23.2. Floral diagram of Calotropsis.

translator may be deposited on the stigmatic surface when the same insect visits another flower. In the subfamily Euasclepiadeae an insect visiting the flower for nectar traps its legs or proboscis between the osmotically elastic anther wings and on withdrawal it carries with it the sutured corpusculum with the pollinia. The retinaculae are elastic and hygroscopic. When in the flower, they are like an extended spring, but as the pollinia are withdrawn the spring closes and the two pollinia quickly cross each other and hold tightly on the insects' leg or proboscis. When this insect visits another flower, in the process of catching leg or proboscis, the pollinia are transferred to the receptive surface of the stigma which is beneath the anther wings.

The *seeds* with a terminal tuft of hairs are distributed by wind. In others they are dispersed by. water or by the human agency.

Economic Importance

The family is of little economic value. Some plants are ornamental while certain others are of medicinal value. A list of few important is given below:

1. *Gymnema sylvestre*; Verna—*Gurmar*—This is a large climbing shrub. The flowers are found to be arranged in umbellate cymes. The roots are used as antidote for snake-bite. The leaves are used as remedy for diabetes.
2. *Daemia extensa*; Syn. *Pergularia extensa*; Verna—Utsan—A fibre is obtained from its stem. The juice of leaves is used medicinally several ways, e.g., as the remedy of diarrhoea, asthama etc.
3. *Oxyrstelma esculentum*; Verna—*Dudhialata*—The fruits are edible; the plant possesses medicinal properties. The latex is used to wash ulcers. The medicine for scabies is prepared from it.
4. *Cosmostigma racemosum*—This is a climbing shrub, found in the Western Ghats, Mysore and Andhra Pradesh. The leaves are used to cure ulcerous sores.
5. *Sarcostemma acidum*; Verna.—*Somlata*—This is a leafless shrub, commonly found in dry areas of Andhra Pradesh, Madras and Mysore. It is used to destroy white ants from sugarcane fields.
6. *Tylophora indica*; Verna—*Antamul*—This is a herb. Its leaves and roots are used as substitute for ipecacuanha. This is also used as emetic, diaphoretic and expectorant. The plants are commonly found in Bengal, Assam and South India.
7. *Asclepias curassavica*; Eng—Blood flower; Verna—*Kakatundi*—This is a perennial herb or a small shrub. It is cultivated as an ornamental. The plant is of medicinal value. The root is used as a purgative and as a medicine for piles.
8. *Hemidesmus indicus*; Eng—Indian sarsaparilla; Verna—*Magrabu*—This is a twining shrub. The roots are used as a tonic, which is useful in loss of appetite, fever, skin diseases, as blood purifier, rheumatism and snake-bite. The tonic (sarsaparilla) is diuretic diaphoretic and demulcent.
9. *Calotropis gigantea*; Eng.—madar; Verna—Ak—A perennial undershrub. The floss, obtained from the seed, is used for stuffing purpose. The stem yields a fibre. All parts of the plant are used medicinally. The fresh leaves are used in the fomentation for swellings. The gun powder charcoal is prepared from the wood. It is commonly found in Northern India.
10. *Carallama fimbriata*; Vern.—*Makedshingi*—This is a fleshy herb, eaten as a vegetable. It is mainly found in Andhra Pradesh.
11. *Cynanchum arnottianum*—This is a herb whose dried leaves are used to destroy maggots. It is found in Kashmir.

12. *Calotropis procera*; Syn. *Asclepias gigantea*; Eng. Akund; Verna—Safed Ak.—The stem fibre is made into cordage and the floss from the seeds is used as a stuffing material. The stem and leaves are used medicinally several ways. It is commonly found in Andhra Pradesh and the Punjab.
13. *Periploca aphylia*; Verna—*Barri*—The stem yields a fibre for cordage; the latex is applied to swellings and tumours, it is used as a purgative; sometimes the flowers are eaten.
14. *Ceropegia lawii*; Verna—*Kharpundi*—The tubers are edible. It is a succulent herb. Found in South India.
15. *Leptadenia pyrotechnica*; Syn *L. spartium*—This is a shrub found in the Punjab, Uttar Pradesh, Rajasthan and Gujrat. The slimy fruits and young twigs are eaten as vegetable.
16. *Ceropegia hirsuta*; Verna—*Khantodi*—This is a climbing shrub. The tubers are edible. It is commonly found in Madras, Travancore are Andhra Pradesh.
17. *Cryptostegia grandiflora*; Verna—*Vilayati-vaknandi*—This is a beautiful climbing shrub with red-rose flowers, grown as ornamental.
18. *Pentatropis cynanchoides*; Verna—*Kanathodi*—The tuberous roots are edible.
19. *Dischidia rafflesiana*; Verna—*Bandikuri*—This is an epiphytic undershrub. The roots are chewed as a remedy for coughs.
20. *Ceropegia oculata*; Verna—*Khantodi*—The tubers are edible. It is a herb, found in South India.
21. *Ceropegia tuberosa*; Verna—*Patala tumbi*—The tubers are edible. It is commonly found in Madras and Andhra Pradesh.
22. *Cryptolepis buchanani*—The plant produces a floss which is used as a stuffing material.

24

AMARANTHACEAE

This family includes 65 genera and 850 species widely distributed in the tropical and temperate regions of the world. Tropical America and India are the chief centres of distribution. In India the family is represented by 17 genera and over 50 species occurring mostly in the warmer parts. The familiar examples are Amaranth (*Amaranthus blitum* L.), Cockscomb (*Celosia argentea* L.) and Globe amaranth (*Gomphrena globosa* L.).

Salient Features

Plants mostly herbs. Inflorescence compound and dense. Flowers bisexual, usually with dry and membranous bracts, bracteoles and perianth. Stamens opposite the perianth lobes. Gynoecium bicarpellary, syncarpous, unilocular with one basal ovule. Fruit generally dry. Seeds with mealy endosperm.

Distribution

The plants are well represented in tropical regions, especially tropical American Africa and India. Some of them also grow in temperate regions. In India the plants grow throughout warmer parts or rarely at a low altitude of Himalayas either as weeds or ornamentals.

Indian Representatives

Achyranthes, *Amaranthus*, *Alternanthera*, *Celosia*, *Cyathula*, *Deeringia*, *Digera*, *Gompherina*, *Bosia*, *Pupalia* etc.

General Habit

Usually annual or perennial herbs, rarely shrubs or trees; annual herb (*Amaranthus*, *Colosia*, *Gompherina*), perennial herbs (*Bosia*), undershrubs (*Deeringia*) or shrubs (*Ptilotus*).

Vegetative Characters

Habit

Usually they are annual or perennial herbs or shrubs and rarely trees.

Stem

Herbaceous or woody, cylindrical, solid with vascular bundles arranged in several concentric rings or irregularly arranged.

Leaves

Opposite or alternate, simple, exstipulate, usually entire with a thick coat or woolly hairs.

Inflorescence

Inflorescence may be either a simple or a branched spike or a raceme. Each flower possesses a pair of usually large membraneous persistent bracteoles which are sometimes sterile. Sometimes the branching occurs in the axils of the bracteoles forming dichasia which may be borne on simple, racemosely or cymosely branched main axis. Sometimes the flowers on the lateral sides are sterile and are often replaced by prickles or hairy tufts which persist in fruit and help in dissimination.

Flowers

The *flowers* are minute and often densely crowded in the inflorescence to give it a very attractive appearance e.g., *Amarantus* (Foxtail), *Celosia* (Cockscomb) etc. The structure of the flowers is similar to that of *Chenopodiaceae*. They are usually hermaphrodite, rarely unisexual by abortion, actinomorphic and hypogynous. Bracts and bracteoles are present.

Perianth

The *perianth* is uniseriate and usually consists of 4-5 perianth leaves which may be free or variously united. They are dry and membraneous, sometimes, white or even coloured, and are often hairy. They thus differe from *Chenopodiaceae* is not having green and herbaceous perianth leaves. The *petals* are absent.

The stamens are five and they lie opposite the sepals. Rarely, as in *Nothosaerva*, there are only one or two stamens. The filaments are sometimes free or nearly so as in *Digera* and *Amaranthus* but they are usually united at the base into a membranous tube which may bear linear, lacerate or., two-fid outgrowths between the stamens representing the staminodes as in *Cyathula*. The anthers are usually dithecous or

Fig. 24.1. Amaranthus spinosus L. A—Flowering branch; B—Pistillate flower; C—Staminate flower; D—Fruit with persistent bracts and calyx; E—Seed.

sometimes as in *Gomphrena* monothecous, introrse and opening by longitudinal slits.

Gynoecium

Monocarpellary usually, may be 2-3 carpels, syncarpous. Ovary superior, unilocular with 1 or more (*Celosia*) ovule. Style sometimes simple or obsolete and with a capitate or small stigma or 2-3 fid with

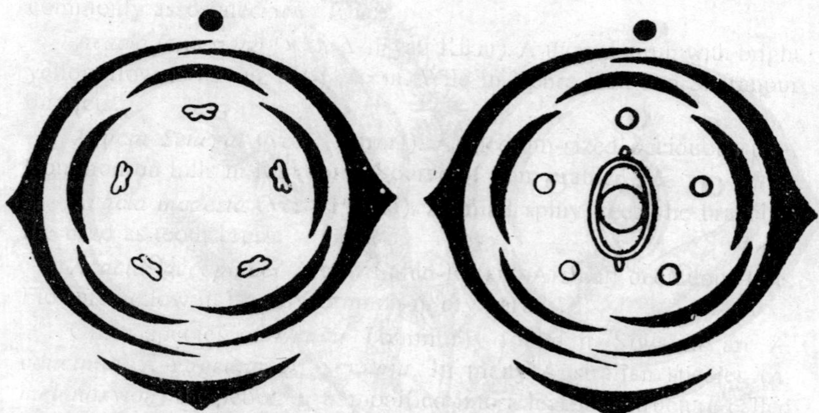

Fig. 24.2. Floral diagram of Amaranthus spinosus, staminate flower.

Fig. 24.3. Floral diagram of Amaranthus spinosus, pistillate flower.

acute stigmas or styles 2-3, erect or recurved and stigmatic on lower face.

Fruits and Seeds

The fruit is usually an utricle which is dehiscent or indehiscent but sometimes it is a berry (*Deeringia*), circumscissile capsule (*Celosia*) or nut (*Digera*). It is enclosed in or seated on the persistent calyx. The *seeds* are compressed or ellipsoid with a crustaceous testa and horse-shoe shaped or annular embryo surrounding a mealy endosperm.

Pollination and Dispersal

The aggregation of small flowers into dense and showy inflorescences and presence of nectar in some favour insect pollination. Some genera, such as *Gomphrena*, with winged fruits are dispersed by wind. Many genera, such as *Achyranthes*, *Cyathula* and *Pupalia* develop hooks on the sepals with which they adhere to passing animals or human beings.

Economic Importance

1. A few plants are grown as ornamentals in the garden e.g. *Amaranthus* (Fox tail), *Celosia cristata* (Cock's comb), *Gompherina* (Globe amaranthus) etc. *Alternanthera* is grown as hedge plant.
2. The plants of this family contribute to the food products. Seeds of *Amaranthus caudatus* are edible. Leaves of some plants (*Amaranthus gangeticus* and *gracilis* are used as vegetables. *A. Cruentus* and *A. Frumentaceous* are raised as cereals by primitive tribes in tropical Asia.

25

ORCHIDACEAE

A fairly large family spread over 735 genera and about 20,000 species (Willis. 1966). Lesile A. Garay (1960) included 800 genera and 30,000 species in this family. According to Pease (1963) there occur approximately 35,000 species in the family Orchidaceae. In India the family is represented by about 1,700 species (Puri, 1965).

Plants usually perennial herbs, terrestrial, climbing or epiphytic, sometimes saprophytic. Roots adventitious, aerial or tuberous, roots of epiphytic plants provided with velamen tissue. Stem generally sympodial, often modified to form a pseudo-bulb, or a tuber. Leaf simple. Inflorescence racemose. Flowers large, brightly coloured, bisexual, epigynous, medinaly zygomorphic. Perianth biseriate, tepals 6 in two distinct whorls, the median tepal of the inner whorl is large and variously modified to form the labellum. Stamen generally one, sometimes two, fertile, with two staminodes. Gynoecium tricarpellary, syncorpous, inferior, unilocular, ovules numerous, parietal placentation, stigma 3 but generally only two respective and the third is sterile and forming the rostellum. The stamens and stigmas form a columnar structure above the ovary called the gynostegium. Fruit a capsule. Seeds numerous, small covered by a thin membrane, endosperm absent.

Distribution

The plants are of world-wide distribution but are abundant in the tropics where they behave as epiphytes. The genera inhabiting temperate regions are usually terrestrial. The chief centres of their distribution are Indo-Malaya, tropical America and Himalayan regions. Orchids are found in almost every part of the globe except for polar regions. Their fossil history dates back to Eocence with the discovery of *Protoorchis monorchis* from Mont Bolcha.

Vegetative Characters

Habit

The plants are generally herbs of various habit (rarely shrubby) of two principal forms: (i) Terrestrial tuberous rooted herb with annual herbaceous simple stems and solitary or spicate or recemose flowers. (ii) Epiphytes with perennial stems or branches variously thickened and

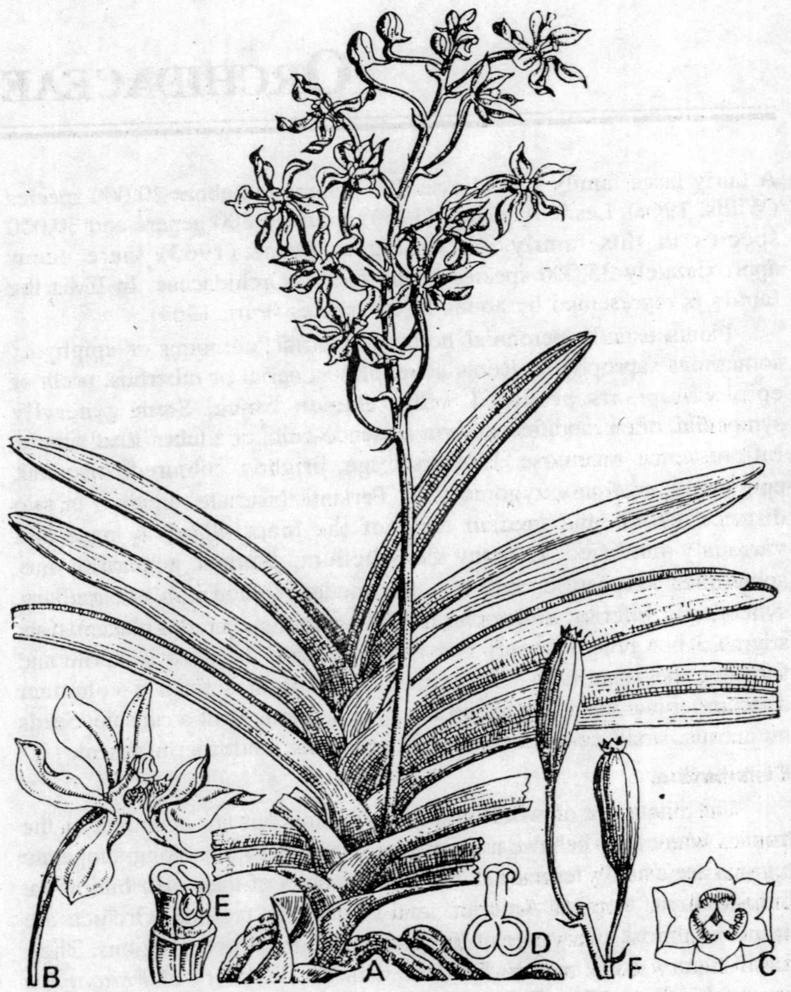

Fig. 25.1. Vanda parviflora Lindl. A—Flowering plant; B—Flower; C—Ovary transverse section; D—Pollinium; E—Gynandrium; F—Fruits.

often forming a pseudobulb, flowering from the top side or base of the pseudobulb. Bracts are usually present. Some of the terrestrial plants live as saprophytes and derive their nourishment from the humus. Associated with the roots or the rhizome is present with an endotropic micorrhiza (*Neottia*, *Corallorhiza*, *Galeola*).

Roots

The development of the main root is always absent and its place is always taken by adventitious roots which arise from the nodes often showing as regular arrangement as the leaves. There are 3 kinds of roots: (i) Normal cylindrical earht roots, (ii) Tuberous roots, serving as stores of reserve material, (iii) Air roots characterised by development of *Velamen*—a tissue consisting of several layers of short tracheids. The sponge-like tissue absorbs dew and rain, and pass it on to the internal tissue.

Schimper in epiphytic orchids distinguished 3 kinds of aerial roots: (i) Clasping or climbing roots which are negatively heliotropic, moving away from the light in search of support and creeping close to the substratum. (ii) Absorptive roots, which are negatively geotropic penetrating the humus which collects between plant and its support. (iii) True aerial roots, which hang down in the air, absorbing water through the velamen lying above the epidermis, that acts as sponge absorbing dew and rain water. Internal tissue is green and carry photosynthesis. In orchids with fleshy leaves (*Vanilla*) velamen is poorly developed.

Stem

Usually erect, but sometimes climbing or trailing, annual terrestrial forms, perennial in epiphytes often thickened into rhizome or pseudobulb bearing aerial root. Axis of the plant is either monopodial (epiphyte) or sympodial. Sometimes the nodes and leaves are very much short and the plant lies flat on the substratum (*Dichaea*), in other cases long internodes may be enclosed in leafsheaths and the blades are flat, short or long and cylindrical. In terrestrial forms the stem is usually slender having 1, 2 or more green leaves terminating in one or several flowered inflorescence (*Spiranthes*, *Cypripedium*, *Habenaria* etc.), below the ground there may be either a fleshy rhizome bearing roots or a tuber.

Leaves

Alternate, rarely opposite or whorled, simple usually linear, often distichous with parallel venation, leaf-base often fleshy and sheathing, marging entire, apex denticulate or symmetrically cut, venation as a rule parallel. Leaves sometimes reduced to scales in saprophytes and parasites, or many become membranous or succulent. Leaf either decay on the stem or in epiphytes separate by distinct joints. Sometimes the leaves fall off during dry season and the plants perennate by means of pseudobulbs.

Foliage leaves may be entirely absent, the shoot is reduced to a scaly bulb or the internodes are more or less elongated with scales at the nodes. In such cases photosynthesis is carried by green surface of the stem (as in *Dendrophylax*). In many genera (as *Pleurothallis*) numerous shoots bear a single leaf at right angles to the axis or apparently forms a continuation of it. Where the axis is short (as in *Masdevallia*) leaves appear to rise from the substratum. The venation of the leaf is plicate or convolute often complicated by longitudinal folding. In transverse section the leaf shows the developed cuticle and presence of water storing cells below the upper epidermis, which is contrivacne for storing water and checking rapid transpiration.

Inflorescence

Inflorescence is mostly of racemose type, either a raceme which in some cases e. g., *Renanthera lowii* reaches a length of 13 feet (Rendle), or a spike or a panicle. Sometimes the flowers are solitary e.g., *Cypripedium*. In *Oberonia*, the flowers are stalked and crowded, the inflorescence resembling that of *Musa*. Usually the inflorescence is only short-lived but sometimes it becomes perennial, producing new flowers throughout the year.

Flowers

Flowers are mostly hermaphrodite. Rarely they may be unisexual, when plants are either monoecious or dioecious. They are zygomorphic, epigynous always bracteate, but may be pedicellate or sessile.

Perianth

Perianth usually consists of six perianth leaves arranged in two whorls of three each. The outer whorl of three perianth leaves (sepals) are either green or petaloid. Either they are all of the same size and shape or the median sepal becomes more conspicuous in size or colour. In *Disa*, the odd sepal is spurred and in *Heamaria* it unites with the lateral petals to form a hood-like structure. Out of the inner whorl of three perianth leaves (so called petals), the posterior one is usually larges and highly modified and is known as the *labellum* or the *lip*. In most of the orchids, the flower is turned or twisted through 180° (resupination) and thus the *labellum* or the lip comes round to the anterior side of the open flower and forms the landing place for the insect at the time of pollination. In *Corysanthes*, the lateral petals are often missing, while in *Epicranthes*, they are filiform. In *Cypripedium caudatum*, the petals are, ribbon-like and about a yard long. In *Disa* the labellum is often small and narrow.

Androecium

The *androecium* is represented by usually one or sometimes two sessile anthers. In monandrous forms there is only single fertile stamen, terminal on the column. This stamen is considered to represent the anterior member of the outer whorl, the two other (lateral) stamens of this whorl and also all the three of the inner whorl are entirely absent. The two lateral members of the outer whorl are sometimes represented by staminodes as in *Epipactis*. In diandrous forms (*Cypripedium* and allied genera) there are two fertile stamens belonging to the inner whorl and they are lateral on the column. The median member of the outer whorl (which is fertile in monandrous forms) is represented by a large staminode. The other staminal members are completely absent. The anthers are dithecous, introrse and opening by longitudinal slits. The pollen grains are either granular or more commonly agglutinated into mealy, waxy or bony masses, the pollinia. One end of the pollinium is extended into a sterile structure, the caudicle. There are two to eight pollinia per anther and they are free or more or less loosely united.

Gynoecium

The *gynoecium* is tricarpellary and syncarpous with an inferior ovary which is unilocular with three parietal- placentae. Sometimes, as in *Apostasia* the ovary is trilocular with axile placentation. There are numerous ovules on each placenta. In monandrous forms the column has two fertile stigmas and a specialized organ, the rostellum, which represents the third stigma. Sometimes (e.g. *Babenaria*) a portion of rostellum is modified into a viscid disk (viscidia) to which pollinia are attached. In diandrous forms there is no *rostellum* and all the three stigmas are simple and fertile.

Fig. 25.2. Floral diagram of a monandrous orchid.

Fruit

Capsule, cylindrical or ovoid, generally ribbed, opening by 3-6 longitudinal slits, minute and light numerous seeds, easily dispersed by wind, non-endospermic.

Pollination

Most of the orchids are epiphytes, growing on trees, some grow on rocks and on ground. The fact that peculiar structure of the flower renders them in most cases entirely dependent on insect agency for pollination has been known for more than a century, but it was not until 1862 when Darwin published *The various contrivances by which orchids are fertilised by insects*, our knowledge of the subject became accurate and complete. The flowers, if not pollinated, remain for a long time without withering. Insects are attracted to the flowers by their size, colour and sweet-smell. The nectar is secreted in the spur or cup-like development of the lip or sometimes from the surface or from the ridges developed on the lip.

Fig. 25.3. Floral diagram of a diandrous orchid.

By the torsion of the flower through 180° the lip provides a landing stage for the insect. The viscid drop or gland placed directly in the path of an insect seeking to obtain honey, glues the pollinia so firmly that they are dragged out of the anther and are carried in such a position that the pollen comes in contact with the stigma of the next flower. The structural details by which this transference of pollen is facilitated differs in different genera and in some cases they are of wonderful mechanical complexity. The method for pollination in *Cypripedium* (= *Paphiopedilum*) is different from other orchids. The three fertile stigma form a convex dry surface and bears minute rigid hairs pointing forwards. The two fertile anthers stand behind and above the stigma, one on each side of the column forms a shield-like body above stigma; and fertile anthers from the top of the column.

The pollengrains are granular and covered with a glutinous mass. The basal portion of the lip surrounds the column, leaving an opening on each side of the column above anthers. Rest of the lip forms a conspicuous structure by which insects are attracted. The insects fly into the lip on the floor which is hairy and secrete honey. The insect in search of honey rub its back first to the stigma and then to anther, so that stigma can be only pollinated by pollens brought previously by the insect.

ECONOMIC IMPORTANCE

1. Many tropical orchids are cultivated in the green houses for their peculiar beautiful flowers. Growing orchid is also a widespread hobby. Some common ornamental orchids include *Cautleya*, *Dendrobnium*, *Epidendrum*, *Cymbidium*, *Cypripedium*, *Hobenaria*, *Bletellia* etc.
2. The family is chiefly valued for medicinal and ornamental plants.
3. The leaves of *Calanthe veratrifolia* are rich in a glycoside 'indican' which on hydrolysis yields 'indigo blue'. The pods of *Vanilla planifolia* yield the "vanilla" a flavouring agent widely used in confectionary. The fibres from the stem cortex of *Dendrobium crumenatum* are used as a braiding material for hats. The tubers of *Eulophia epidendra* are used as a vermifuge. The rootstocks of *Geodorum densiflora* are crushed and rubbed on cattle to kill flies. The dried leaves of *Jumellea fragrans* constitute 'faham tea'.
4. The roots of *Vanda roxburghii* are used as a cure for rheumatism and antidote for scorpion stings.

26

LILIACEAE

The family is spread over 250 genera and 3,700 species (Willis, 1966). However, the number of genera and species included in this family varies according to different authors: Lawrence, 1951 (2409 genera, 4000 species); Baily, 1947 (75 genera, 1200 species): de Wit, 1968 (220 genera, 3500 species); Cronquist (4,200 species). *Genus Asparagus* alone has about 300 species. In India the family is represented by a quite large number of species.

Salient Features

Plants generally rhizomatous bulbous perennial herbs. Leaves radical or cauline, alternate or whorled. Inflorescence generally racemose. Flowers bisexual regular, homochlamydous, trimerous and pentacyclic. Perianth 6, biseriate. Stamens generally 6, in two whorls of 3 each, epiphyllous. Gynoecium tricarpellary, syncarpous, ovary superior, trilocular, ovules several, palcentation axile. Fruits generally a capsule. Seeds with small embryo, endosperm fleshy or horny.

Distribution

The plants are cosmopolitan, distributed all over the world chiefly in warm temperate and tropical regions. The plants occur throughout India and are distributed both in the hills and plains.

Indian Representatives

Allium, *Aloe*, *Agave*, *Asparagus*, *Chlorophytum*, *Colchicum*, *Cordyline*, *Dracaena*, *Fritillaria*, *Lilium*, *Polygonatum*, *Sensevieria*, *Scilla*, *Asphodelus*, *Urginea*, *Yucca*, *Tulipa*, *Haemerocallis*, *Eremurus*, *Aspidistra*, *Clintonia*, *Haworthia* etc.

Vegetative Characters

Habit

The members of this family show great variation in habit. Most of them are herbs perennating by means of rhizome (*Polygonatum*), bulb (*Tulip*, *Lily*, *Allium*) or by corms (*Colchicum*). Some of the plants are adapted to xerophytic mode of climate (*Yucca*, *Dracaena*, *Aloe*,

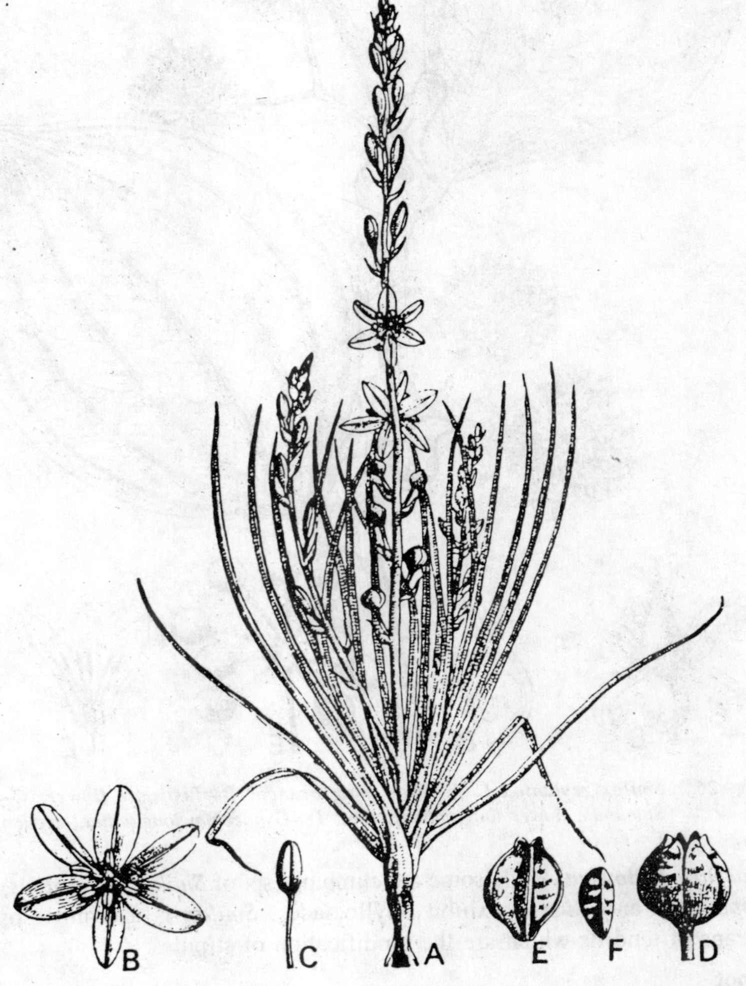

Fig. 26.1. Asphodelus tenuifolius Cav. A—Flowering plant; B—Flower; C—Stamen; D—Fruit; E—Fruit longitudinally cut; F—Seed.

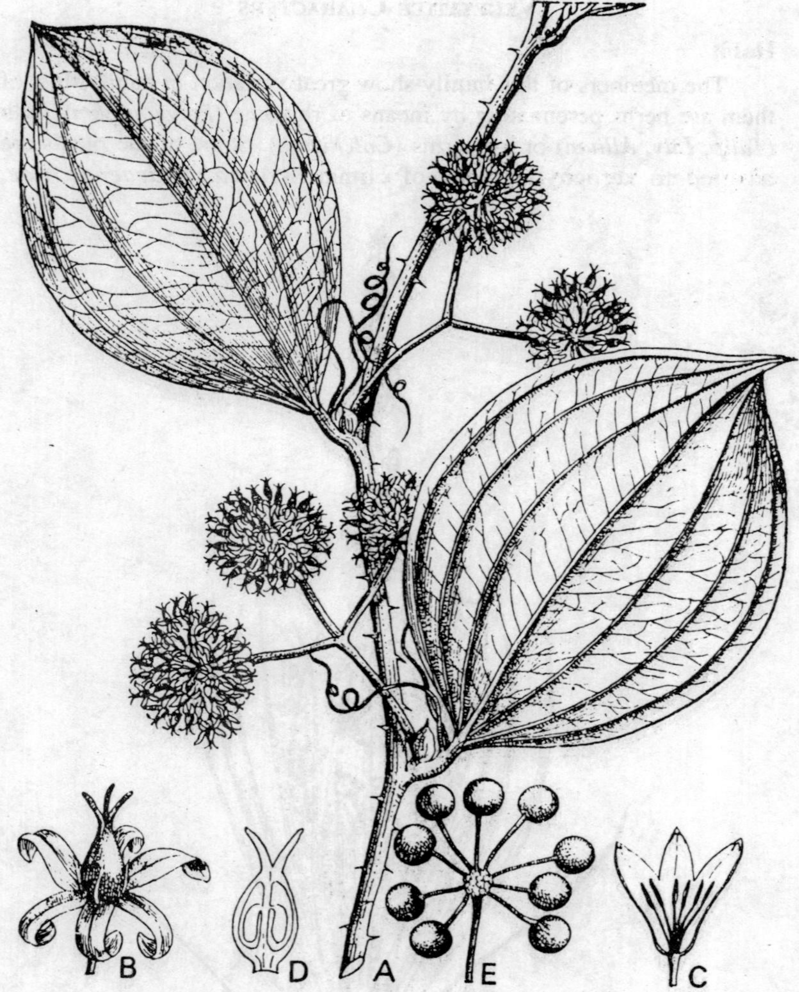

Fig. 26.2. Smilax zeylanica L. A—Flowering branch; B—Pistillate flower; C—Staminate flower longitudinally cut; D—Gynoecium longitudinal section; E—Fruit.

Dasylirion, *Bowiea*) while some are climbing (sp. of *Smilax*, *Asparagus*). *Asparagus* and *Ruscus* exhibit phylloclades. *Smilax* is a climber by means of tendrils which are the modification of stipules.

Root

Adventitious root system. Tuberous roots in *Asparagus*.

Stem

Herbaceous (*Asparagus*) or woody (*Dracaena, Yucca*) or modified in bulb (*Allium* species), rhizome (*Polygonatum*) and corms (*Colchicum*), cylindrical (*Asparagus, Dracaena*) glabrous or hairy, branched and solid. The stem of *Dracaena, Yacca* and *Aloe* show anomalous secondary growth.

Leaves

Alternate, opposite or whorled, radical or cauline, petiolate or sessile, exstipulate or stipulate, (Stipules modified into tendrils in *Smilax*), simple, leaves minute and scale like in *Asparagus*, leaf apex modified into tendrils in *Gloriosa*, unicostate parallel venation. Leaves caducous in *Asparagus*.

Floral Characters

Inflorescene

It is variable, usually racemose types as raceme, umbel (*Smilax*), spadix (*Rhodea*), panicle (*Yucca*); or solitary terminal (*Tulip, Lilium, Fritillaria*) or axillary (sp. of *Gloriosa*). Apparent Umbel in species of *Allium* has been shown to consist of a monochasial cyme with shortened internode. The flowers are with bracts and sometimes with bracteoles. When the bracteoles are present, further branching takes place in its axil (*Hemerocallis*).

Flower

Regular, actinomorphic (sometimes a slightly zygomorphic in *Lilium* and *Hemerocallis*), hermaphrodite (unisexual in *Smilax* and *Ruscus*), hypogynous, complete or incomplete in unisexual flowers). Usually trimerous, rarely 2, 4 (*Aspidistra*) and 5-merous.

Perianth

6, segments in two whorls of 3 + 3 (rarely 8-12) usually similar and petal like, undifferentiated into calyx and corolla. The segments are also termed as tepals. The segments may be free or rarely partially united, outer valvate or connate into a tube, usually imbricate in bud. In some species of *Kingia* and *Xanthorrhoea* perianth may be scarious or membranous.

Androecium

The *androecium* usually consists of six stamens arranged in two whorls of three each and situated opposite the perianth leaves (rarely the number may be up to 12 or reduced to 3 only). They may be free or united to the perianth leaves. The missing inner or the outer whorl may

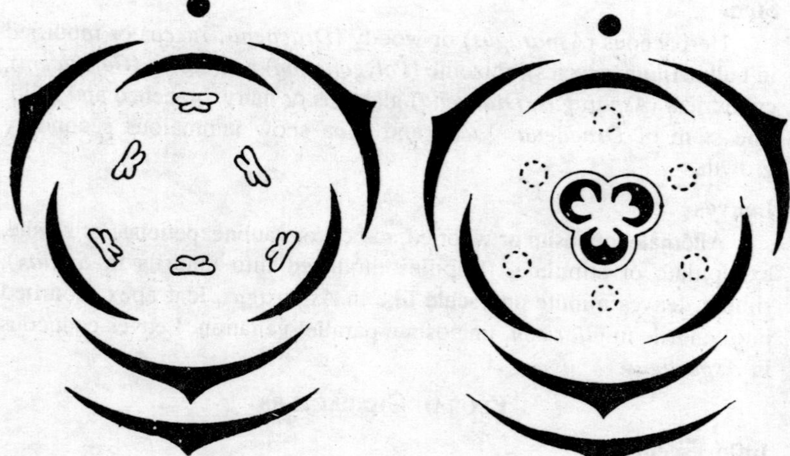

Fig. 26.3. Floral diagram of Smilax zeylanica, staminate flower.

Fig. 26.4. Floral diagram of Smilax zeylanica, pistillate flower.

be represented by staminodes or may be entirely suppressed. The filaments may be free or variously connate. The anthers are bilocular, mostly introrse, opening by a slit lengthwise or rarely by a terminal pore.

Gynoecium

The *gynoecium* consists of three carpels which are syncarpous. The ovary is mostly superior, vary rarely more or less adnate to the base of the perianth tube and then semi-inferior. Mostly it is trilocular with axile placentation, but rarely unilocular with parietal placentation. The style may be entire or divided. Rarely the style may be free. The ovules are usually numerous and mostly 2-seriate in each locule, rarely solitary.

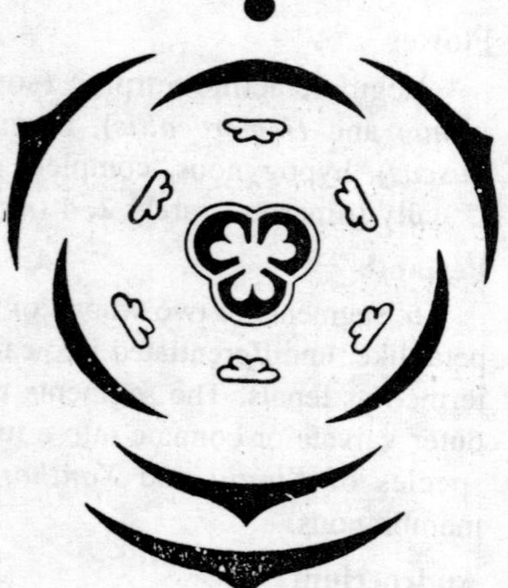

Fig. 26.5. Floral diagram of Asphodelus tenuifolius

Fruits and Seeds

The *fruit* is usually a loculicidal or septicidal capsule or sometimes a berry as in *Smilax* and *Polygonatum*. The *seeds* are endospermic with a straight or curved embryo.

Pollination and Dispersal

The flowers are usually insect pollinated. They are attracted by the showy and often fragrant flowers and the nectar which is secreted by the septal nectaries in most of the genera. Sometimes, as in *Asparagus*, the nectaries are present at the base of the tepals. *Yucca* has an unique pollination mechanism which is intimately associated with the life history of a moth, *Pronuba yuccasella*. This is a rare example of mutual dependence and adaptation of a single insect and a single flower. The flowers are fully expanded and emit their perfume especially at night and are then visited by the moths.

The female moth which has a long ovipositor collects a load of pollen from one flower. She then visits another flower and deposits some eggs in the ovary singly just below the ovules by piercing the ovary wall with her ovipositor. After depositing eggs the moth climbs to the top and thrusts the pollen mass into the stigmatic opening. The ovules are thus fertilized and they develop simultaneously with the larvae of the moth. The larvae feed on some of the numerous ovules. The moth remains dormant in its cocoon until the next flowering season when the same cycle is repeated again.

Economic Importance

The family is of considerable economic importance for food, fibre, medicine and ornamentals.

1. Several ornamental plants included in this family are *Asparagus*, *Dracaena*, *Hemerocallis*, *Brodiaea*, *Yucca*, *Hyacinth*, *Scilla*, *Gloriosa*, *Aloe*, *Sensiveria*, *Lilium*, *Tulipa* etc.
2. From *Aloe africana* and *A. barbedensis* an important drug 'Aloin' is obtained. Other plants of medicinal importance include *Gloriasa*, *Colchicum*, *Scilla*, *Urgentia*, *Veratrum*, *Asparagus* etc.
3. Several plants of this family are edible e.g., (*Allium cepa*—Onion, *A. asealonium*—shallot, *A. sativum*—garlic and *A. porrum*—leek). The young shoots of *Asparagus* are edible. The roots of *Chlorophytum* and *Asparagus* are also edible. *Lapagyria rosea* is valued for its edible fruits.
4. *Colchicum luteum* provides a laboratory reagent 'Colchicine'. *Dracaena* yields a resinous juice, the dragon's blood. *Hyacinthus orientalis* is cultivated for a popular perfume 'hyacinth' in France. The bulbs of *Scilla* and *Urgenia* are valued for rodent control. The leaves of *Xanthorrheoea* provide an acrid resin used for making sealing wax, gold size etc.

27

PALMAE

Mainly trees, with stout unbranched stems ending in a crown of leaves. Stem erect or scandent, sometimes armed. Leaves usually large, alternate, pennatisect or pinnatifid. Inflorescence at first enclosed in a coriaceous spathe. Flowers usually small, 1-sexual (monoecious or dioecious). Perianth segments 6 in two series, persistent, never petaloid. Stamens 6 on the base of perinath, anthers usually versatile. Ovary 1-3 celled, superior. Carpels 3, distinct, free of variously united. Fruit a berry, drupe or a nut.

Distribution

The family includes about 210 genera and 400 species (*Hutchinson*), according to *Rendle* the number of genera and species is 140 and 1200 respectively, especially abundant in warmer and sandy parts of the world. They may also be found in the warmer parts of the temperate zone.

Indian Representative

Areca, *Cocos*, *Phoenix*, *Ptychoraphis*, *Bentinckia*, *Pinanga*, *Arenga*, *Calamus*, *Caryota*, *Wallichia*, *Borassus*, *Didymosperma*, *Livistonia*, *Licuala*, *Trachycorpus*, *Corypha*, *Daemonorops*, *Zalacca*, *Plectcomia* etc.

General Habit

The plants range from a small herbaceous form (*Eleutheropetalum*, *Reinhardia*) to usually large unbranched trees 100-200 feet high (*Borassus*, *Cocos* etc.), sometimes woody, branched shrubs or rarely climbing (*Calamus*). In *Iriartea* thorns occur all over leaves, stem and even roots.

VEGETATIVE CHARACTERS

Roots

The roots are adventitious. The primary root soon dies off and is replaced by adventitious roots emerging from the base of the stem.

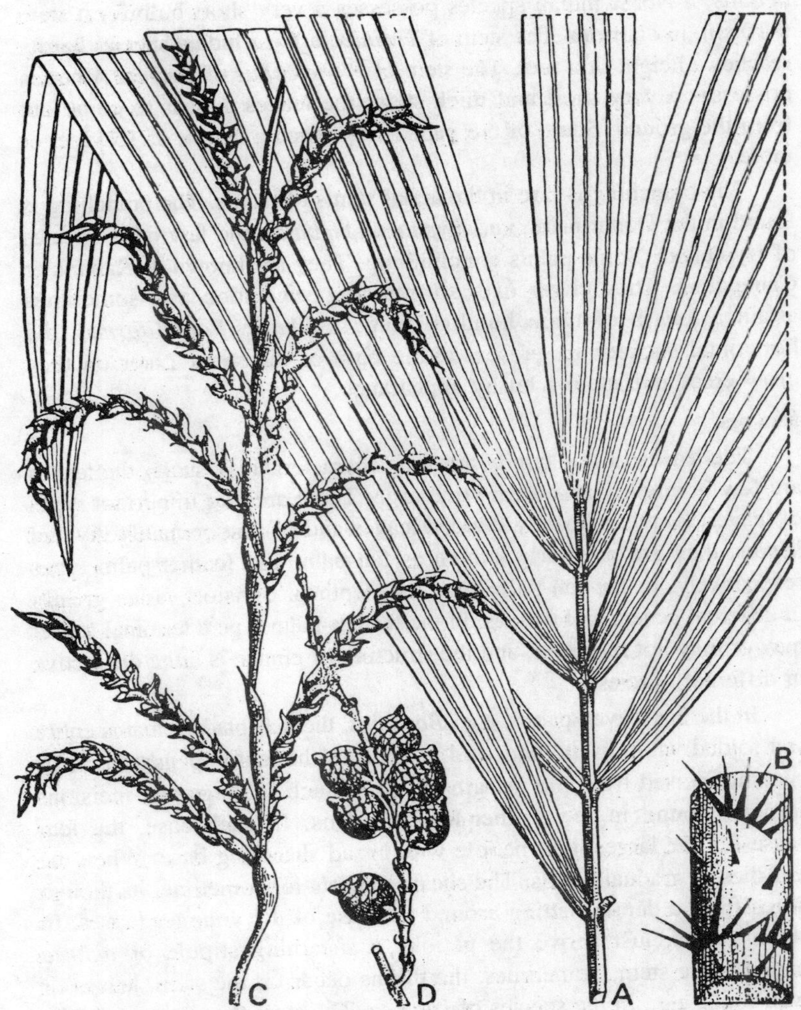

Fig. 27.1. Calamus pseudo-tenuis Becc. A—Vegetative branch; B—Part of the stem showing thorn; C—Male inflorescences; D—Female inflorescences.

Stem

Usually the plants possess an upright tall and unbranched stem which bears a crown of leaves at its top. The stem of date-palm, i.e., *Phoenix dactylifera* reaches a height of sixty feet or more than that and remains covered with remnants of persistent leaf bases. *Phoenix acaulis*, a North Indian species possesses a very short bulbiform stem 6-10 cm. in diameter. The stem of *Thrinax*, a West Indian species hardly reaches a height of a foot. The stem of *Phytelephus* of tropical America possesses a very short and thick stem; the leaves appear to come out from the ground. Some of the palms may reach a height of 150 feet or more.

The branching is rare in the aerial stems of palms. The branching is found in the Doum-palm, i.e., *Hyphaene thebaica* and few other species of *Hyphaene*. Some palms are climbing. They are known as Rattans or Canepalms. They climb over surrounding vegetation and sometimes attain a length of three hundred feet. In *Rhapis flabelliformis*, the horizontal suckers are present at the base of the stem. Later on they grow erect and attain a bushy appearance.

Leaves

The leaves of this family are very characteristic. Usually the leaves are few in number and often very large. There are two important types of leaves, (i) the palmately divided leaves and (ii) the pinnately divided leaves; they are popularly known as fan-palm and feather-palm types respectively. In the fan-palm type the depth of division varies greatly in different genera and species. In the feather-palm type a terminal leaflet may or may not represent, and the structure of pinnae is quite distinctive in different species.

In the fan-leaved palms, e.g., *Borassus*, the leaf blade remains entire and folded while in the bud, and as soon as the leaf expands the folds become incised from the margin inwards. Such foldings and incisions are also found in the feather-leaved palms. In each case, the leaf possesses the large, stout petiole with broad sheathing base. When the leaf dies, it gradually falls. The sheath persists for sometime, its though fibres form a dense matting around the base of the younger leaves. In rattan palms, just above the petiole, a sheathing stipule or *ochrea* encircles the stem. Sometimes, the thorns occur on the stem, leaves or even roots, e.g., in the species of *Iriartea*. The stem thorns may also be present within the leaf-sheath. In rattanpalms, many curved spines occur on the stems and leaves, which help them in climbing.

Inflorescence

Inflorescence is usually of large size and profusely branched. The plants may be monocarpic, i.e., flower only once in the life-time e.g., *Corypha* (Talipot palm) or polycarpic, bearing axillary inflorescences on maturity. Most of the palms are of the latter type, and they flower throughout their lives. In the former case the plant grows vegetatively for a large number of years producing very strong woody trunk and a cluster of leaves after which the growing point stops producing vegetative leaves and develops into a huge inflorescence, sometimes containing several million flowers. This production of the inflorescence so completely exhausts the food reserves of the plant that it ultimately dies after fruiting. The inflorescence may be in the axil of the current leaves or it may be lower down on the stem.

The branching of the inflorescence is of the racemose type and may be either a simple (e.g., Borassus) or compound (e.g., Cocos) spike or even a profusely branched panicle (e.g., *Daemonorops*); but in all cases either the entire inflorescence or each branch separately (in case of panicle) is enclosed by a fleshy spathe consisting of one or more leaves. The inflorescence thus usually termed a spadix. (A massive fleshy spike containing usually unisexual flowers and protected or enclosed by a large enveloping and usually petaloid bract known as spathe). As the floral axis grows and the flowers mature, the spathe gets torn and the whole structure emerges out of it.

Flowers

As a rule the flowers are unisexual. They are actinomorphic, incomplete and hypogynous. They are sessile and rarely possess short pedicels. The flowers are very small in size and present in very large numbers in the inflorescence. The majority of plants are monoecious, i.e., the male and the female flowers occur on the same plant, e.g., *Cocos*. In some cases the plants are dioecious, i.e., the male and female flowers occur on two separate plants, e.g., *Phoenix*.

In the monoecious plants, the male and female flowers occupy the different positions in the inflorescence in the different genera. In many cases a few female flowers occur at the base of the branches, whereas the male ones are crowded in the upper part. In *Raphia ruffia*, the branches of the spike bear female flowers in the lower half, and the male flowers in the upper half. There may be considerable difference in size between the two kinds of flowers, e.g., in *Borassus*, the female flowers are much larger than the male ones which are very minute in size.

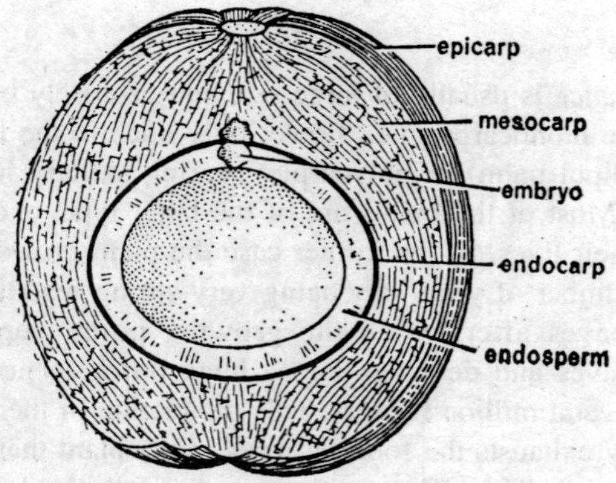

Fig. 27.2. Cocos nucifera, drupe longitudinally cut.

Perianth

It consists of six perianth leaves (tepals) arranged in two whorls of three each. They are polyphyllous and persistent. In *Areca catechu* the perianth leaves of the outer whorl are much smaller than the perianth leaves of the inner whorl. They are green to yellow or white in colour.

Androecium

The *androecium* is of usually six stamens arranged in two series of three each opposite the perianth segments. Sometimes there is a single whorl of three stamens as in *Nypa* or there are numerous stamens as in *Caryota*. Filaments are free and often short, and the anthers are dithecous, introrse and dehiscing by longitudinal slits. A conical pistillode is often present in the male flower.

Gynoecium

The *gynoecium* is tricarpellary and syncarpous but sometimes the three carpels are partly (e.g. *Nypa*) or completely (e.g. *Phoenix*) free. The ovary is superior and usually trilocular with a single ovule in each locule from the inner angle of the locule. The placentation is axile. Sometimes the ovary is unilocular with parietal (e.g. *Oncosperma*) or basal (e.g. *Areca*) ovules. When the gynoecium is of three free carpels, the ovary of each carpet is unilocular with a single ovule. Sometimes, as in *Cocos* and *Phoenix*, two of the three carpels abort during maturation. There are usually three sessile stigmas. The staminodes are usually present in the female flowers in several genera.

Fruits and Seeds

The *fruit* is a berry or drupe, the latter often develops from one of the three carpels, the other two degenerate. The exocarp is often fibrous

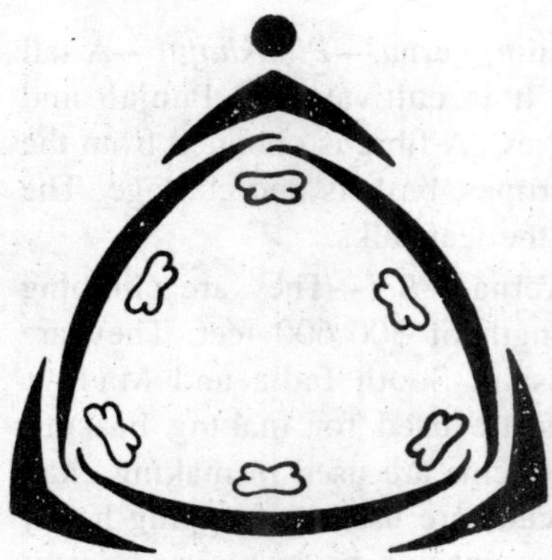

Fig. 27.3. Floral diagram of Calamus pseudo-tenuis, staminate flower.

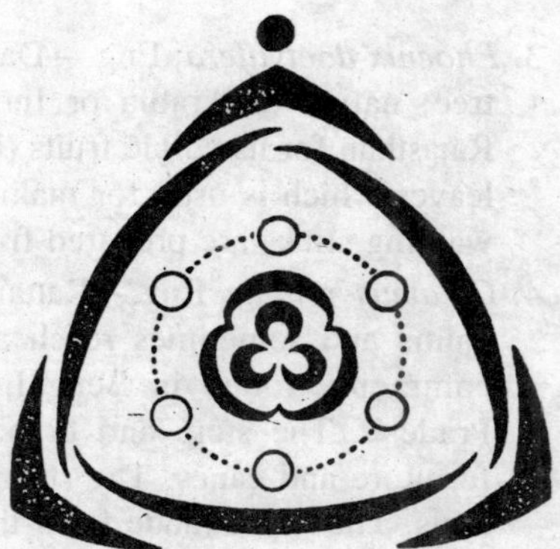

Fig. 27.4. Floral diagram of Calamus pseudo-tenuis, pistillate flower.

and the endocarp is usually united to the seed. The *seeds* are endospermic, the endosperm is large, sometimes very hard (*Phoenix*) or ruminate (*Areca*). The embryo is small.

Pollination

The pollination takes place with the aid of wind as the pollen grains are produced in large numbers. It is a cross pollination which takes place in the plants of this family, though male and the female may be situated on the same plants; it is on account of marked protandry. Thephrastus in his famous book "Enquiry into Plants" has mentioned the method of pollination in the date-palm. According to him, Arabs held special ceremoney at an appropriate time of the year. During this function a man used to climb a male palm, got the inflorescence and gave it to a priest who touched the female flowers with it resulting in pollination. Sometimes pollination by insects may also take place.

Economic Importance

1. *Lodoicea maldivica*; Eng.—Double coconut; Verna—*Daryaka nariyal*—A tall palm—The plant bears one of the largest known fruits which takes ten years to ripen. The fruit is used to check diarrhoea and vomiting. The leaves are used for making that ches and hats. The shells of the large fruits are used as water containers.
2. *Phoenix acaulis*; Verna.—*Khajeria*—A stemless palm. found in Bihar, Uttar Pradesh and North Bengal. The young buds are used as vegetable. The fruits are edible.

3. *Phoenix dactylifera*; Eng —Datepalm; Verna.—*Pindkhajur*—A tall tree; native of Arabia or India. It is cultivated in Punjab and Rajasthan for its edible fruits (berries). A fibre is obtained from the leaves which is used for making ropes, baskets and cordage. The walking sticks are prepared from the leafstalks.
4. *Calamus rotang*; Eng.—Rattan; Verna.—*Bet*—They are climbing palms and sometimes reach a length of 500-600 feet. They are commonly found in Bengal, Assam, South India and Madhya Pradesh. The stem and branches are used for making baskets, furniture and canes. The stripped stems are used in making chair seats. The ropes made from this cane are used in dragging heavy loads. It is also used in making suspension bridges over streams and rivers in the hills. The root is given in chronic fever.
5. *Phoenix sylvestris*; Eng.—Wild date; Verna.—*Khajur*—This is a tall palm. It is native of India. It is cultivated largely in Andhra Pradesh, Madhya Pradesh and Mysore. The fruits are edible.
6. *Areca catechu*; Eng.—Betelnut palm; Verna.—Supari—This is a beautiful slender stemmed palm; native of Malaya, now grown along the coasts of Mysore, Madras, West Bengal, Assam and Maharashtra. The nuts are chewed along with betel leaves (Pan), Katha (from *Acacia catechu*), *lime*, *spices* and tobacco. Besides, they are used as a vermifuge for animals and for the preparation of denitrifices. The husks are used for making hard board and brown wrapping paper. Galls, caused by insects are used as tanning material. The nuts are roasted, powdered and then made into tooth powder by adding some other substances. The beads and other fancy articles are made from the nuts. The trunk is used as water pipe. The leaves are used for thatching and woven into mats.
7. *Metroxylon sagu*; Syn. *M. rumphii*; Eng.—Sago palm; Verna.—*Sago*—The plants flower once in their life time. The fruit matures in three years. The sago (*sabudana*) is obtained from the pith of the old trunk by crushing and washing. This is used as food.
8. *Caryota urens*; Eng.—Toddy palm; Verna.—*Mari*—It is a palm with large bipinnate leaves. They are found in Orissa. North Bengal and Assam. The plant yields toddy by simply tapping the stem. The fresh toddy is extremely rich in vitamins and recommended for tuberculosis patients. The plant yields juice which is used for making sugar. The trunks are used as water pipes. The fibre obtained from the leaf sheaths, is used for ropes, baskets and soft brushes. The pith of stem yields sago.

9. *Daemonorops jenkinsianus*; Syn. *Calamus jenkinsianus*—Verna.—*Gola bet*—A tall palm found in Assam, the Khasia hills and Sikkim. The baskets are manufactured from its long, soft stems.
10. *Elaeis guineensis*; Eng.—Oil palm—This is native of West Africa but now grown in Travancore. The oil, obtained from the nuts, is used in making soap, margarine and as a fuel for diesel motors. The oil is also used in rheumatism.
11. *Borassus flabellifer*; Syn. *B. flabelliformis*; Eng.—Palmyra palm; Verna.—*Tar*—This is a beautiful tall palm which yields several products of commercial importance, such as, toddy, nira, palm jaggery, fruit and fibre. The fibre is used for making brushes and brooms and the leaves for making fans. umbrellas, baskets and mats. The leaf stalks are used for making writing and printing papers. These palms are found along the coastal areas of Bengal, Bihar and the Western and Eastern Peninsulas. In old times the leaves were used for writing purposes. The wood of the trunk is hard and durable and resists saline water. The trunks of female plants are strongest. They are used in making poles, pillars and posts. The fruit is roasted and eaten. The bows of Tar were popular in ancient India.
12. *Cocos nucifera*; Eng.—Coconut; Verna—*Nariyal*—A tall palm, cultivated chiefly in Kerala, Madras and Mysore. An edible oil, obtained from the nut, is used in the manufacture of soap and candles. *Copra*, obtained by cutting the nut, is used in confectionery, soap and margarine. The copra residue is used as fodder and as manure. The fibrous rind of the nut yields the coir, which is used for making carpets and ropes. The endosperm yields the oil, which is used as hair oil and for several other purposes. The coconut pith and husk are used in the preparation of thermal insulating boards. The inflorescence is tapped when young for toddy, which when fermented gives an alcoholic drink. The root is astringent and is used in uterine diseases.
13. *Roystonea regia*; Syn. *Oreodoxa regia*; Eng.—bottle palm—It is a beautiful palm and largely grown in avenues.
14. *Copernicia*—It is found in tropical America. The Carnauba palm wax is obtained from it, which is used in making gramophone records, candles etc. The wood is also used several ways.
15. *Calamus acanthospathus*—This is a common climber of North-East India. It is extensively used for making tea-baskets. It is also used in making temporary bridges, walking sticks and chairs.

16. *Corypha umbraculifera*; Eng.—Talipot palm—This is a palm reaching upto 80 feet height. Found in South India. It flowers once in its life time and thereafter dies. The leaves are used as umbrellas. The mats, fans and baskets are made from the leaves.
17. *Calamus tennis*—They are found in Assam, sub-Himalayan tract and Dehradun. The walking sticks are made from stems. The mats, screens and baskets are also made.
18. *Caryota mitis*—This is a tall palm of Andaman islands. The young leaves are used as vegetable. The seed is used as a masticatory.
19. *Calamus latifolius*; Eng.—East India cane; Verna—*Danagribet*—A climbing palm found in Bengal and Assam. The stems are used for walking sticks and baskets.

28

GRAMINEAE

Herb, shrubs or tree, generally with terete jointed stem. Leaves alternate, usually narrow, sheath shlit up in front up to the base, distinct from the blade; often with a ligule. Flowers 1-2 sexual, subtended by several distichous imbricate glumes. Perianth of 2-3 minute hypogynous scales (lodicules) or none. Stamens in most cases in whorls of 3, hypogynous. Anthers versatile Styles 2-3, stigmatose throughout. Fruit terete and caryopsis in most cases.

Salient Features

Usually herbs, sometimes shrubs or trees. Stem often with terete jointed. Leaves alternate, usually narrow, sheath split up in front upto the base, distinct from the blade, ligule present. Inflorescence compound spike. Flower bisexual or unisexual, zygomorphic, subtended by several distichous imbricate glumes. Perianth represented by 2-3 minute, hypogynous, scaly, lodicules, sometimes absent. Stamens 3, anthers versatile. Carpels with a single basal ovule, style 2-3, stigmose throughout. Fruit caryopsis.

Distribution

The members of this family are widely distributed all over the globe and are seen in every vegetation type dominating areas where the rainfall is neither high to support trees nor so low that only arid plants may grow. These areas include prairies and plains of North America, desert regions of India and Australia Savannahs of South America, Steppes of Eurasia, Veldt of Aftica etc. Many of them grow in arctic and antarctic zones beyond tree limit in the alpine regions where there is perpetual snow. The grasses are found to occur right from sea level to the highest altitude where vegetation can exist. The plants of this family represent all the three ecological types hydrophytes, mesophytes, and xerophytes.

In India the plants of this family grow under varied habitats: many are cultivated as cereal crops, as ornamentals, as plants of medicinal and industrial importance or growing wild under various conditions. A large number of grasses are grown in the lawns or grow as hydrophytes or xerophytes.

Indian Representatives

Triticum, *Zea*, *Oryza*, *Avena*, *Hordeum*, *Saccharum*, *Pannisetum*, *Sorghum*, *Pannicum*, *Secale*, *Arundodonax*, *Veteveria*, *Bambusa*, *Cynodon*, *Andropogon*, *Cymbopogan*, *Paspalidum*, *Hygroryza*, *Cenchrus*, *Dichanthium*, *Lasiurus*, *Eragrostis*, *Melica*, *Stipa*, *Dendrocalamus*, *Melocalamus*, *Arundinaria*, *Ochlandra*, *Setaria*, etc.

Vegetative Characters

Habit

Mostly the plants are annual, biennial or perennial herbs or shrubs. The largest woody species are the bamboos, which may grow to more than 100 feet in height and several inches in diameter. Most of the plants e.g., grasses, are wild while some are cultivated for valuable food.

Root

Generally adventitious, fibrous, fascicled, stilt (e.g., *Zea mays*, *Sorghum vulgare* etc.).

Stem

The stem may be erect, prostrate or even creeping. Usually fistular (hollow), rarely solid (e.g., *Zea*, *Saccharum* etc.). The internodes are long and nodes are conspicuous. In many grasses the runners and suckers are also developed. Among the perennial grasses, usually the rhizomes and root stocks are formed, which help in vegetative propagation. Small tubers and corms are also found in so many species.

In this family the aerial stem is commonly known as 'culm' which is usually terete (cylindrical) but in few cases flattened. In most grasses the main axis only develops lateral branches from the basal buds. These branches are '*tillers*'.

Leaves

In most of the members of this family, the foliage leaf can easily be divided into two parts, the sheath and the blade (lamina). The leaf sheath usually covers the internode completely or partially. The inner surface

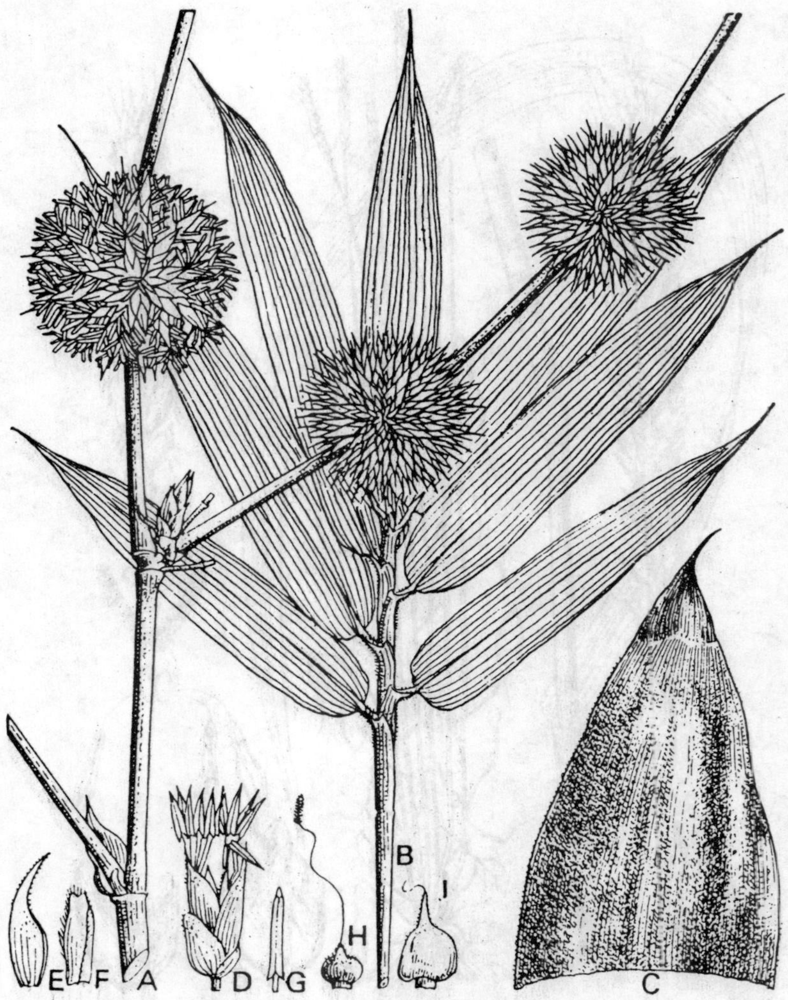

Fig. 28.1. Dendrocalamus strictus Nees. A—Flowering branch; B—Vegetative branch with leaves; C—Tip of a leaf; D—Spikelet; E—Glume; F—Lemma; G—Stamen; H—Gynoecium; I—Fruit.

of leaf sheath is usually glabrous (smooth) while the upper surface is variously grooved and either glabrous or hairy.

At the junction of the leaf sheath and leaf blade there occurs a thin membranous outgrowth, the *ligule*. In some species the ligule is represented by only few hairs. In many species there are two lateral,

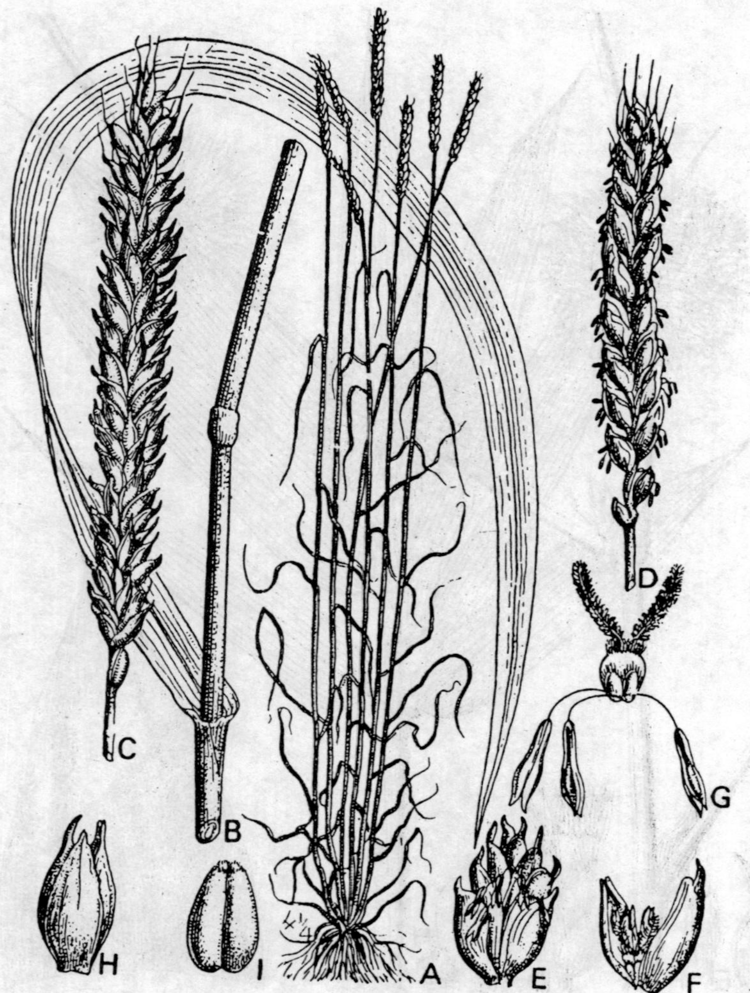

Fig. 28.2. Triticum aestivum L. A—Flowering plant; B—Culm with leaf; C, D—Spikes; E—spikelet; F—A flower enclosed by lemma and palaea; G—Flower; H—Caryopsis enclosed in glumes; I—Caryopsis.

pointed outgrowths above the ligule are known as *auricles*. They are well developed in *Hordeum* and *Oryza*. The leaf blade is long, narrow, acuminate and with parallel venation. The leaves are usually sessile very rarely petiolate (e.g., in *Panicum sagittaefolium*), alternate, lanceolate, entire or hairy margins, surface smooth of glabrous.

Inflorescence

The flowers are bisexual, may be unisexual and the unit of the inflorescence is the spikelet which are arranged in various ways on the main axis forming compound inflorescences. It may be spike of spikeltet (*Triticum*) or panicle of spikelet (Oat). The branches of the panicle may be often spreading or contracted to form a compact or even cylindrical panicle (*Pennisetum*).

The axis of the whole inflorescence is called the *rachis* and that of a spikelet *rachilla*; the usual character of the spikelet being as follows: Each spikelet bears a number of (1-7) flowers arranged alternately on the opposite of its central axis. At the base of spikelet is usually a pair of boat-shaped glumes (sterile) arising slightly one above the other on the opposite side of the rachilla and usually enclose all the flowers till they are ready to open. In a few genera there are more than two empty glumes.

Each flower usually consist on its anterior side a boat shaped flowering glume—*Lemma* or *inferior palae* which may bear a bristle-like *awn*, at its base or tip which is the prolongation of the midrib. Arising above, opposite to, but slightly above, and fitting into inferior palae in the true flower. At the base on the anterior side are usually 2, rarely more, small scale-like hygroscopic hypogynous ciliate scales called lodicules. When the flower is ripe and about to open the lodicules expand by absorbing the moisture and push apart the two palae.

Flowers

The flowers are small; inconspicuous, bisexual or sometimes unisexual as in *Zea* and related genera, zygomorphic and hypogynous.

Perianth

The *perianth* is highly modified and much reduced and is represented usually by two minute or *fleshy* and hyaline structures, known as lodicules. The two lodicules are present antero-laterally. Sometimes, as in *Bambusa*, the third posterior lodicule is also present.

Androecium

The *androecium* consists of usually three stamens in a whorl, the odd stamen being anterior. In *Bambusa* and *Oryza* there are two alternate trimerous whorls of stamens. Sometimes, as in *Uniola*, only a single stamen is present which is usually anterior. Rarely there are numerous stamens as in *Pariana*. The filaments are free and delicate, and the anthers are basifixed (apparently and functionally versatile), dithecous, introrse and opening usually by a longitudinal slit.

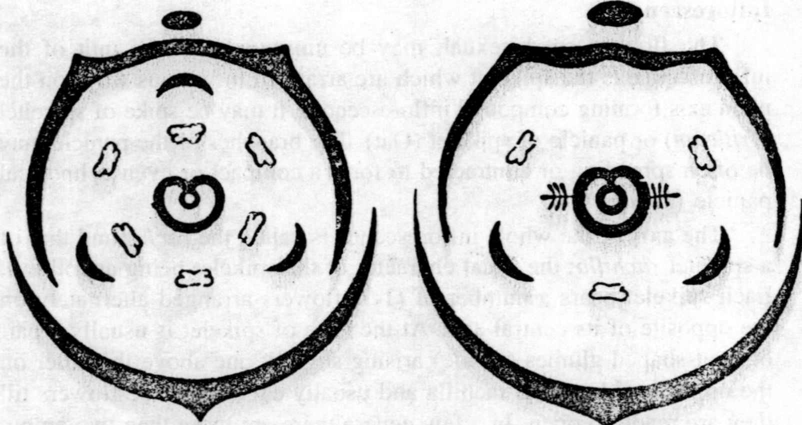

Fig. 28.3. Floral diagram of Dendrocalamus strictus.

Fig. 28.4. Floral diagram of Triticum aestivum.

Gynoecium

The *gynoecium* is monocarpellary with a superior unilocular ovary having a single anatropous ovule adnate to the adaxial side of the ovary. The placentation is basal. There are usually two styles but a third style is also present in *Bambusa*. The stigmas are generally plumose.

Fruits and Seeds

The fruit is mostly a one-seeded caryopsis in which the thin ovary wall closely adheres to the seed. Sometimes the fruit is a nut or berry (as in some *Bambusa*) or a utricle (*Eleusine*). The endosperm is starchy and the embryo is straight.

Pollination and Dispersal

The grasses are self or wind pollinated. The inconspicuous flowers, long flexible filaments, abundant powdery pollen, and large feathery stigmas are characteristic of wind pollinated plants and so also of grasses. The 'seeds' which are very small and light are usually distributed by wind. In others where the glumes enclosing the seeds are barbed, bristly or awned, dispersal takes place by animals to which the seeds adhere.

Economic Importance

From the stand point of its economic value, this family is most important. It contains all the., cereals, grasses, bamboos, canes and other plants. A list of important plants is given below:

1. *Heteropogon contortus*; Syn. *Andropogon contortus*; Verna.—*Kumeria*—This is a common perennial grass used as a fodder and the fibre is made into mats. It is also used to check soil erosion and in the manufacture of paper and board.
2. *Imperata cylindrica*; Syn. *I. arundinacea*; Verna—*Ulu*.—This is a perennial grass, the stem is used for thatching and a raw material for paper making. The pillows and cushions are stuffed with its fruits. It is also used in soil conservation.
3. *Pennisetum typhoides*; Syn. *P. typhoideum*; Eng.—Pearl millet; Verna.—*Bajra*.—It is grown for its edible grains. The straw is used as a fodder. It is cultivated in Uttar Pradesh, Punjab, Andhra Pradesh, Rajasthan and Maharashtra,
4. *Pennisetum purpureum*; Eng.—Elephant grass.—It is grown in Assam, Bengal, Bihar, Uttar Pradesh and Punjab. It is used as a fodder.
5. *Avena sterilis*; Eng.—Indian oat, Red oat; Verna.—*Jai*. This is a herb used as a fodder. A hybrid has been raised between this and *A. sativa*. It is grown in Uttar Pradesh, Punjab, Bihar, Bengal, Orissa and in the hills of Madras.
6. *Panicum miliare*; Eng.—Little millet; Verna—*Sawa*, *Kutki*.—The grains are edible. The stems and leaves are used as fodder. It is cultivated in Uttar Pradesh and Madhya Pradesh.
7. *Panicum maximum*; Eng.—Guinea grass; Verna—*Ginighas*.—It is grown in Maharashtra, Madras and Northern India for fodder.
8. *Setaria sphecelata*.—It is introduced from Brazil and grown for fodder.
9. *Setaria verticillata*; Eng.—Bristly foxtail Verna.—*Laptuna*.—The grains are used as food. It is grown in Madras and Andhra Pradesh.
10. *Eleusine coracana*; Eng.—Finger millet, Ragi; Verna.—*Mandal*, *Mandua*.—This is a herb, cultivated as a food crop. This is the main food of the farmers in South India; Maharashtra and of hill-men in Northern India. This is made into cakes, porridge and sweetmeats. A beer is brewed from the grain by the hill tribes. The enzymic extract of the germinated grain can be used for unhairing hides and skins. It is chiefly cultivated in Andhra Pradesh, Madras, Mysore, Orissa, Bihar, -Uttar Pradesh and Maharashtra.
11. *Triticum aestivum*; Syn. *T. vulgare*; Eng.—Wheat; Verna.—*Gehu*.—This is a herb, cultivated as a food crop. The wheat straw is used as cattle fodder and in the manufacture of paper. It is mainly grown in

Uttar Pradesh, Punjab, Madhya Pradesh, Maharashtra, Bihar and Rajasthan.

12. *Sacchaum officinarum*; Eng.—Sugarcane; Verna.—*Ganna, Ikh*—The stems yield the cane syrup. The by-products of sugar industry are commercially used for various purposes, e.g., molasses is used in cooking, also for rum, industrial alcohol, synthetic rubber. A mixture of refuge canes and molasses called molas-cuit is used as fodder. Cane refuge is also utilized for manufacturing bagpaper, wrapping-writing and printing papers, cardboard and as fuel. It is chiefly grown in Uttar Pradesh, Bihar and Punjab.
13. *Triticum dicoccum*; Syn. *T. dicoccum*; Eng.—Emmer; Verna.—*Gehu.*—It is cultivated as a food crop.
14. *Erianthus arundinaceus*; Syn. *Saccharum arundinaceum*; Eng.—Pin reed grass; Verna.—*Ramsar, Sarkanda.*—This is a perennial grass. The leaf sheaths yield a fibre which is used for making ropes, twines and paper, and the stems are used for making chairs, stools, tables, baskets and screen.
15. *Erianthus munja*; Syn. *Saccharum munja*; Verna.—*Munj sentha, Sarkanda, Munja*—The stem fibre is used for making baskets, mats and cordage. The leaves are used for thatching, and are also the source of paper. It is also used as soil-binder. It is commonly found in the Punjab and Uttar Pradesh.
16. *Pseudostachyum polymorphum.*—This is a shrubby bamboo found in Sikkim, Assam. It is used by tea planters for making baskets. It is also used for making umbrella handles and walking sticks.
17. *Urochloa panicoides*—A herb used as a fodder. The gains are also edible.
18. *Urochloa reptans*—A herb used as a fodder. The grains are also eaten.
19. *Echinochloa crus-galli*; Syn. *Panicurn crus-galli*; Eng.—Baryard millet; Verna.—*Samak*—It is used as a fodder and food crop. It is chiefly cultivated in Maharashtra and Andhra Pradesh.
20. *Echinochloa frumentacea*; Syn. *Panicum crus galli*; var. *frumentacea*; Eng.—Japanese millet; Verna.—*Sanwa, Sawa.*—It is used as a fodder and food crop. It is used as a basis of a very potent beer. It is mainly grown in Madhya Pradesh and Uttar Pradesh.
21. *Setaria italica*; Syn. *Panicum italicum*; *Chaetochloa italica*; Eng.—Italian millet; *Ksme, Kangni, Kangu, Kakun.*—It is chiefly cultivated in Andhra Pradesh and Madras for fodder and food. It is

used as diuretic and astringent and also used externally for rheumatism.

22. *Triticum durum*; Eng.—Durum wheat; Verna.—*Gehu.*—This is cultivated mainly in Maharashtra as a cereal crops.
23. *Triticum compaclum*; Eng.—Club *wheat.*—This is grown as a fodder crop in South India.
24. *Triticum sphaerococcum*; Eng.—Indian dwarf wheat; Verna.—*Gehu.*—This is grown as a grain crop mainly in the Punjab.
25. *Hordeum vulgare*; Syn. *H. sativum*; Eng.—Barley; Verna.—Jau.—This is a herb, cultivated as a food crop mainly in Uttar Pradesh, Punjab, Rajasthan, Madhya Pradesh, Bihar and West Bengal. The leaves and stems are used as fodder. The addition of I or 2 ppm. of gibberellic acid to barley is claimed to improve maltin.
26. *Oryza sativa*; Eng.—Rice; Verna.—*Chaval*, *Dhan.*—This is a herb, grown as a food crop. The rice straw is used for making strawboards, paper, mats. Rice-bran oil is used for soaps and cosmetics, and as an anticorrosion oil. It is grown all over India.
27. *Avena sativa*; Eng.—Oats; Verna.—*Jai.*—It is grown to a limited extent in North-Western Himalayas. The grains are used by poor people as food.
28. *Avena ludoviciana*—It is introduced from Uruguay in India and grown for fodder.
29. *Avena stigosa*—This has also been introduced from Uruguay in India and grown for fodder.
30. *Sorghum vulgare*; Syn. *Andropogon sorghum*; *Holcus sorghum*; Eng.—Sorghum; Verna,—*Jower.*—The grains are used as food and the stem and leaves are consumed as a cattle fodder. A spirit is distilled from the grain. It is cultivated mainly in Uttar Pradesh, Punjab, Madhya Pradesh, Andhra Pradesh, Maharashtra and Rajasthan.
31. *Sorghum vulgare* var. *durra*; Syn. *S. durra*; Eng.—Durra, Kaffir-corn.—The grains are used as food and the straw makes good fodder.
32. *Sorghum vulgare* var. *saccharum*; Syn. *S. saccharatum* Eng.—Sugar Sorghum; Verna.—*Deodhan*—The grain is used as food. A sweet juice obtained from the stem and leaves is given as fodder.
33. *Sorghum sudanense*; Eng.—Sudan grass.—This is an annual: grass found in Maharashtra, Assam and Punjab. It is used as a fodder.
34. *Sorghum halpnense*; Syn. *Holcus halepensis*; Eng.—Johnson grass; Verna.—*Baru Kala*, *mucha.*—This is a perennial grass, used as fooder.

35. *Sorghum purpureo-sericeum*; Syn. *Andropogon purpurcasericeus.* —It is used as fodder, chiefly cultivated in Bombay.
36. *Pennisetum nervosum*; Eng.—Bent spike Pennisetum—It is used as a fodder. It is introduced from Uruguay. It appears to be a promising drought-resistant grass.
37. *Pennisetum clandestinum*; Eng.—Kikuyu grass.—It is used as a fodder and a soil-binder. It is grown in the Nilgiris and Assam.
38. *Panicum miliaceum*; Verna.—*Chin, Morha, Anu.*—The grains are edible and the straw is used as a fodder. It is grown mainly in Uttar Pradesh.
39. *Panicum antidotale*; Eng.—Blue paniccum; Verna—*Bansi, Gunara.*—This is a tall grass, used as a fodder and also for the fixation and reclamation of sand dunes. It is found in Punjab, Uttar Pradish and the Western Peninsula.
40. *Arundo donax* var *versicolor.*—This is an ornamental grass with variegated leaves.
41. *Bambusa arundinacea*; Syn. *B. spinosa*; Eng.—Thorny bamboo; Verna.—*Bans.*—This is a tall woody grass found throughout our country, particularly along river valleys and in moist situations. The culms are used for making paper, thatches etc. The young buds and grains are consumed as food.
42. *Lolium perenne*—Perennial rye grass, used as fodder.
43. *Phragmites karka*; Syn. *P. maxima*; Vernu.—*Narkul.*—This is a perennial grass, the stems are made into pipes, and are also used for making writing and printing papers. They are also used for making chairs. baskets and mats.
44. *Phyllstachys bambusoides.*—It is used for walking sticks.
45. *Arundinaria racemosa.*—It is found in Bhutan and Western Himalayas. The culms are used for making mats and roofs of houses.
46. *Arundo donax*; Eng.—Giant reed; Verna.—*Baranal, Narhal.*—This is a tall perennial grass. The stems are used in the manufacture of baskets, mats, pipes, writing and printing papers. It is chiefly found in Kashmir, Assam and the Nilgiris.
47. *Bambusa tulda*; Eng.—Bamboo; Verna.—*Peka*—The mature culms are used for manufacturing paper, baskets and fans. The young culms are eaten as vegetable. It is found in Assam and Bengal.
48. *Bambusa balcooa*; Eng.—Plains bamboo; Verna.—*Bhaluka.*—It is used for building purposes. It is cultivated in Assam, Bengal, Bihar and Uttar Pradesh.

49. *Aira caryophyllea.*—This is an ornamental grass.
50. *Apluda mutica*; Syn. *A. aristata*; Verna.—*Bhanjura*—This is a perennial grass, used as a fodder. This is mainly grown in Uttar Pradesh.
51. *Aristida depressa.*—The panicles are made into brooms.
52. *Aristida setacea.*—The panicles are used for making brooms and brushes.
53. *Eriochloa procera*—This is a grass, used as fodder.
54. *Festuca gigantea*—It is used as fodder.
55. *Festuca rubra*—It is used as fodder.
56. *Festuca ovina*—It is used as fodder.
57. *Festuca elatior*—It is a perennial grass, used as fodder.
58. *Halopyrum mucronatum*—It is used for the reclamation of sand dunes.
59. *Panicum turgidum.*—It is used for the fixation and reclamation of sand dunes.
60. *Zea mays*; Eng.—Maize, Corn, Indian corn; Verna.—*Makai, Makka, Bhutta*. This is a tall annual herb. It is grown as a food crop mainly in Uttar Pradesh, the Punjab, Madhya Pradesh, Bihar, Andhra Pradesh, Jammu and Kashmir. The immature cobs are largely eaten after roasting. The grains are also used in making corn starch and industrial alcohol. It is native of South America.
61. *Setaria glauca*; Syn. *Panicum glaucum*; Eng.—Cat tail millet; Verna.—*Bandra.*—This is a herb, used as a fodder. The grains are also eaten as food.
62. *Setaria pallide fusca*; Eng.—Kavatta grass.—It is used as a fodder. The grains are also edible.
63. *Setaria palmifolia*; Eng.—Palm grass; Verna.—*Bhapatia*. It is grown for fodder.
64. *Eleusine indica*; Eng.—goose grass; Verna.—*Balraja.*—It is found throughout India and used as a fodder.
65. *Eragrostis lehmanniana.*—This has also been introduced from United States and grown for erosion control and fodder.
66. *Eragrostis tef.*; Syn. *E. abessinica.*—This is an introduced annual grown in Uttar Pradesh, Maharashtra and Baroda as fodder. The grain is eaten in times of scarcity.
67. *Eragrostis tremula.*—It is used as fodder.

68. *Erianthus ravennae*; Syn. *Andropogon ravennae*; *Saccharuin ravennae*; Verna.—*Moonj*—The stem fibre is used for making chairs, muddas, chhappars and ropes.
69. *Indocalamus wightianus*; Syn. *Arundinaria wightiana*.—It is used for making baskets, mats and walking sticks.
70. *Saccharum spontaneum*; Eng.—Thatch grass; Verna.—*Kans*.—It is used as a sand binder and also for paper pulp. It is found throughout India. The stems and leaves are used for making thatches.
71. *Saccharum fuscum*.—This grass is used for making cheaper grades of papers, wrapping-boards, and other cellulose products.
72. *Agropyron repens*; Syn. *Triticum repens*.—The decoction of the rhizome is a demulcent and diuretic and used in the treatment of catarrhal diseases of the urino-genital tract.
73. *Arundinaria falcata*; Syn. *Bambusa falcata*; Eng.—Himalayan bamboo.—The stems are used for making baskets, arrows, fishing rods and the lining of roofs of houses. It is found in Western Himalayas and Sikkim.
74. *Arundinaria spathiflora*.—This is a small bamboo growing the undergrowth of the fir, oak and deodar forests of the Western Himalayas at an elevation of 6,500 feet to 9,500 feet. The culms are used for pipe stems and baskets.
75. *Dendrocalamus strictus*; Eng.—Solid bamboo; Verna.—*Bans kaban*.—It is used in the manufacture of paper, baskets and brushes. It is found in Northern India. Madhya Pradesh and South India.
76. *Echinochloa colonum*; Syn. *Panicum colonum*; Eng.—Shama millet; Verna.—*Samak*—The grains are eaten in times of scarcity.
77. *Echinochloa stagnina*; Verna.—*Banti*—It is used as a fodder. The grains are edible. It is found throughout India.
78. *Eragrostis curvula*.—This is an introduced grass from United States. It is grown for erosion control and fodder.
79. *Bambusa vulgaris*; Eng.—Feathery bamboo; Verna.—*Basini bans*.—The split culms are made into mats and baskets and the young buds are consumed as vegetable. It is cultivated in the hotter parts of our country.
80. *Bambusa polymorpha*.—It is used for making paper, in Bengal and Assam.
81. *Cephalostachyum pergracile*.—This is a tufted bamboo and frequently found in association with teak. The culms are used for building and mat-making, and cooked rice is often carried in the internodes on a journey.

82. *Chloris gayana*; Eng.—Rhodes grass. A grass used as fodder. This has been introduced from Brazil.
83. *Chrysopogon aciculatus.*—This is a natural lawn grass, grown in the areas of high rainfall.
84. *Chrysopogon fulvus*; Syn. *C. montanus*; Verna.—*Goria.*—A perennial herb, used as fodder.
85. *Cymbopogon jwarancusa.*—The oil smells like peppermint but the yield is low. The roots are used in the treatment of fevers.
86. *Cymbopogon caesius*; Syn. *Andropogon caesius*; Eng.—Ginger grass.—This is a perennial grass. The leaves yield an essential oil, which is used in soap manufacture. The grass is grown in Madras and Tranvancore.
87. *Cynodon dactylon*; Eng.—Bermuda-grass; Verna.—*Doob.*—This is a perennial grass used as fodder. This is the commonest used for tennis lawns.
88. *Dendrocalamus giganteus.*—This is the largest of the Indian bamboos. It produces culms upto twenty five cm. in diameter. They are used for water-buckets and boxes. It is cultivated in Assam, Bengal and Malabar.
89. *Dendrocalamus hamiltonii*; Verna.—*Kag zi bans.*—This is a tall tufted grass, used in the manufacture of paper, baskets and mats. It is mainly cultivated in Dehradun.
90. *Cymbopogon nardus*; Syn. *Andropogon nardus*; Eng.—Citronella grass; Verna.—*Ganjni.*—The oil known as Ceylon type, obtained from the leaves ; is used in perfumery and cosmetics. It is also used in mosquito-repellent preparations. Also used in an infusion as a stomachic and carminative.
91. *Cymbopogon schoenanthus*; Syn. *Andropogon schoenanthus*; Eng.—Camel hay; Verna.—*Rousaghas.*—The leaves yield an essential oil which is used in perfumery.
92. *Cymbopogon winterianus*; Eng.—Citronella grass. The oil commercially known as Java type contains high proportions of geraniol, citronella and citronollol and hence used in the synthesis of organic compounds such as synthetic menthol.
93. *Cymbopogon flexuosus*; Syn. *Andropogon flexuosus*; Eng.—East Indian lemon grass.—The leaves yield an oil which is used in perfumery and cosmetics.
94. *Chrysopogon gryllus*; Syn. *Andropogon gryllus*; Verna.—*Saluni.*—It is used as a fodder.

95. *Cymbopogon citratus*; Syn. *Andropogon citratus*; Eng.—West Indian lemon grass; Verna.—*Sugandh rohisha.*—The leaves yield an aromatic oil which is used in perfumery, cosmetics and as a flavouring substance. It is chiefly grown in Travancore and Malabar.
96. *Cymbopogon martini*; Syn. *Andropogon martini*; Eng.—Ginger grass; Verna.—Gandhejghas.—The leaves are the source of an aromatic oil, which is used in perfumery and cosmetics. This occurs in two varieties known as the *Motia* and. *Sofia*. The essential oil, obtained from the Motia. variety, is of a superior quality and is referred to as palmarosa oil, while the oil from the Sofia variety is inferior in quality and is known as Gingergrass oil. Next to Sandal wood and lemon-grass oils, palmarosa oil is the important of essential oils produced in our country.

INDEX